Reservoir Simulation

CBMS-NSF REGIONAL CONFERENCE SERIES IN APPLIED MATHEMATICS

A series of lectures on topics of current research interest in applied mathematics under the direction of the Conference Board of the Mathematical Sciences, supported by the National Science Foundation and published by SIAM.

GARRETT BIRKHOFF, *The Numerical Solution of Elliptic Equations*
D. V. LINDLEY, *Bayesian Statistics, A Review*
R. S. VARGA, *Functional Analysis and Approximation Theory in Numerical Analysis*
R. R. BAHADUR, *Some Limit Theorems in Statistics*
PATRICK BILLINGSLEY, *Weak Convergence of Measures: Applications in Probability*
J. L. LIONS, *Some Aspects of the Optimal Control of Distributed Parameter Systems*
ROGER PENROSE, *Techniques of Differential Topology in Relativity*
HERMAN CHERNOFF, *Sequential Analysis and Optimal Design*
J. DURBIN, *Distribution Theory for Tests Based on the Sample Distribution Function*
SOL I. RUBINOW, *Mathematical Problems in the Biological Sciences*
P. D. LAX, *Hyperbolic Systems of Conservation Laws and the Mathematical Theory of Shock Waves*
I. J. SCHOENBERG, *Cardinal Spline Interpolation*
IVAN SINGER, *The Theory of Best Approximation and Functional Analysis*
WERNER C. RHEINBOLDT, *Methods of Solving Systems of Nonlinear Equations*
HANS F. WEINBERGER, *Variational Methods for Eigenvalue Approximation*
R. TYRRELL ROCKAFELLAR, *Conjugate Duality and Optimization*
SIR JAMES LIGHTHILL, *Mathematical Biofluiddynamics*
GERARD SALTON, *Theory of Indexing*
CATHLEEN S. MORAWETZ, *Notes on Time Decay and Scattering for Some Hyperbolic Problems*
F. HOPPENSTEADT, *Mathematical Theories of Populations: Demographics, Genetics and Epidemics*
RICHARD ASKEY, *Orthogonal Polynomials and Special Functions*
L. E. PAYNE, *Improperly Posed Problems in Partial Differential Equations*
S. ROSEN, *Lectures on the Measurement and Evaluation of the Performance of Computing Systems*
HERBERT B. KELLER, *Numerical Solution of Two Point Boundary Value Problems*
J. P. LASALLE, *The Stability of Dynamical Systems*
D. GOTTLIEB AND S. A. ORSZAG, *Numerical Analysis of Spectral Methods: Theory and Applications*
PETER J. HUBER, *Robust Statistical Procedures*
HERBERT SOLOMON, *Geometric Probability*
FRED S. ROBERTS, *Graph Theory and Its Applications to Problems of Society*
JURIS HARTMANIS, *Feasible Computations and Provable Complexity Properties*
ZOHAR MANNA, *Lectures on the Logic of Computer Programming*
ELLIS L. JOHNSON, *Integer Programming: Facets, Subadditivity, and Duality for Group and Semi-Group Problems*
SHMUEL WINOGRAD, *Arithmetic Complexity of Computations*
J. F. C. KINGMAN, *Mathematics of Genetic Diversity*
MORTON E. GURTIN, *Topics in Finite Elasticity*
THOMAS G. KURTZ, *Approximation of Population Processes*
JERROLD E. MARSDEN, *Lectures on Geometric Methods in Mathematical Physics*
BRADLEY EFRON, *The Jackknife, the Bootstrap, and Other Resampling Plans*
M. WOODROOFE, *Nonlinear Renewal Theory in Sequential Analysis*
D. H. SATTINGER, *Branching in the Presence of Symmetry*
R. TEMAM, *Navier–Stokes Equations and Nonlinear Functional Analysis*

Miklós Csörgo, *Quantile Processes with Statistical Applications*
J. D. Buckmaster and G. S. S. Ludford, *Lectures on Mathematical Combustion*
R. E. Tarjan, *Data Structures and Network Algorithms*
Paul Waltman, *Competition Models in Population Biology*
S. R. S. Varadhan, *Large Deviations and Applications*
Kiyosi Itô, *Foundations of Stochastic Differential Equations in Infinite Dimensional Spaces*
Alan C. Newell, *Solitons in Mathematics and Physics*
Pranab Kumar Sen, *Theory and Applications of Sequential Nonparametrics*
László Lovász, *An Algorithmic Theory of Numbers, Graphs and Convexity*
E. W. Cheney, *Multivariate Approximation Theory: Selected Topics*
Joel Spencer, *Ten Lectures on the Probabilistic Method*
Paul C. Fife, *Dynamics of Internal Layers and Diffusive Interfaces*
Charles K. Chui, *Multivariate Splines*
Herbert S. Wilf, *Combinatorial Algorithms: An Update*
Henry C. Tuckwell, *Stochastic Processes in the Neurosciences*
Frank H. Clarke, *Methods of Dynamic and Nonsmooth Optimization*
Robert B. Gardner, *The Method of Equivalence and Its Applications*
Grace Wahba, *Spline Models for Observational Data*
Richard S. Varga, *Scientific Computation on Mathematical Problems and Conjectures*
Ingrid Daubechies, *Ten Lectures on Wavelets*
Stephen F. McCormick, *Multilevel Projection Methods for Partial Differential Equations*
Harald Niederreiter, *Random Number Generation and Quasi-Monte Carlo Methods*
Joel Spencer, *Ten Lectures on the Probabilistic Method, Second Edition*
Charles A. Micchelli, *Mathematical Aspects of Geometric Modeling*
Roger Temam, *Navier–Stokes Equations and Nonlinear Functional Analysis, Second Edition*
Glenn Shafer, *Probabilistic Expert Systems*
Peter J. Huber, *Robust Statistical Procedures, Second Edition*
J. Michael Steele, *Probability Theory and Combinatorial Optimization*
Werner C. Rheinboldt, *Methods for Solving Systems of Nonlinear Equations, Second Edition*
J. M. Cushing, *An Introduction to Structured Population Dynamics*
Tai-Ping Liu, *Hyperbolic and Viscous Conservation Laws*
Michael Renardy, *Mathematical Analysis of Viscoelastic Flows*
Gérard Cornuéjols, *Combinatorial Optimization: Packing and Covering*
Irena Lasiecka, *Mathematical Control Theory of Coupled PDEs*
J. K. Shaw, *Mathematical Principles of Optical Fiber Communications*
Zhangxin Chen, *Reservoir Simulation: Mathematical Techniques in Oil Recovery*
Athanassios S. Fokas, *A Unified Approach to Boundary Value Problems*
Margaret Cheney and Brett Borden, *Fundamentals of Radar Imaging*
Fioralba Cakoni, David Colton, and Peter Monk, *The Linear Sampling Method in Inverse Electromagnetic Scattering*

ZHANGXIN CHEN
University of Calgary
Calgary, Alberta, Canada

Reservoir Simulation

Mathematical Techniques in Oil Recovery

siam.
SOCIETY FOR INDUSTRIAL AND APPLIED MATHEMATICS
PHILADELPHIA

10 9 8 7 6 5 4 3

Library of Congress Cataloging-in-Publication

Chen, Zhangxin, 1962–

Reservoir simulation : mathematical techniques in oil recovery / Zhangxin Chen.

p. cm. – (CBMS-NSF regional conference series in applied mathematics ; 77)

Includes bibliographical references and index.

ISBN 978-0-898716-40-5 (alk. paper)

1. Oil reservoir engineering–Mathematical models. 2. Oil reservoir engineering–Simulation methods. 3. Porous materials–Permeability–Mathematical models. 4. Transport theory–Mathematical models. I. Title II. Series.

TN871.C465 2007
622'.3382015118–dc22

2007061749

THIS BOOK IS DEDICATED

TO MY PARENTS, WIFE, AND CHILDREN.

Contents

List of Figures

List of Tables

List of Notation

Symbol	Quantity	Unit
$\mathcal{A}$	Function in simplified gas state	dimensionless
A	Adjustable parameter in tortuosity (Chap. 7)	dimensionless
A	Cross-sectional area normal to flow (Chaps. 2, 3, 5)	L^2, ft^2 (m^2)
A_i	Cross section normal to x_i-direction	L^2, ft^2 (m^2)
a	Generic coefficient (Chaps. 3, 5)	dimensionless
a	Empirical or experimental coefficient (Chap. 4)	dimensionless
a_{num}	Numerical dispersion	
$\mathcal{B}$	Function in simplified gas state	dimensionless
B	Formation volume factor (Chaps. 2, 3)	L^3/L^3
B	Neighborhood of origin (Chap. 4)	L^2, ft^2 (m^2)
B	Adjustable parameter in tortuosity (Chap. 7)	dimensionless
B_α	Formation volume factor of phase α	L^3/L^3
B_g	Formation volume factor of gas	L^3/L^3, RB/SCF (m^3/std m^3)
B_o	Formation volume factor of oil	L^3/L^3, RB/STB (m^3/std m^3)
B_{ob}	Oil formation volume factor at p_b	L^3/L^3, RB/STB (m^3/std m^3)
B_w	Formation volume factor of water	L^3/L^3, RB/BBL (m^3/std m^3)
B_{wi}	Initial water formation volume factor	L^3/L^3, RB/BBL (m^3/std m^3)
b	Generic coefficient	dimensionless
$\mathcal{C}$	Function in simplified gas state	dimensionless
C	Generic constant (Chaps. 3, 4)	dimensionless
C	Adjustable parameter in tortuosity (Chaps. 6, 7)	dimensionless
c_b	Bulk specific heat capacity	$L^2/(Tt^2)$, Btu/(lbm-R) (J/(kg-K))
c_f	Fluid compressibility	Lt^2/M, psi^{-1} (kPa^{-1})
c_g	Gas compressibility	Lt^2/M, psi^{-1} (kPa^{-1})
C_{Go}	Mass fraction of gas component in oil	fraction
C_{Oo}	Mass fraction of oil component in oil	fraction
$C_{p,ob}$	Overburn heat capacity (fixed pressure)	$L^2/(Tt^2)$, Btu/(lbm-R) (J/(kg-K))
$C_{p\alpha}$	Heat capacity of phase α (fixed pressure)	$L^2/(Tt^2)$, Btu/(lbm-R) (J/(kg-K))

c_p	Fluid heat capacity	$L^2/(Tt^2)$, Btu/(lbm-R) (J/(kg-K))
C_s	Specific heat capacity	$L^2/(Tt^2)$
c_s	Solid specific heat capacity	$L^2/(Tt^2)$, Btu/(lbm-R) (J/(kg-K))
$C_{p,ub}$	Underburn heat capacity (fixed pressure)	$L^2/(Tt^2)$, Btu/(lbm-R) (J/(kg-K))
$C_{V\alpha}$	Heat capacity of phase α (fixed volume)	$L^2/(Tt^2)$, Btu/(lbm-R) (J/(kg-K))
c_V	Fluid specific heat capacity	$L^2/(Tt^2)$, Btu/(lbm-R) (J/(kg-K))
c_o	Oil compressibility	Lt^2/M, psi^{-1} (kPa^{-1})
c_w	Water compressibility	Lt^2/M, psi^{-1} (kPa^{-1})
c_R	Rock compressibility	Lt^2/M, psi^{-1} (kPa^{-1})
c_t	Total compressibility	Lt^2/M, psi^{-1} (kPa^{-1})
c_μ	Oil viscosity compressibility	Lt^2/M, psi^{-1} (kPa^{-1})
$c(p)$	Compressibility	Lt^2/M, psi^{-1} (kPa^{-1})
D	Inner diameter of tubing	L, ft (m)
DS_{max}	Max variation of saturation	fraction
$\mathbf{D}_{ij}^M$	Molecular diffusion coefficient	L^2/t, ft^2/D (m^2/d)
$\mathbf{D}_i^p$	Pressure diffusion coefficient	L^3t/M, $ft^2/(D\ psi)$ ($m^2/(d\ kPa)$)
$\mathbf{D}_i^T$	Thermal diffusion coefficient	$L^2/(tT)$, $ft^2/(D\ R)$ ($m^2/(d\ k)$)
$\mathbf{D}_i$	Diffusion/dispersion of component i	L^2/t, ft^2/D (m^2/d)
$D_{m,ii}$	Diffusion of component m in x_i-direction	L^2/t, ft^2/D (m^2/d)
$\mathcal{D}_m$	Diffusion coefficient of component m (numerical)	M/t, lbm/D (kg·mole/d)
d_m	Molecular diffusion	L^2/t, ft^2/D (m^2/d)
d'_m	Scaled molecular diffusion	L^2/t, ft^2/D (m^2/d)
d_l	Longitudinal dispersion	L, ft (m)
d_t	Transverse dispersion	L, ft (m)
$E_i(\cdot)$	Exponential integral function	
$\mathbf{E}(\mathbf{u})$	Orthogonal projection along $\mathbf{u}$	dimensionless
$\mathbf{E}^\perp(\mathbf{u})$	Complement of $\mathbf{E}(\mathbf{u})$: $\mathbf{I}-\mathbf{E}(\mathbf{u})$	dimensionless
F	Formation resistivity factor (Chap. 7)	dimensionless
F	Total mass variable (Chap. 8)	M/L^3, lbm/ft^3 (kg/m^3)
F_m	Total mass of component m	M/L^3, lbm/ft^3 (kg/m^3)
F_w	distribution function of saturation	fraction
f	Generic function	
f_i	Fugacity function of component i	$M/(Lt^2)$, $lbm/(ft\ D^2)$ ($kg/(m\ d^2)$)
$f_{i\alpha}$	Fugacity of component i in phase α	$M/(Lt^2)$, $lbm/(ft\text{-}D^2)$ ($kg/(m\text{-}d^2)$)
f_m	Friction factor	dimensionless

f_w	Water cut	fraction
f_α	Fractional flow function of phase α	fraction
$\mathcal{F}$	Right-hand side function	
G_i^n	Right-hand side vector	
g_i	Boundary data ($i = 1, 2, 3$)	
g	Boundary datum	
g_u	Boundary parameter	
g_p	Boundary parameter	
$g_{\alpha,i}$	Boundary data ($i = 1, 2, 3$)	
$g_{\alpha,p}$	Boundary datum	
$g_{\alpha,u}$	Boundary datum	
H	Reservoir thickness	L, ft (m)
H_α	Enthalpy of α-phase	L^2/t^2, Btu/lbm (J/kg)
h	Grid step size	L, ft (m)
h_i	Grid step size in x_i-direction	L, ft (m)
h'	Spatial step size	L, ft (m)
h''	Spatial step size	L, ft (m)
$\mathbf{I}$	Identity tensor	
$\mathbf{J}$	Diffusion/dispersion	$M/(L^2t)$, lbm/(ft^{2}D) kg·mole/(m^{2}d)
$\mathbf{J}_i$	Diffusion/dispersion of component i	M/L^2t, lbm/(ft^{2}D) kg·mole/(m^{2}d)
J	Time interval of interest $(0, T]$	t, D (d)
J^n	Subinterval in time $(t^{n-1}, t^n]$	t, D (t)
K_h	Domain partition	dimensionless
K_m	Equilibrium K-value of component m	dimensionless
$K_{m\alpha}$	Equilibrium value of component m in phase α	dimensionless
$\mathcal{K}$	Empirical or experimental constant	dimensionless
$\mathbf{k}$	Permeability tensor	L^2, darcy (μm^2)
k	Permeability	L^2, darcy (μm^2)
k_H	Horizontal permeability	L^2, darcy (μm^2)
k_V	Vertical permeability	L^2, darcy (μm^2)
$\mathbf{k}_\alpha$	Effective permeability of phase α	L^2, darcy (μm^2)
k_{ii}	Permeability in x_i-direction	L^2, darcy (μm^2)
k_{ij}	Binary interaction parameter	dimensionless
$k_{r\alpha}$	Relative permeability of phase α	dimensionless
k_{row}	Relative permeability of oil-water	dimensionless
k_{rog}	Relative permeability of oil-gas	dimensionless
k_{rc}	Value of k_{row} at S_{wc}: $k_{row}(S_{wc})$	dimensionless
$\bar{\mathbf{k}}$	Certain average of $\mathbf{k}$	L^2, darcy (μm^2)
k_{ob}	Thermal conductivity of overburden	$ML/(Tt^3)$, Btu/(ft-D-R) (J/(m-d-K))
k_{ub}	Thermal conductivity of underburden	$ML/(Tt^3)$, Btu/(ft-D-R) (J/(m-d-K))
k_T	Bulk thermal conductivity	$ML/(Tt^3)$, Btu/(ft-D-R) (J/(m-d-K))

k_f	Fluid thermal conductivity	ML/(Tt^3), Btu/(ft-D-R) (J/(m-d-K))
k_s	Solid thermal conductivity	ML/(Tt^3), Btu/(ft-D-R) (J/(m-d-K))
L	Length of domain Ω in x_1-direction (Chap. 5)	L, ft (m)
L	Mass fraction of oil (Chap. 8)	fraction
L_i^n	Difference operator	
L_{ik}	Phenomenological coefficient	dimensionless
$\mathcal{L}$	Differential operator	
$M_{w\nu}$	Number of perforated zones of νth well	dimensionless
m	Parameter in tortuosity	dimensionless
N_c	Number of components	dimensionless
N_w	Number of wells	dimensionless
n	Parameter in tortuosity	dimensionless
n	Total number of moles	mole
n_i	Number of moles of component i	mole
$n_{m\alpha}$	Number of moles of component i in phase α	mole
n_α	Number of moles of phase α	mole
P_i^n	Approximate pressure	M/(Lt^2), psi (kPa)
p	Pressure	M/(Lt^2), psi (kPa)
$\bar{p}$	Averaged pressure	M/(Lt^2), psi (kPa)
p_α	Pressure of phase α	M/(Lt^2), psi (kPa)
p_b	Bubble point pressure	M/(Lt^2), psi (kPa)
p_{bh}	Pressure at z_{bh}	M/(Lt^2), psi (kPa)
p_c	Capillary pressure	M/(Lt^2), psi (kPa)
p_{cb}	Threshold pressure	M/(Lt^2), psi (kPa)
p_{cm}	Critical pressure of component m	M/(Lt^2), psi (kPa)
p_{cow}	Capillary pressure $p_o - p_w$	M/(Lt^2), psi (kPa)
p_{cgo}	Capillary pressure $p_g - p_o$	M/(Lt^2), psi (kPa)
p_i	Pressure at node i	M/(Lt^2), psi (kPa)
$p_{i,j,k}$	Pressure at block (i, j, k)	M/(Lt^2), psi (kPa)
p^o	Reference pressure	M/(Lt^2), psi (kPa)
p^0	Initial pressure	M/(Lt^2), psi (kPa)
p_s	Formation pressure at std	M/(Lt^2), psi (kPa)
p_{sp}	Specified surface pressure	M/(Lt^2), psi (kPa)
p_w	Pressure of wellbore	M/(Lt^2), psi (kPa)
p_{wh}	Pressure of wellhead	M/(Lt^2), psi (kPa)
p_h^0	Pressure at z_d	M/(Lt^2), psi (kPa)
p_α^0	Initial pressure of phase α	M/(Lt^2), psi (kPa)
Q_c	Heat source rate	ML^2/t^3, Btu/D (J/d)
Q_L	Heat loss rate	ML^2/t^3, Btu/D (J/d)
Q	Production/injection rate	L^3/t, RB/D (m^3/d)
Q_α	Source/sink of phase α	M/t, RB/D (m^3/d)
Q_i	Flow rate of component i	M/t, RB/D (m^3/d)
$Q_{i,j,k}$	Production/injection rate at (i, j, k)	M/t, lbm/D (kg/d)
$\tilde{Q}_{\alpha s}$	Source/sink rate	L^3/t, RB/D (m^3/d)

$\tilde{Q}_{i,j,k}$	Production/injection rate at (i, j, k)	L^3/t, RB/D (m^3/d)
$\tilde{Q}_{w,i,j,k}$	Water source/sink at (i, j, k)	L^3/t, RB/D (m^3/d)
q	Source/sink	$M/(L^3t)$, $lbm/(ft^3D)$ ($kg/(m^3d)$)
q_i	Source/sink of component i	$M/(L^3t)$, $lbm/(ft^3D)$ kg·mole/(m^3d))
q_T	Heat source/sink	$M/(Lt^3)$, $Btu/(ft^3D)$ ($J/(m^3d)$)
$\tilde{q}$	Total well flow rate	t^{-1}, D^{-1} (d^{-1})
q'	Perturbated source/sink	$M/(L^3t)$, $lbm/(ft^3D)$ ($kg/(m^3d)$)
$q^{(i)}$	Production/injection rate at well i	L^3/t, ft^3/D (m^3/d)
q_α	Source/sink of phase α	$M/(L^3t)$, $lbm/(ft^3D)$ ($kg/(m^3d)$
$\tilde{q}_\alpha$	Source/sink of phase α	t^{-1}, D^{-1} (d^{-1})
$q_{\alpha s}$	Source/sink of phase α at std	1/t, 1/D (1/d)
$q_{spe}^{(\nu)}$	Specified source/sink	$M/(L^3t)$, $lbm/(ft^3D)$
$\tilde{q}_{\alpha s}$	Source/sink rate at std	t^{-1}, D^{-1} (d^{-1})
q_c	Heat source	$M/(Lt^3)$, $Btu/(ft^3\text{-}D)$ ($J/(m^3\text{-}d)$)
q_L	Heat loss	$M/(Lt^3)$, $Btu/(ft^3\text{-}D)$ ($J/(m^3\text{-}d)$)
R	Universal gas constant	
R_{gl}	Gas/liquid ratio	L^3/L^3, SCF/STB (std m^3/std m^3)
R_{go}	Gas/oil ratio	L^3/L^3, SCF/STB (std m^3/std m^3)
R_{so}	Dissolved gas/oil ratio	L^3/L^3, SCF/STB (std m^3/std m^3)
R_i^n	Truncation error	
$R_{i,j,k}^l$	Residual of Newton–Raphson iteration	
r	Radial coordinate	L, ft (m)
r_e	Equivalent radius	L, ft (m)
r_w	Radius of well	L, ft (m)
r^o	Reference radius	L, ft (m)
S_α	Saturation of phase α	fraction
S_{gmax}	Maximum gas saturation	fraction
S_{iw}	Irreducible water saturation	fraction
S_{gc}	Critical gas saturation	fraction
$S_{n\alpha}$	Normalized saturation of phase α	fraction
S_{nc}	Residual saturation	fraction
S_{or}	Residual oil saturation	fraction
S_{rg}	Residual gas saturation	fraction
S_{wc}	Critical water saturation	fraction
S_{we}	Saturation of water at $x_1 = L$	fraction
S_{wf}	Water saturation at water front	fraction

S_{wr}	Residual water saturation	fraction
S_{wmax}	Maximum water saturation	fraction
S_w^0	Initial saturation of water	fraction
s	Skin factor	dimensionless
T	Temperature	T, R (K)
T_i	Transmissibility in x_i-direction (Chap. 3)	Lt, ft-D (m-d)
T_i	Transmissibility in x_i-direction (Chap. 5)	L^4t/M, STB/(D-psi) (std m^3/(d·kPa))
T_s	Formation temperature at std	T, R (K)
T_w	Transmissibility of water	L^4t/M, STB/(D-psi) (std m^3/(d·kPa))
$\mathbf{T}$	Transmissibility of fluid	Lt, ft-D (m-d)
$\mathbf{T}_m$	Transmissibility of component m	Lt, ft-D (m-d)
$\mathbf{T}_\alpha$	Transmissibility of α phase (Chap. 9)	Lt, ft-D (m-d)
$\mathbf{T}_\alpha$	Transmissibility of α phase (Chaps. 6, 8)	L^3t/M, ft^3D/lbm (std m^3d/kg)
$\mathbf{T}_\alpha$	Transmissibility (numerical) (Chap. 6)	L^4t/M, STB/(D-psi) (std m^3/(d·kPa))
$\mathbf{T}_{m\alpha}$	Transmissibility of component m in phase α (Chap. 9)	Lt, ft-D (m-d)
$\mathbf{T}_{m\alpha}$	Transmissibility of component m in phase α (Chap. 8)	L^4t/M, STB/(D-psi)
T_{cm}	Critical temperature of component m	T, R (K)
T_{rm}	Reduced temperature of component i	dimensionless
T_{ob}	Temperature of overburden	T, R (K)
T_{ub}	Temperature of underburden	T, R (K)
t	Time	t, days (d)
t_B	Water breakthrough time	t, D (d)
t_0	Initial time	t, days (d)
t^n	Time level n	t, days (d)
U_α	Specific internal energy of phase α	L^2/t^2, Btu/lbm (J/kg)
$\mathbf{u}$	Darcy velocity (u_1, u_2, u_3)	L/t, ft/D (m/d)
$\mathbf{u}_\alpha$	Darcy's velocity of phase α	L/t, ft/D (m/d)
$u(r)$	Velocity in r-direction	L/t, ft/D (m/d)
V	Volume (Chaps. 2, 3, 4, 5, 6, 8, 9)	L^3, ft^3 (m^3)
V	Molar volume of fluid (Chap. 7)	L^3, ft^3 (m^3)
V	Mass fraction of gas (Chap. 8)	fraction
V_{ci}	Critical volume of component i	L^3, ft^3 (m^3)
V_{Gs}	Volume of gas component at std	L^3, SCF (std m^3)
V_g	Volume of gas phase	L^3, ft^3 (m^3)
$V_{i,j,k}$	Volume of block (i, j, k)	L^3, ft^3 (m^3)
V_{Os}	Volume of oil component at std	L^3, STB (std m^3)
V_o	Volume of oil phase	L^3, ft^3 (m^3)
V_s	Volume at std	L^3, std m^3
V_w	Water volume	L^3, ft^3 (m^3)
$V(t)$	Cumulative liquid production	L^3, ft^3 (m^3)
$\bar{V}$	Nondimensional $V(t)$	fraction
$V_w(t)$	Cumulative water production	L^3, ft^3 (m^3)
$\bar{V}_w$	Nondimensional $V_w(t)$	fraction

$V_o(t)$	Cumulative oil production	L^3, ft^3 (m^3)
$\bar{V}_o$	Nondimensional $V_o(t)$	fraction
v	Interstitial velocity (Chap. 5)	L/t, ft/D (m/d)
v	Pore velocity (Chap. 3)	L/t, ft/D (m/d)
v_i	Molar volume of component i	L^3, ft^3 (m^3)
W	Molecular weight	M/mole, lbm/mole
W_i	Molecular weight of component i	M/mole
WI	Well index	L^3, ft^3 (m^3)
W_G	Molecular weight of gas	M/mole, lbm/mole
W_O	Molecular weight of oil	M/mole, lbm/mole
W_t	Total well productivity	L^3, ft^3 (m^3)
$\mathbf{x}$	Spatial variable (x_1, x_2, x_3) (Chaps. 3, 4, 5, 6, 7, 8, 9)	L, ft (m)
$\mathbf{x}$	Mole vector (Chap. 7)	mole
x_i	Mole fraction of component i	fraction
$x_{i\alpha}$	Mole fraction of component i in phase α	fraction
x_m	Total mole fraction of component m	mole/mole
$\mathbf{x}^{(i)}$	Location of well	L, ft (m)
$\mathbf{x}^{(\nu)}$	Location of well	L, ft (m)
$\mathbf{x}_0$	Location of well	L, ft (m)
$\mathbf{x}^{\star}$	Spatial variable	L, ft (m)
$\mathbf{x}^{\star\star}$	Spatial variable	L, ft (m)
Y_G	Raw gas gravity	dimensionless
Z	Gas compressibility	dimensionless
Z_α	Compressibility factor of phase α	dimensionless
z	Depth	L, ft (m)
z_{bh}	Datum level depth	L, ft (m)
z_d	Datum level depth	L, ft (m)
z_m	Total mole fraction of component m	fraction
α_{Ti}	Thermal diffusion factor of component i	dimensionless
γ	Fluid gravity	M/(L^{2}t^2), psi/ft (kPa/m)
γ_α	α-phase gravity	M/(L^{2}t^2), psi/ft (kPa/m)
$\bar{\gamma}$	Weighted fluid gravity	L/t, ft/D (m/d)
μ	Viscosity	M/(Lt), cp (Pa·s)
μ_w	Viscosity of water	M/(Lt), cp (Pa·s)
μ_α	Viscosity of phase α	M/(Lt), cp (Pa·s)
μ_{ob}	Oil viscosity at p_b	M/(Lt), cp (Pa·s)
μ_{ig}	Viscosity of ith component in gas	M/(Lt), cp (Pa·s)
μ_{io}	Viscosity of ith component in oil	M/(Lt), cp (Pa·s)
μ_m^*	Viscosity of component m at low pressure	M/(Lt), cp (Pa·s)
μ^*	Viscosity of fluid at low pressure	M/(Lt), cp (Pa·s)
μ_i	Chemical potential of component i	ML^2/t^2, lbm-ft^2/D^2 (kg-m^2/d^2)
ρ	Density	M/L^3, lbm/ft^3 (kg/m^3)
$\rho^{(i)}$	Density of injected fluid	M/L^3, lbm/ft^3 (kg/m^3)
ρ^o	Reference density	M/L^3, lbm/ft^3 (kg/m^3)
ρ_{air}	Air density	M/L^3, lbm/ft^3 (kg/m^3)
ρ_{Gs}	Density of gas at std	M/L^3, lbm/ft^3 (kg/m^3)

ρ_{Go}	Partial density of gas component in oil	M/L^3, lbm/ft^3 (kg/m^3)
ρ_{Oo}	Partial density of oil component in oil	M/L^3, lbm/ft^3 (kg/m^3)
ρ_{Os}	Density of oil at std	M/L^3, lbm/ft^3 (kg/m^3)
ρ_{Ws}	Density of water at std	M/L^3, lbm/ft^3 (kg/m^3)
ρ_α	Density of α phase	M/L^3, lbm/ft^3 (kg/m^3)
ρ_b	Bulk density	M/L^3, lbm/ft^3 (kg/m^3)
ρ_o	Density of oil	M/L^3, lbm/ft^3 (kg/m^3)
ρ_s	Density at std	M/L^3, lbm/ft^3 (kg/m^3)
ρ_s	Mass density of solid	M/L^3, lbm/ft^3 (kg/m^3)
ρ_{ob}	Density of overburden	M/L^3, lbm/ft^3 (kg/m^3)
ρ_{ub}	Density of underburden	M/L^3, lbm/ft^3 (kg/m^3)
Φ	Potential	M/(Lt^2), lbm/(ft D^2) (kg/(m d^2))
Φ'	Pseudopotential	L, ft (m)
ϕ	Porosity	fraction
ϕ^o	Reference porosity	fraction
ψ	Pseudopressure	M/(Lt^3), lbm/(ft D^3) (kg/(m d^3))
$\varphi_{i\alpha}$	Fugacity coefficient of component i in phase α	dimensionless
σ	Surface tension	M/t^2, lbm/D^2 (kg/d^2)
θ	Contact angle (Chap. 2)	rad (rad)
θ	Polar coordinate (Chaps. 3, 4)	rad (rad)
λ_α	Mobility of phase α	Lt/M, ft D/lbm (m d/kg)
λ	Total mobility	Lt/M, ft D/lbm (m d/kg)
β_g	Parameter in Stone's permeability	dimensionless
β_o	Parameter in Stone's permeability	dimensionless
$\frac{\partial}{\partial t}$	Time derivative	t^{-1}, D^{-1} (d^{-1})
$\frac{\partial}{\partial x_i}$	Spatial derivative	L^{-1}, ft^{-1} (m^{-1})
$\frac{\partial^2}{\partial x_i^2}$	Second spatial derivative	L^{-2}, ft^{-2} (m^{-2})
∇	Gradient operator	L^{-1}, ft^{-1} (m^{-1})
$\nabla\cdot$	Divergence operator	L^{-1}, ft^{-1} (m^{-1})
$\delta(\mathbf{x})$	Dirac delta function	1/L^3, 1/ft^3 (1/m^3)
δ	Constrictivity	dimensionless
δ_{jk}	Kronecker symbol	dimensionless
Δ	Laplacian operator	L^{-2}, ft^{-2} (m^{-2})
Δh	Well length in a gridblock	L, ft (m)
Δl	Length step size in well direction	L, ft (m)
δp	Pressure increment	M/(Lt^2), psi (kPa)
δp_{bh}	Bottom hole pressure increment	M/(Lt^2), psi (kPa)
Δp_{a}	Acceleration pressure drop	M/(Lt^2), psi (kPa)
Δp_{f}	Friction pressure drop	M/(Lt^2), psi (kPa)
Δp_{g}	Hydrostatic pressure drop	M/(Lt^2), psi (kPa)
Δp_{w}	External work pressure drop	M/(Lt^2), psi (kPa)
δS	Saturation increment	fraction

Δt	Time step size	t, days (d)
Δt^n	Time step size at level n	t, days (t)
Δt_p	Time step size for pressure	t, days (d)
Δt_S	Time step size for saturation	t, days (d)
Δx_i	Spatial step size in x_i-direction	L, ft (m)
$\Delta_{x_i}^2$	Second difference quotient	
Δz	Length step size in vertical direction	L, ft (m)
$\Delta\Phi_\alpha$	Potential difference of phase α	M/(Lt^2), psi (kPa)
$\wp$	Gravitational acceleration	L/t^2, ft/D^2 (m/d^2)
ϵ	Small parameter	dimensionless
Γ	Boundary of Ω ($\partial\Omega$)	L^2, ft^2 (m^2)
Ω	Solution domain	L^3, ft^3 (m^3)
ξ	Molar density of fluid	mole/L^3, mole/ft^3 (mole/m^3)
ξ_i	Molar density of component i	mole/L^3, mole/ft^3 (mole/m^3)
ξ_r	Reduced density	dimensionless
$\xi_{i\alpha}$	Molar density of component i in phase α	mole/L^3, mole/ft^3 (mole/m^3)
ξ_α	Molar density of phase α	mole/L^3
Ω_{ma}	Equations of state parameter of components m	dimensionless
Ω_{mb}	Equations of state parameter of components m	dimensionless
ω_m	Acentric factor of components m	dimensionless
τ	Tortuosity	dimensionless
$\boldsymbol{\nu}$	Outward unit normal	dimensionless

Subscripts

Symbol	Quantity
G	Gas component
g	Gas phase
i	Component number
l	Newton–Raphson iteration level
m	Layer, well, or perforation number (Chaps. 4, 5, 6, 8, 9)
m	Component number (Chap. 7)
N	Final time level
n	Time iteration level
O	Oil component
o	Oil phase
R	Rock
s	Solid (Chaps. 2, 6, 7, 8, 9)
s	Standard conditions (Chaps. 2, 3, 6)
w	Water phase (Chaps. 4, 5, 6, 8, 9)
w	Well (Chaps. 4, 5, 6, 7, 8, 9)
α	Phase index
ν	Well number

Preface

This book evolved from the lectures I presented at the U.S. NSF-CBMS Regional Research Conference, University of Nevada, Las Vegas (UNLV), May 23–27, 2006, on the subject of multiphase flows in porous media and simulation. It can be used as a text for senior undergraduate students and first-year graduate students in geology, petroleum engineering, and applied mathematics. It can also serve as a reference book for geologists, petroleum engineers, applied mathematicians, and scientists in the area of petroleum reservoirs. In addition, it can be used as a handbook for employees in the oil industry who need a basic grasp of modeling and computational method concepts. Calculus, basic physics, and some familiarity with partial differential equations and matrix algebra are necessary prerequisites.

The ten chapters correspond to the ten respective lectures I presented at UNLV. In Chapter 1, an overview of classical reservoir engineering and basic reservoir simulation methods is presented. In Chapter 2, a glossary of terms used in petroleum engineering and their units are reviewed. In Chapters 3 and 5–9, governing partial differential equations and their numerical solutions are given for, respectively, single-phase, two-phase, black oil (three-phase), single-phase with multicomponents, compositional, and thermal flows. For each of these flows, (1) basic flow and transport equations are first given; (2) the corresponding rock and fluid properties are stated; (3) peculiar features of these equations are discussed; (4) the procedure to obtain their numerical solution is described in detail; and (5) difficulties and practical issues in the solution are addressed. Particularly, the treatment of rock, fluid, and rock/fluid properties at the internal boundaries of gridblocks is studied in great detail. Well representations used in numerical simulation of these flows are described in Chapter 4. Some practical aspects of reservoir simulation, such as data gathering and analysis, selection of a simulation model, history matching, and reservoir performance prediction, are summarized in Chapter 10. In the numerical solution, as an example, the discretization procedure is carried out in detail for the finite difference method; finite volume and finite element discretizations can be found in the book by Chen, Huan, and Ma (2006). Numerical benchmark examples have been presented for all the flows under consideration by Chen et al. (2006) except for the single-phase, multicomponent flow, for which a couple of numerical examples are given.

I take this opportunity to thank many people who have made this book possible. I thank Professor Jichun Li and Professor Yitung Chen, who spent incredible time and energy in organizing this NSF-CBMS Regional Research Conference. I also thank the U.S. NSF for their generous support of the conference. In addition, I thank the editors of the CBMS book series in the Society for Industrial and Applied Mathematics, who are enthusiastic in publishing this book. Furthermore, I thank all the conference participants whose attendance and encouragement made the conference enjoyable and successful. Finally, I want to acknowledge financial support from Foundation CMG (Computer Modelling Group).

Zhangxin Chen
Calgary, Alberta, Canada
Professor of Mathematics and Engineering
NSERC/AERI/Foundation CMG Chair in Reservoir Simulation
iCORE Industrial Chair in Reservoir Modeling

May 5, 2007

Chapter 1

Introduction

1.1 Petroleum Reservoir Simulation

A *petroleum reservoir* is a porous medium that contains hydrocarbons. The major goal of *reservoir simulation* is to predict future performance of the reservoir and find ways and means of optimizing the recovery of some of the hydrocarbons under various operating conditions. It involves four main interrelated *modeling stages*—establishment of physical models, development of mathematical models, discretization of these models, and design of computer algorithms—and requires a combination of skills of physicists, mathematicians, reservoir engineers, and computer scientists. This book is devoted to the development of mathematical models and their numerical solutions for a wide variety of fluid flows, including single-phase, two-phase, black oil, compositional, and thermal. To show the role of reservoir simulation in optimizing the management and production of hydrocarbon resources, this chapter briefly touches on classical *reservoir engineering methods* used in hydrocarbon recovery and their relations to the reservoir simulation.

1.2 Classical Reservoir Engineering Methods

Classical methods of predicting reservoir performance include analogical, experimental, and mathematical methods. The *analogical methods* utilize features of mature reservoirs that are analogous to the target reservoir in an attempt to forecast the performance of this target zone or reservoir. The *experimental methods* measure physical properties, such as pressure, saturation, and/or rates, in laboratory cores and then scale them up to the whole hydrocarbon accumulation. Finally, the *mathematical methods* use model equations to forecast reservoir performance. These methods are described in detail in reservoir engineering books (Ahmed, 2006). Here we briefly mention the mathematical methods because of their direct relationships with reservoir simulation.

1.2.1 Material Balance Methods

Mathematical methods are the most widely used classical reservoir engineering methods in the petroleum industry in predicting reservoir performance. These methods include material balance, decline curve, statistical, and analytical methods. In general, graphical or hand computations suffice when these methods are applied. With the advent of personal computers, software for these methods is now available.

The classical *material balance methods* use a mathematical representation of a reservoir or drainage volume. Their basic principle is based on mass conservation; i.e., the amount of mass for water, oil, or gas remaining in the reservoir after a production period equals the difference of the amount of mass originally in place and that was removed from the reservoir due to production, plus the amount of mass added due to injection and encroachment. The literature is rich in the material balance methods (Ahmed, 2006).

1.2.2 Decline Curve Methods

The classical *decline curve methods* use one of three mathematical declines (exponential, hyperbolic, and harmonic) to describe the rate of oil production decline. A decline curve has the general form

$$Cq^b = -\frac{1}{q}\frac{dq}{dt}, \tag{1.1}$$

where C is a decline rate parameter, q is the production rate (m^3/d), and t is time (d). The cases $b = 0$, $0 < b < 1$, and $b = 1$ correspond to the exponential, hyperbolic, and harmonic declines, respectively.

The decline curve methods match historical production data to select a proper form of the rate equation (1.1). After the form is chosen, historical data are matched by choosing the parameters C and b that minimize the error (often using the least squares error) between the data and the equation. Extrapolating the historical data into the future predicts reservoir performance using the matched equation. A major assumption of any extrapolating method is that all processes occurring in the past will continue in the future. The decline curve methods cannot be used for "what-if" scenarios.

1.2.3 Statistical Methods

Statistical methods employ empirical correlations that are statistically obtained using the past performance of some reservoirs to forecast the future performance of others. They are a generalization of the analogical methods. A correlation is developed with data from mature reservoirs in the same region, with the same lithology (e.g., carbonate or sandstone) and under the same operating conditions (e.g., solution gas drive or water flood). For the reservoir engineer to be confident in using an empirical correlation model, reservoir properties must be within the limit of the regression database used to develop such a model. Predicting errors with the statistical methods can be as high as 20 to 50%.

1.2.4 Analytical Methods

Analytical methods, such as pressure-transient and Buckley–Leverett methods, use the analytical solution of a mathematical model. The model consists of a set of differential equations that describe the flow and transport of fluids in a petroleum reservoir, together with an appropriate set of boundary and/or initial conditions. To solve these equations exactly, simplifying assumptions must be made to reduce the complexity of the model.

In general, these assumptions are very restrictive. For example, in the *pressure-transient method*, the assumptions require single-phase flow in a horizontal reservoir with uniform thickness and small pressure differences under the laminar flow condition. The *Buckley–Leverett method* for two-phase flow ignores gravity and capillary forces under the incompressibility condition. However, since much of the physics of a problem is preserved, the analytical methods are often used to determine how various parameters influence reservoir performance. Furthermore, these methods can be used to validate reservoir simulators. Advanced reservoir engineering includes other sophisticated analytical methods such as separation of variables, Laplace transform, and integral methods.

1.3 Reservoir Simulation Methods

Reservoir simulation has become a standard predictive tool in the oil industry. It can be used to obtain accurate performance predictions for a hydrocarbon reservoir under different operating conditions. A hydrocarbon recovery project usually involves a capital investment of hundreds of millions of dollars, and the risk associated with its selected development and production strategies must be assessed and minimized. This risk includes such important factors as the complexity of a petroleum reservoir and the fluids filling it, the complexity of hydrocarbon recovery mechanisms, and the applicability of predictive methods. These complexities can be taken into account in reservoir simulation through data input into the simulation model, and this applicability can be estimated through sound engineering practices and accurate reservoir simulation.

1.3.1 Reservoir Simulation Stages

Reservoir simulation involves four major interrelated stages (Fig. 1.1). First, a physical model of relevant processes is developed incorporating as much physics as is deemed necessary to describe the essential features of the underlying physical phenomena. Second, a set of coupled systems of time-dependent nonlinear partial differential equations is developed and analyzed for existence, uniqueness, stability, and regularity. Third, a numerical model with the basic properties of both the physical and mathematical models is derived and

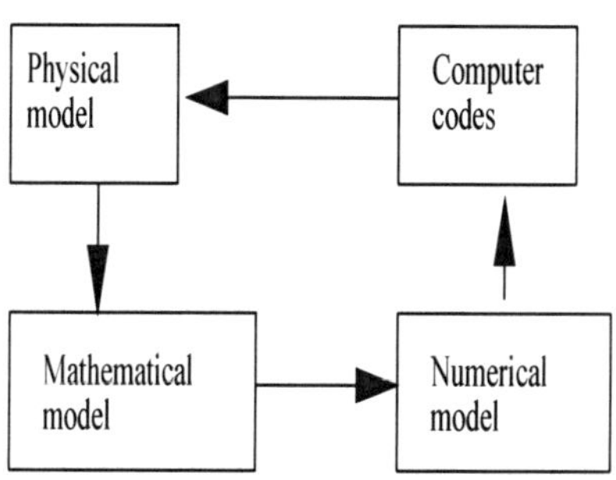

Figure 1.1. *Reservoir simulation stages.*

analyzed. Fourth, computer algorithms (and codes) are developed to solve efficiently the systems of linear and nonlinear algebraic equations arising from the numerical discretization. Each of these stages is essential to reservoir simulation, and a number of iterations among these stages are sometimes necessary to adjust the physical, mathematical, and numerical models and computer algorithms so that accurate reservoir performance forecast can be obtained. The widespread acceptance of reservoir simulation can be attributed to the advances in computing facilities, mathematical modeling, numerical methods, solver techniques, and visualization tools.

1.3.2 Reservoir Simulator Classifications

Reservoir simulators can be classified in different approaches. The most common approaches are based on the type of reservoir fluids being studied and the recovery processes being modeled. Other approaches include the number of dimensions (1D, 2D, and 3D), the number of phases (single-phase, two-phase, and three-phase), and the coordinate system used in the model (rectangular, cylindrical, and spherical). The type of reservoir simulators can also be determined by the rock structure or response (ordinary, dual porosity/permeability, and coupled hydraulic/thermal fracturing and flow).

Reservoir simulators based on the classification of the type of reservoir fluids include gas, black oil, and compositional simulators. The black oil simulators are conventional recovery simulators, and are used in cases where recovery processes are not sensitive to compositional changes in the reservoir fluids. Compositional simulators are used when recovery processes are sensitive to compositional changes, and include primary depletion of volatile oil and gas condensate reservoirs, pressure maintenance operations in these reservoirs, and multiple contact miscible processes.

Reservoir simulators classified using recovery processes include conventional recovery (black oil), miscible displacement, thermal recovery, and chemical flood simulators (cf. Fig. 1.2). Primary oil recovery processes, such as water drive, solution gas drive, gas cap expansion, gravity drainage, and capillary imbibition, can be all modeled with the black oil simulators. Moreover, secondary recovery stages, such as water or gas injection, can also be modeled with these simulators. Thermal recovery mechanisms involve steam floods or in situ combustion, and use an equation of energy conservation in addition to the mass conservation law. Chemical flood simulators include alkaline, surfactant, polymer, and/or foam (ASP + foam) flooding and can be used to change the mobility ratio dynamically in the displacement (polymer) or mobilize the residual oil (surfactant). Additional effects such as adsorption onto rock, permeability reduction, and non-Newtonian fluid need to be considered (Chen, Huan, and Ma, 2006).

1.3.3 Reservoir Simulation Applications

Reservoir simulation is usually applied in the following steps (Ertekin, Abou-Kassem, and King, 2001):

- Set simulation study objectives. The first step in any reservoir simulation study is to set clear objectives. These objectives must be achievable and compatible with available reservoir and production data.

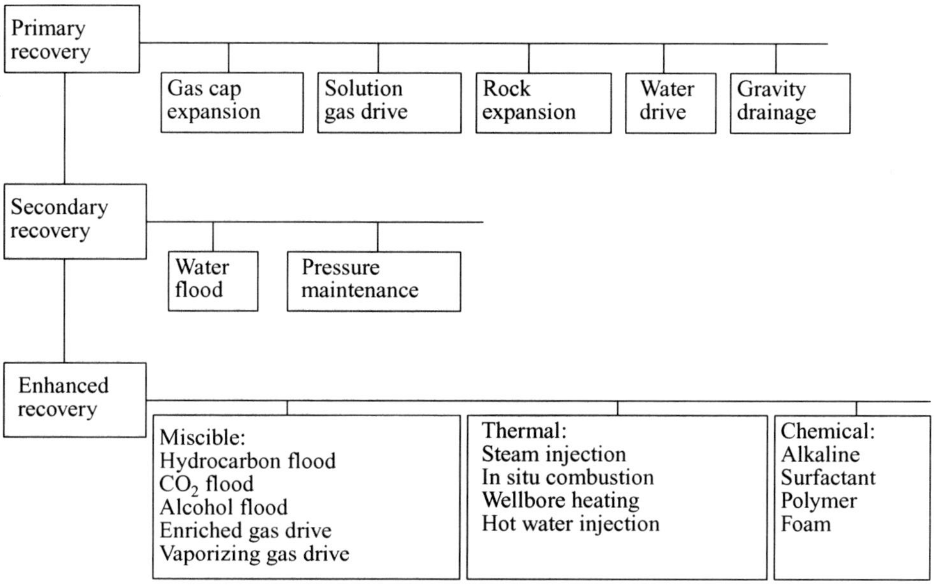

Figure 1.2. *Recovery processes.*

- Gather and validate reservoir data. After the simulation objectives have been set, reservoir and production data are gathered. The data meeting the objectives are incorporated into the simulator.

- Design the reservoir simulator. Once the data are gathered and validated, the simulator is designed. This step involves the four major interrelated stages outlined above: construction of a conceptual physical model, development of mathematical and numerical models, and design of computer codes.

- History match the reservoir simulator. After the reservoir simulator is constructed, it must be tuned, or history matched, with available reservoir and production data since much of the data in a typical simulator needs to be verified.

- Make predictions. In the final application step, various development and production plans are evaluated, and a sensitivity analysis of various reservoir and production parameters is carried out.

The above reservoir simulation steps will be further discussed in Chapter 10. While reservoir simulation is the most comprehensive method, the classical reservoir engineering methods are still in use to predict reservoir performance. These classical methods can be used to generate input data for reservoir simulators. For example, a pressure build-up analysis can be used to obtain formation permeability in reservoir characterization, and material balance methods yield information on water encroachment and aquifer size during history match.

1.4 SI Metric Conversion Factors

The *units* used throughout this book are customary English and metric. The conversion factors are listed below.

API	141.5/(131.5+API)	= g/cm^3
BBL	× 1.589 873E-01	= m^3
cp	× 1.0E-03	= Pa·s
ft	× 3.048-01	= m
ft^2	× 9.290 304E-02	= m^2
ft^3	× 2.831 685E-02	= m^3
lbm	× 4.535 924E-01	= kg
md	× 9.869 233E-04	= μm^2
psi	× 6.894 757E+09	= kPa
psi^{-1}	× 1.450 377E-01	= kPa^{-1}
R	× 5/9	= K
Btu		≈ kJ

Chapter 2

A Glossary of Petroleum Terms

Reservoir simulation requires an understanding of the reservoir rock and the fluids (water, oil, and gas) filling it. Reservoir rock properties of interest reflect the capacity of the rock to transmit and store fluids in its pores, and include porosity and permeability. Fluid properties of interest include those that heavily depend on pressure such as fluid densities, viscosities, formation (i.e., reservoir at high pressure and temperature) volume factors, compressibilities, and gas solubility factor. The rock-fluid interaction properties (capillary pressures and relative permeabilities) depend on saturation. This chapter gives a glossary of these petroleum terms.

2.1 Reservoir Rock Properties

The concepts of pores, porosity, and permeability are described.

Pores and pore throats. *Pores* are the tiny connected passages that exist in a permeable rock, typically of size 1 to 200 μm and easily visible in scanning electron microscopy. They can be lined by diagenetic minerals (e.g., clays). The narrower constrictions between pore bodies are *pore throats*. It is these pore throats that control the capillary entry pressure in a drainage process.

Porosity. *Porosity* is the fraction of a rock that is pore space. There are two types of porosities: total and effective. The total porosity includes both interconnected and isolated pore spaces, while the effective porosity includes only the former. Because only the interconnected pores store and transmit fluids, one is mainly concerned with the effective porosity. Hereafter, the term *porosity* will solely mean the effective porosity. In this sense it measures the capacity of the reservoir to store producible fluids in its pores.

Porosity is commonly denoted by ϕ (fraction) and varies from 0.25 for a fairly permeable rock down to 0.1 for a very low permeable rock. A reservoir rock property, such as porosity, often varies in space. If a property is independent of reservoir location, the reservoir rock is referred to as *homogeneous* with respect to this property. If it varies with location, it is termed *heterogeneous*. Variation of pore volume with pore pressure p can be taken into account by the pressure dependence of porosity. Porosity depends on pressure due to *rock compressibility*, which is often assumed to be constant (typically 10^{-6} to 10^{-7} psi^{-1}) and can be defined as

$$c_R = \frac{1}{\phi}\frac{d\phi}{dp}. \tag{2.1}$$

After integration, it is given by

$$\phi = \phi^o e^{c_R(p-p^o)}, \tag{2.2}$$

where ϕ^o is the porosity at the reference pressure p^o. Using a Taylor series expansion, we see that

$$\phi = \phi^o \left\{1 + c_R(p-p^o) + \frac{1}{2!}c_R^2(p-p^o)^2 + \cdots\right\},$$

so an approximation results:

$$\phi \approx \phi^o\big(1 + c_R(p-p^o)\big). \tag{2.3}$$

The reference pressure p^o is usually the atmospheric pressure or initial reservoir pressure.

Permeability. *Permeability* is the capacity of a rock to conduct fluids through its interconnected pores. This conducting capacity is sometimes referred to as *absolute permeability*. It is commonly indicated by **k**, with dimensions of area and units darcy (d) or milli-darcy (md). To the reservoir engineer, permeability is probably the most important quantity because its distribution dictates connectivity and fluid flow in a reservoir. Typical values of permeability for reservoir rocks are given in Table 2.1.

Permeability often varies with location and, even at the same location, may depend on a flow direction. In many practical situations, it is possible to assume that **k** is a diagonal tensor:

$$\mathbf{k} = \begin{pmatrix} k_{11} & & \\ & k_{22} & \\ & & k_{33} \end{pmatrix} = \mathrm{diag}(k_{11}, k_{22}, k_{33}).$$

Furthermore, it is even possible to assume that $k_H = k_{11} = k_{22}$ in the horizontal plane since directional trend is not apparent in many depositional environments. The vertical permeability $k_V = k_{33}$ is usually different from k_H since even very thin shale stringers significantly influence k_V. The horizontal permeability is generally larger than the vertical permeability. If $k_{11} = k_{22} = k_{33}$, the porous medium is called *isotropic*; otherwise, it is *anisotropic*. Homogeneity, heterogeneity, isotropy, and anisotropy each correspond to a single reservoir property, so these terms are always used in reference to a specific property. For example, a reservoir can be homogeneous with respect to porosity but heterogeneous with respect to thickness.

Table 2.1. *Classification of rock permeabilities.*

Classification	Permeability range (md)
Poor to fair	1–15
Moderate	15–20
Good	50–250
Very good	250–1,000
Excellent	over 1,000

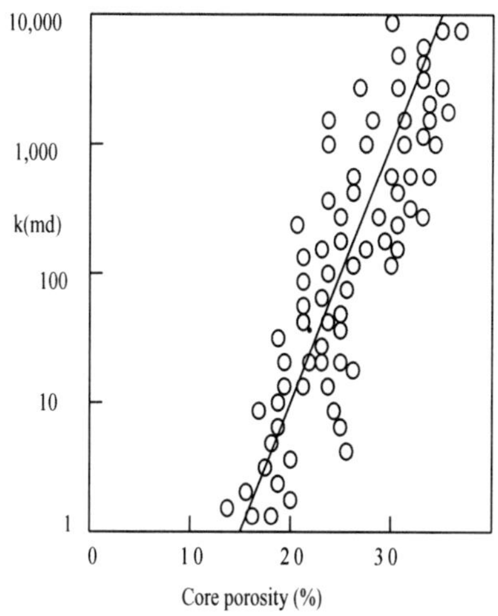

Figure 2.1. *Permeability-porosity correlation.*

Permeability-porosity correlation. In many systems there is an approximate correlation between the permeability **k** and the porosity ϕ. This is not always the case, and much scatter can be seen in a k/ϕ crossplot (cf. Fig. 2.1). In general, the larger the porosity, the higher the permeability.

2.2 Reservoir Fluid Properties

The definitions of fluid densities, viscosities, formation volume factors, compressibilities, and gas solubility factor are now introduced.

Phase. *Phase* refers to a chemically homogeneous region of fluid that is separated from another phase by an interface, e.g., oleic (oil), aqueous (mainly water), or gas. Phases are frequently subscripted by w (water), o (oil), and g (gas). The solid (rock) phase is indicated by s (R).

Component. *Component* is a single chemical species that may be present in a phase. For example, the aqueous phase contains components water (H_2O), sodium chloride (NaCl), and dissolved oxygen (O_2), and the oil phase contains hundreds of components, e.g., C_1, C_2, C_3, etc.

Type of reservoir fluid. In general, water, oil, and gas can exist simultaneously in a hydrocarbon reservoir. These fluids can be classified as incompressible, slightly compressible, or compressible, depending upon how they respond to pressure. An incompressible fluid has zero compressibility; its density is independent of pressure. Water and gas-free (dead) oil can be incompressible. A slightly compressible fluid has a small but constant compressibility that typically ranges from 10^{-5} to 10^{-6} psi^{-1}. At reservoir conditions, water, dead oil, and undersaturated oil behave like slightly compressible fluids. A compressible fluid has a compressibility typically in the range 10^{-3} to 10^{-4} psi^{-1}; its density increases as pressure increases but tends to be stabilized at higher pressures (cf. Fig. 2.2). At reservoir conditions, gas is compressible.

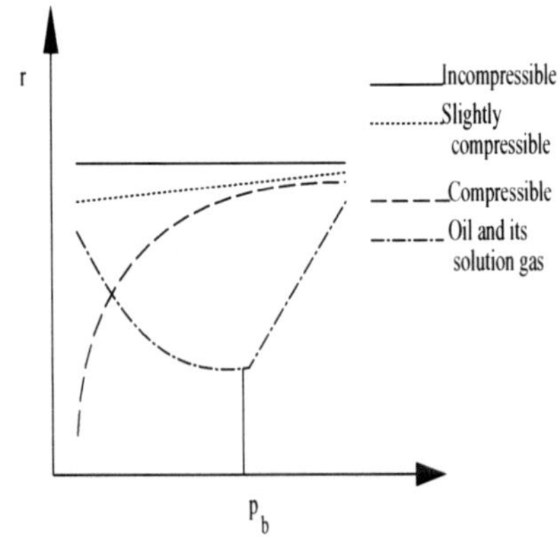

Figure 2.2. *Density-pressure relation.*

In general, in reservoir simulation, water is either incompressible or slightly compressible, and natural gas is compressible. Oil and its solution gas are treated as slightly compressible when the reservoir pressure is higher than the oil bubble point pressure (p_b) and as compressible when the reservoir pressure is lower than p_b.

Compressibility. Compressibility of a fluid can be defined in terms of the volume (V) or density (ρ) change with pressure:

$$c_f = -\frac{1}{V}\frac{\partial V}{\partial p}\bigg|_T = \frac{1}{\rho}\frac{\partial \rho}{\partial p}\bigg|_T \tag{2.4}$$

at a fixed temperature T. After integration, equation (2.4) is expressed as

$$\rho = \rho^o e^{c_f(p-p^o)}, \tag{2.5}$$

where ρ^o is the density at the reference pressure p^o. Using a Taylor series expansion, we see that

$$\rho = \rho^o \left\{ 1 + c_f(p - p^o) + \frac{1}{2!}c_f^2(p - p^o)^2 + \cdots \right\},$$

so an approximation is obtained:

$$\rho \approx \rho^o\big(1 + c_f(p - p^o)\big). \tag{2.6}$$

A different form of equation (2.6) can be derived if we use the real *gas law* (the pressure-volume-temperature (PVT) relation)

$$\rho = \frac{pW}{ZRT}, \tag{2.7}$$

where W is the molecular weight, Z is the gas compressibility factor, and R is the *universal gas constant*. If pressure, temperature, and density are in atm, K, and g/cm^3 (physical unit system), respectively, the value of R is 82.057. For the English units (psia, R, and lbm/ft^3),

$R = 10.73$; for the SI system (N/m^2, K, and kg/m^3), $R = 8{,}314$. Substituting (2.7) into (2.4) gives (with $c_g = c_f$)

$$c_g = \frac{1}{p} - \frac{1}{Z}\frac{\partial Z}{\partial p}\bigg|_T. \tag{2.8}$$

Gas solubility factor. *Gas solubility factor* R_{so} (also called the *dissolved gas/oil ratio*) is the volume of gas (measured at standard conditions) dissolved at a given reservoir pressure and temperature in a unit volume of stock tank oil:

$$R_{so}(p, T) = V_{Gs}/V_{Os}, \tag{2.9}$$

where the subscript s in the volumes denotes standard conditions and the capital letters G and O represent the gas and oil components, respectively. It usually has the unit SCF/STB (standard cubic feet/stock tank barrels). Note that

$$V_{Os} = W_O/\rho_{Os}, \qquad V_{Gs} = W_G/\rho_{Gs}, \tag{2.10}$$

where W_O and W_G are the weights of the oil and gas components, respectively. Then equation (2.9) becomes

$$R_{so} = \frac{W_G \rho_{Os}}{W_O \rho_{Gs}}. \tag{2.11}$$

Formation volume factors. *Formation volume factor* describes the ratio of volume V of a phase (measured at reservoir conditions) to the volume V_s of the phase measured at standard conditions:

$$B(p, T) = V(p, T)/V_s. \tag{2.12}$$

It has the unit RB/STB (reservoir barrels/stock tank barrels) for liquids. For gas, the unit is RB/SCF. For a single phase (water, dead oil, or gas), it can be written in density:

$$B(p, T) = \rho_s/\rho. \tag{2.13}$$

For the black oil model, since the oil phase contains both oil and solution gas,

$$V_o = \frac{W_O + W_G}{\rho_o}. \tag{2.14}$$

Consequently, combining (2.10), (2.12), and (2.14), the oil formation volume factor becomes

$$B_o = \frac{(W_O + W_G)\rho_{Os}}{W_O \rho_o}. \tag{2.15}$$

Fluid density. The density of a single phase (water, dead oil, or gas) can be obtained from equation (2.13):

$$\rho = \rho_s/B. \tag{2.16}$$

Using (2.11) and (2.15), the *mass fractions* of oil and solution gas in the oil phase are, respectively,

$$C_{Oo} = \frac{W_O}{W_O + W_G} = \frac{\rho_{Os}}{B_o \rho_o},$$

$$C_{Go} = \frac{W_G}{W_O + W_G} = \frac{R_{so} \rho_{Gs}}{B_o \rho_o},$$

Table 2.2. *Typical viscosity values of oils.*

Classification	Viscosity range (cp)
Light oil	0.3–1
Medium oil	1–6
Moderate oil	6–50
Very viscous oil	50–1,000
Heavy oil and oil sands	over 1,000

which, together with $C_{Oo} + C_{Go} = 1$, yield the density of the oil phase:

$$\rho_o = \frac{R_{so}\rho_{Gs} + \rho_{Os}}{B_o}. \tag{2.17}$$

Fluid viscosity. The *viscosity* of a fluid is a measure of the (frictional) energy dissipated when it is in motion resisting an applied shearing force, with dimensions of force/area $\cdot$ time and units Pa $\cdot$ s (metric) or poise (customary). The most common unit in oil field practices is cp (centipoise). For a gaseous fluid, the molecules are far apart and have low resistance to flow as a result of their random motion. On the other hand, a dense fluid has high resistance to flow since the molecules are close to each other. The water viscosity at standard conditions is 1 cp. At reservoir conditions (4,000–6,000 psi and 200°F), typical viscosity values of oils are given in Table 2.2. The viscosity of bitumen can be 4,500,000 cp. In general, fluid viscosity depends on pressure, temperature, and its compositions and is commonly denoted by μ.

2.3 Wettability

Wettability of a reservoir rock affects a fluid displacement process, particularly the form of relative permeability and capillary pressure functions.

Wettability. *Wettability* measures the preference of the rock surface to be wetted by a particular phase—oleic, aqueous, or some mixed (intermediate) combination. The wettability of a porous medium determines the form of the relative permeability and capillary pressure functions.

Water wet. *Water wet formation* is where water is the preferred wetting phase. Water occupies the smaller pores and forms a film over all of the rock surface, even in the pores containing oil. Water flood in such a system will be an imbibition process; water spontaneously imbibes into a core containing mobile oil at the residual oil saturation S_{or}, thus displacing the oil.

Oil wet. *Oil wet formation* is where oil is the preferred wetting phase. In the same basic principle as above, oil occupies the smaller pores and forms a film over all of the rock surface, even in the pores containing water. Water flood in such a system will be a drainage process; oil spontaneously imbibes into a core containing mobile water at the residual water saturation S_{wr}, thus displacing the water.

Intermediate wet. An *intermediate wet formation* is where some degree of both water and oil wetness is displayed by the same rock. Various types of intermediately wet systems have been known as mixed or fractionally wet. Both water and oil may spontaneously imbibe into such a system to some extent.

2.4 Fluid Displacement Processes

The choice of a simulation model depends on the fluid displacement process being modeled.

Imbibition. An *imbibition* displacement process occurs when the wetting phase increases. For example, in a water wet porous medium, imbibition will be water displacing oil.

Drainage. A *drainage* displacement process occurs when the nonwetting phase increases. For example, in a water wet porous medium, drainage will be oil displacing water. The imbibition and drainage capillary pressure and relative permeability functions are distinct because these petrophysical functions depend on the saturation history.

Spontaneous imbibition. A *spontaneous imbibition* process is when a wetting phase invades a porous medium in the absence of any external driving force. The wetting phase invades under the action of surface forces. For example, for a water wet core at irreducible water saturation S_{wr}, water may spontaneously imbibe and displace oil.

Oil recovery methods. *Oil recovery methods* include primary depletion, secondary recovery (usually water flood), and tertiary recovery (or enhanced oil recovery). A range of methods that are designed to recover additional oil that cannot be produced by primary and secondary recovery methods includes thermal methods (steam injection or in situ combustion), gas injection (N_2, CO_2, and hydrocarbon gas), chemical flooding (alkaline, surfactant, polymer, and/or foam injection), and microbial methods (using bugs to recover oil).

Process simulation models. Various types of model formulations of the flow and transport equations for multiphase, multicomponent systems are used to simulate the different recovery processes. They include the black oil, compositional, thermal, and chemical models (cf. Fig. 1.2).

2.5 Reservoir Rock/Fluid Properties

Now, we describe the rock and fluid interaction properties.

Fluid saturation. The *saturation* of a phase (water, oil, or gas) is the fraction of the pore space that it occupies. It is commonly indicated by S (fraction). For two-phase flow of water and oil, for example, the saturations S_w and S_o satisfy

$$S_w + S_o = 1,$$

which means that the two fluids jointly fill the voids. For three-phase flow of water, oil, and gas, the fact that the three fluids jointly fill the pore space gives

$$S_w + S_o + S_g = 1.$$

Multiphase flow functions such as capillary pressure and relative permeability depend strongly on saturations.

Residual saturation. The *residual saturation* of a phase is the amount of that phase (fraction of pore space) that is trapped or is irreducible. For example, after many pore volumes of water displace oil from a reservoir, the *residual oil saturation* S_{or} is reached; the corresponding *connate* (irreducible) water level is S_{wc}, and the associated residual gas saturation is S_{rg}. At the residual saturation of a phase, the corresponding relative permeability of that phase is zero. Strictly, a phase should be referred to as a wetting or nonwetting

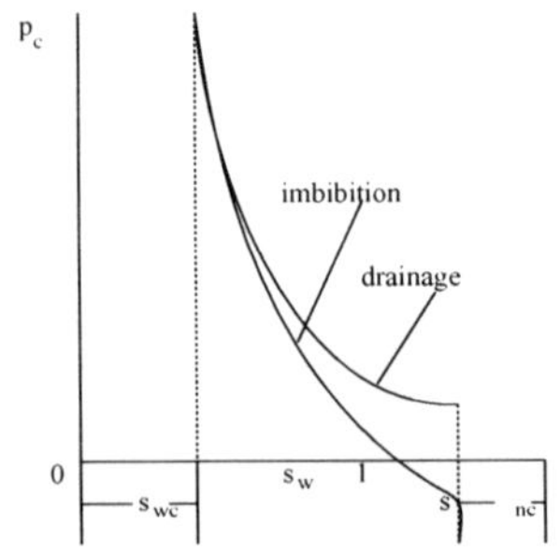

Figure 2.3. *Typical capillary pressure curve.*

phase. The residual nonwetting phase is trapped in the pores by capillary forces. In a moderately water wet sandstone, S_{or} typically ranges from 0.2 to 0.35. The amount of residual or trapped phase depends on the permeability and wettability of the rock.

Capillary pressure. In two-phase flow, a discontinuity in fluid pressure occurs across an interface between any two immiscible fluids (e.g., water and oil). This is a consequence of the interfacial tension that exists at the interface. The discontinuity between the pressure in the nonwetting phase (say, oil), p_o, and that in the wetting phase (say, water), p_w, is referred to as the *capillary pressure*, p_c:

$$p_c = p_o - p_w, \tag{2.18}$$

where the phase pressures at the interface are taken from their respective sides. A typical curve of the capillary pressure is shown in Fig. 2.3. The capillary pressure depends on the wetting phase saturation S_w and the direction of saturation change (imbibition or drainage). The phenomenon of dependence of the curve on the history of saturation is called *hysteresis*. While it is possible to develop a model that takes into account the hysteresis resulting from the saturation history (Mualem, 1976; Bedrikovetsky, Marchesin, and Ballin, 1996), in most cases the direction of flow can be predicted and only a set of capillary pressures is needed. Various curves describing a drainage or imbibition cycle can be found in Brooks and Corey (1964), van Genuchten (1980), and Corey (1986).

The value p_{cb} that is necessary to start displacement is termed a *threshold pressure* (Bear, 1972). The capillary pressure curve has an asymptote at whose value the pressure gradient remains continuous in both phases. This can be observed by considering vertical gravity equilibrium. When the value of the irreducible saturation of the nonwetting phase is approached, an analogous situation occurs at the other end of the curve during the imbibition process (Calhoun, Lewis, and Newman, 1949; Morrow, 1970).

In the discussion so far, the capillary pressure has been assumed to depend only on the saturation of the wetting phase and its history. In general, however, it also depends on the surface tension σ, porosity ϕ, permeability k, and the contact angle θ with the rock surface of the wetting phase, which, in turn, depend on the temperature and fluid compositions (Poston et al., 1970; Bear and Bachmat, 1991):

$$J(S_w) = \frac{p_c}{\sigma \cos\theta}\sqrt{\frac{k}{\phi}},$$

which is the *J-function*. If the contact angle is ignored, this function becomes

$$J = \frac{p_c}{\sigma}\sqrt{\frac{k}{\phi}}.$$

Using the J-function, typical curves for p_c can be obtained from experiments. This function is also the basis for some theoretical methods of measuring permeability k (Ashford, 1969).

For three-phase flow, two *capillary pressures* are needed:

$$p_{cow} = p_o - p_w, \qquad p_{cgo} = p_g - p_o. \tag{2.19}$$

Note that the third capillary pressure, p_{cgw}, can be found using p_{cow} and p_{cgo}:

$$p_{cgw} = p_g - p_w = p_{cow} + p_{cgo}.$$

The capillary pressures p_{cow} and p_{cgo} are usually assumed to take the forms (Leverett and Lewis, 1941)

$$p_{cow} = p_{cow}(S_w), \qquad p_{cgo} = p_{cgo}(S_g), \tag{2.20}$$

where S_w and S_g are the phase saturations of water and gas, respectively. These forms remain in wide use, though revised forms have been proposed (Shutler, 1969).

Relative permeability. *Relative permeability* is a quantity (fraction) that describes the amount of impairment to flow of one phase on another. In two-phase flow, it is a function of the phase saturation; in three-phase, it may depend on the saturation of another phase. The relative permeabilities to the water, oil, and gas phases are, respectively, denoted by k_{rw}, k_{ro}, and k_{rg}.

Mobility. The *mobility* of a phase is defined as the ratio of the relative permeability and viscosity of that phase. For example, the mobilities of the water, oil, and gas phases are $\lambda_w = k_{rw}/\mu_w$, $\lambda_o = k_{ro}/\mu_o$, and $\lambda_g = k_{rg}/\mu_g$, respectively.

Fractional flow. *Fractional flow* is a quantity (fraction) that determines the fractional volumetric flow rate of a phase under a given pressure gradient in the presence of another phase. Symbols for water and oil in a two-phase flow system are $f_w = \lambda_w/\lambda$ and $f_o = \lambda_o/\lambda$, where $\lambda = \lambda_w + \lambda_o$ is the total mobility.

2.5.1 Two-Phase Relative Permeability

Measurements on *relative permeabilities* have been made mostly for two-phase flow. Typical curves suitable for an oil-water system with water displacing oil are presented in Fig. 2.4. The value of S_w at which water starts to flow is termed the *critical saturation*, S_{wc}, and the value at which oil ceases to flow, S_{nc}, is called the *residual saturation*. Analogously, during a drainage cycle S_{nc} and S_{wc} are referred to as the critical and residual saturations, respectively.

The slopes of capillary pressure curves at irresidual saturations must be finite in numerical simulation, so these curves themselves cannot be utilized to define the saturation value at which the displaced phase becomes immobile. This saturation value is found using the residual saturation at which the relative permeability of this phase is zero. Darcy's law implies that the phase stops flowing because the mobility becomes zero (not because the

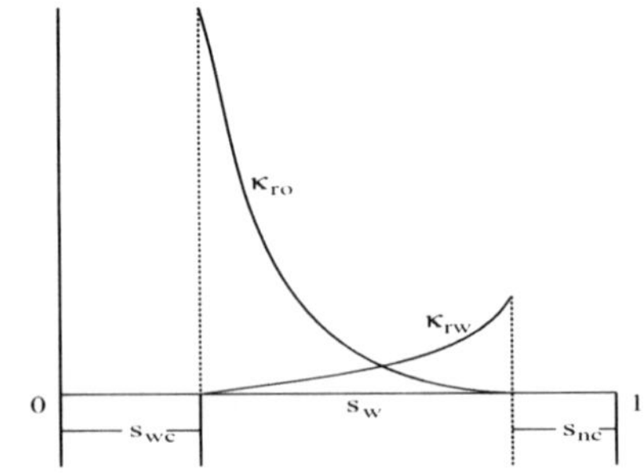

Figure 2.4. *Typical relative permeability curves.*

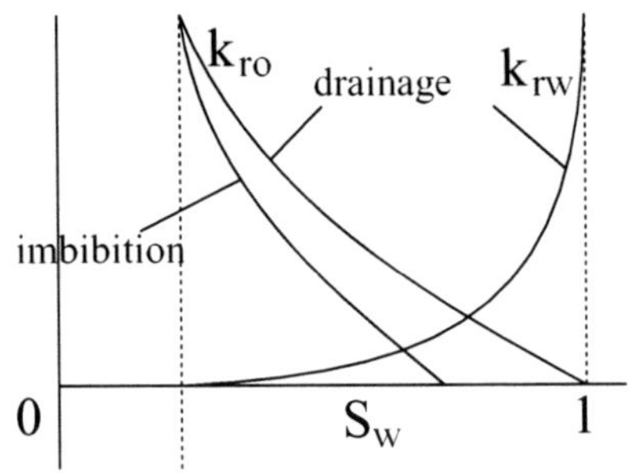

Figure 2.5. *Hysteresis in relative permeability curves.*

external force becomes zero). As a result, it is not necessary to distinguish the critical and residual saturations.

As for capillary pressures, relative permeabilities depend not only on the wetting phase saturation S_w, but also on the direction of saturation change (drainage or imbibition). Fig. 2.5 shows the phenomenon of dependence of a relative permeability for the nonwetting phase on the history of saturation. Note that the curve in imbibition is always lower than that of drainage. For the wetting phase, the relative permeability does not depend on the history of saturation.

Wettability of the rock also strongly influences relative permeabilities (Owens and Archer, 1971). Because of this, reservoir fluids should be employed for experiments instead of refined fluids.

Relative permeabilities must be determined empirically or experimentally for each particular porous medium of interest. However, the literature is rich on analytical expressions for the relationship between relative permeabilities and the saturation of the wetting phase (Corey, 1954; Naar and Henderson, 1961). These expressions were usually obtained from simplified porous media models (e.g., bundle of capillary tubes and capillary tube networks).

Corey's two-phase relative permeability model. Corey's model applies to the drainage process in a consolidated rock. The normalized wetting phase saturation is

$$S_{nw} = \frac{S_w - S_{wc}}{1 - S_{wc}},$$

and its relative permeability is given by

$$k_{rw} = S_{nw}^4. \tag{2.21}$$

The relative permeability of the nonwetting phase is

$$k_{ro} = (1 - S_{nw})^2(1 - S_{nw}^2). \tag{2.22}$$

Naar and Henderson's relative permeability model. Naar and Henderson's model is applicable to a water-oil system for the imbibition process. The water phase relative permeability is the same as (2.21),

$$k_{rw} = S_{nw}^4, \tag{2.23}$$

while the oil phase relative permeability is

$$k_{ro} = (1 - 2S_{nw})^{3/2}\left(2 - (1 - 2S_{nw})^{1/2}\right). \tag{2.24}$$

Note that $k_{ro} = 0$ for all values of $S_{nw} \geq 0.5$.

2.5.2 Three-Phase Relative Permeability

In contrast, the determination of *relative permeabilities* for three-phase flow is rather difficult. From experiments, a *ternary diagram* for the relationship between the relative permeabilities and saturations can be shown as in Fig. 2.6. This diagram is based on the level curve of the relative permeability being equal to 1% for each phase. From it we can figure out where single-, two-, or three-phase flow occurs under different combinations of saturations. In the triangular region bounded by the three level curves, for example, three fluids flow simultaneously.

Starting from Leverett and Lewis (1941), most of the measurements on three-phase relative permeabilities have been experimental. These measurements have indicated that the relative permeabilities for the wetting and nonwetting phases in a three-phase system are functions of their respective saturations as they are in a two-phase system (Corey et al., 1956; Snell, 1962):

$$k_{rw} = k_{rw}(S_w), \quad k_{rg} = k_{rg}(S_g). \tag{2.25}$$

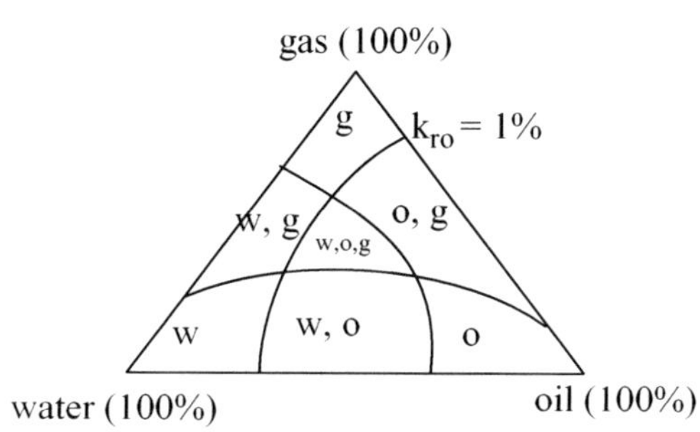

Figure 2.6. *A three-phase ternary diagram.*

The relative permeability for the intermediate wetting phase is a function of the two independent saturations:

$$k_{ro} = k_{ro}(S_w, S_g). \tag{2.26}$$

The functional form in (2.26) is rarely known. In practice, the estimation of three-phase relative permeabilities is based on two sets of two-phase data: the relative permeability in an intermediate and wetting system,

$$k_{row} = k_{row}(S_w), \tag{2.27}$$

and that in an intermediate and nonwetting system,

$$k_{rog} = k_{rog}(S_g). \tag{2.28}$$

The underlying concept is that for the wetting phase, both the intermediate and nonwetting phases act like a single nonwetting phase, while for the nonwetting phase, both the intermediate and wetting phases behave as a single wetting phase. Fig. 2.7 illustrates typical relative permeability curves for a water, oil, and gas system in an isotropic porous medium. The point where $k_{row} = 0$ indicates the maximum water saturation rather than the critical oil saturation since the oil saturation can be further reduced by increasing the gas saturation. It has been experimentally observed, however, that a nonzero residual (or minimal) oil saturation S_{or} exists when oil is displaced simultaneously by water and gas. The earlier remark on hysteresis of the relative permeability for the nonwetting phase also applies to the three-phase system.

The simplest procedure to determine k_{ro} is

$$k_{ro} = k_{row} k_{rog}. \tag{2.29}$$

Other models were suggested by Corey et al. (1956), Naar and Wygal (1961), Stone (1970, 1973), and Delshad and Pope (1989). As an example, we describe Naar and Wygal's model and two of Stone's models, model I and model II.

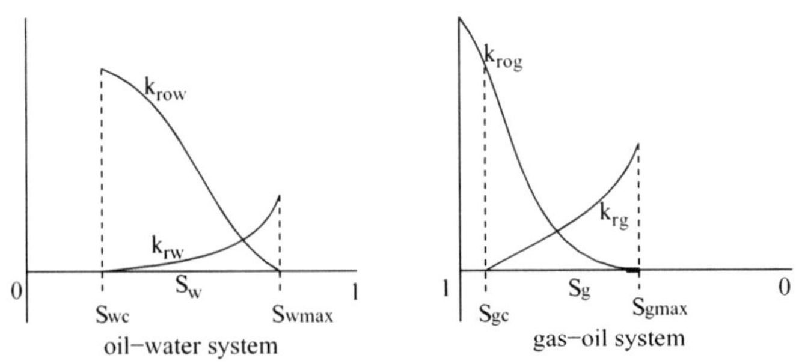

Figure 2.7. *Relative permeability curves in a three-phase system.*

Naar and Wygal's model. The relative permeabilities of the water, oil, and gas phases are

$$\begin{aligned} k_{rw} &= \left(\frac{S_w - S_{wc}}{1 - S_{wc}}\right)^4, \\ k_{ro} &= \frac{S_o^3(1 - S_g + 2S_w - 3S_{wc})}{(1 - S_{wc})^4}, \\ k_{rg} &= \frac{S_g^3(2 - S_g - 2S_{wc})}{(1 - S_{wc})^4}. \end{aligned} \tag{2.30}$$

This model does not have the effect of S_{or} and S_{gc}, so the following constraints must be imposed on the second and third equations in (2.30):

$$\begin{aligned} k_{ro} &= 0 \qquad \text{if } S_o \leq S_{or}, \\ k_{rg} &= 0 \qquad \text{if } S_g \leq S_{gc}. \end{aligned} \tag{2.31}$$

Stone's model I. The saturations are normalized as follows:

$$\begin{aligned} S_{no} &= \frac{S_o - S_{or}}{1 - S_{wc} - S_{or}}, \qquad S_o \geq S_{or}, \\ S_{nw} &= \frac{S_w - S_{wc}}{1 - S_{wc} - S_{or}}, \qquad S_w \geq S_{wc}, \\ S_{ng} &= \frac{S_g}{1 - S_{wc} - S_{or}}. \end{aligned}$$

Note that

$$S_{no} + S_{nw} + S_{ng} = 1.$$

The relative permeability of oil is defined by

$$k_{ro} = S_{no}\beta_w\beta_g. \tag{2.32}$$

To determine β_w, we take $S_g = S_{ng} = 0$; i.e., the three-phase system reduces to a water and oil system. In this case, $\beta_g = 1$ and $k_{ro} = k_{row}$, which, together with equation (2.32), gives

$$\beta_w = \frac{k_{row}(S_w)}{1 - S_{nw}}. \tag{2.33}$$

Similarly, to determine β_g, we choose $S_w = S_{wc}$ so that $\beta_w = 1$ and $k_{ro} = k_{rog}$. Then using equation (2.32) yields

$$\beta_g = \frac{k_{rog}(S_g)}{1 - S_{ng}}. \tag{2.34}$$

Substituting (2.33) and (2.34) into (2.32) gives the expression of k_{ro} for *Stone's model I*.

This model reduces exactly to two-phase data only if the following condition is satisfied:

$$k_{row}(S_{wc}) = k_{rog}(S_g = 0) = 1. \tag{2.35}$$

Otherwise, the relative permeability $k_{ro}(S_w, S_g)$ provides only an approximation of the two-phase data. A model that does not have this limitation can be obtained if the oil-gas data are measured in the presence of irreducible water. In this case, a water-oil system at S_{wc} and a gas-oil system at $S_g = 0$ are physically identical; i.e., both systems satisfy $S_w = S_{wc}$ and $S_o = 1 - S_{wc}$. Hence equation (2.35) is equivalent to the definition of the absolute permeability being the effective permeability of oil in the presence of S_{wc}.

Set

$$k_{row}(S_{wc}) = k_{rog}(S_g = 0) = k_{rc}.$$

Then Stone's model I can be modified as follows:

$$k_{ro} = k_{rc} S_{no} \beta_w \beta_g, \tag{2.36}$$

where

$$\beta_w = \frac{k_{row}(S_w)}{(1 - S_{nw})k_{rc}}, \qquad \beta_g = \frac{k_{rog}(1 - S_g)}{(1 - S_{ng})k_{rc}}.$$

Stone's model II. In the definition of Stone's model I, S_{or} must be specified. In fact, this value can be predicted from an equation derived from *channel-flow* considerations:

$$k_{ro} = (k_{row} + k_{rw})(k_{rog} + k_{rg}) - (k_{rw} + k_{rg}), \tag{2.37}$$

where $k_{ro} \geq 0$ is required (i.e., negative values of k_{ro} mean immobile oil). As for Stone's model I, to satisfy equation (2.35), *model II* can be altered as follows:

$$k_{ro} = k_{rc} \left\{ (k_{row}/k_{rc} + k_{rw})(k_{rog}/k_{rc} + k_{rg}) - (k_{rw} + k_{rg}) \right\}. \tag{2.38}$$

2.6 Terms Used in Numerical Simulation

Finally, we briefly review some of the terminologies used in numerical reservoir simulation. More will be given in each of the subsequent chapters as needed.

Numerical method. A *numerical method* for solving a differential equation problem involves discretizing this problem, which has infinitely many degrees of freedom, to produce a *discrete problem*, which has finitely many degrees of freedom and can be solved using a computer. For a collection of numerical methods including finite difference, finite volume, and finite element methods, the reader should refer to the books by Chen (2005) and Chen, Huan, and Ma (2006).

Grid structure. A *grid structure* is the geometry of a computational grid that is used for the numerical simulation of a reservoir. This grid can be Cartesian, radial, or distorted and 1D, 2D, or 3D.

2D areal grid. A *2D areal grid* is a 2D grid structure that is imposed from looking down onto the reservoir. For a Cartesian coordinate system, it is a division of the reservoir in the x_1- and x_2-directions using spatial steps h_1 and h_2 (cf. Fig. 2.8).

2D cross-sectional model. A *2D cross-sectional model* is a 2D grid structure that is imposed on a vertical slice down through the reservoir. For a Cartesian system, it is a division of the reservoir in the x_1- and x_3-directions using spatial steps h_1 and h_3 (cf. Fig. 2.9). Cross-sectional models are used to asses the effect of vertical stratification in the reservoir and to yield pseudofunctions in upscaling.

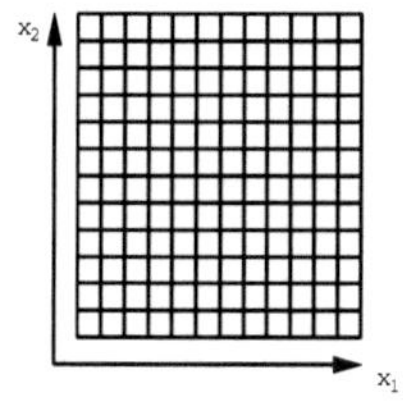

Figure 2.8. *2D areal grid.*

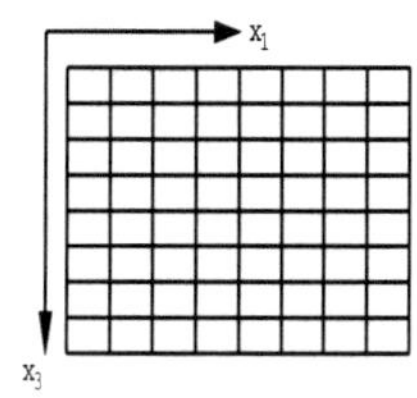

Figure 2.9. *2D cross-sectional grid.*

Transmissibility. The *transmissibility* between two adjacent gridblocks measures how easily fluids flow between them. For example, for two-phase flow, the transmissibility at the interface of two blocks for water is

$$T_w = \left(\frac{kA}{h}\right)_{\text{av}} \left(\frac{k_{rw}}{\mu_w B_w}\right)_{\text{av}}, \tag{2.39}$$

where A is the cross-sectional area of the interface. This quantity consists of two parts, each of which is an average between the blocks: the single-phase part $(kA/h)_{\text{av}}$ and the two-phase part $(k_{rw}/(\mu_w B_w))_{\text{av}}$. The single-phase part average will be a *harmonic average* between blocks (cf. Chapter 3). The two-phase part average is more complex. An *upstream weighting* will be used for the averaged relative permeability, and an *arithmetic average* between blocks will be used for the viscosity and volume formation factor (cf. Chapter 5).

Spatial discretization. *Spatial discretization* refers to a process of dividing the reservoir domain in space into small subdomains with spatial steps h_1, h_2, and h_3 and then modeling the flow by a numerical method. In numerical reservoir simulation, one always divides the reservoir into gridblocks and then models the flow between blocks.

Temporal discretization. *Temporal discretization* refers to a process of dividing a time interval of interest into subintervals with temporal step Δt and advancing the simulation in time.

Grid orientation. A grid orientation effect exists when fluids flow both oriented with the principal grid direction and diagonally across the grid. Simulation results differ for each of the fluid paths through this grid. This problem stems from the use of a five-point (respectively, seven-point) difference scheme in a 2D (respectively, 3D) spatial discretization. It may be alleviated by using more accurate discretization schemes such as a 2D nine-point scheme (respectively, a 3D 27-point scheme) or other numerical methods (Chen, Huan, and Ma, 2006).

Numerical dispersion. *Numerical dispersion* is the spreading of a flood front in a displacement process such as a water flood. This phenomenon is due to numerical effects. Specifically, it comes from the spatial and temporal discretization or a truncation error that arises from gridding. This front spreading tends to lead to early water breakthrough and other errors in recovery. How bad the error is depends on the actual fluid recovery process being simulated (e.g., water flood and water-alternating-gas flood), spatial and temporal steps, and numerical methods used.

Mass conservation. *Mass conservation* is a general principle used in checking the accuracy of a numerical method in reservoir simulation. It is simply stated as follows:

$$\begin{aligned}&(\text{mass into a block}) - (\text{mass out of the black})\\&= \text{mass accumulation within the block.}\end{aligned}$$

Reservoir simulation models are basically composed of mass conservation and Darcy's law relating a fluid velocity to a pressure (or potential) gradient (cf. Chapter 3). In thermal methods, energy conservation is added. *Material balance* is an engineering terminology for mass conservation over a fixed volume, which is usually the hydrocarbon reservoir.

Chapter 3

Single-Phase Flow and Numerical Solution

The basic differential equations that govern the flow of a single phase through a reservoir are described. They include a mass conservation equation, Darcy's law, and an equation of state relating the fluid pressure to its density. The cases of incompressible, slightly compressible, and compressible fluids are considered. Then an analytic solution for a 1D radial flow is obtained, which is usually used to estimate the pressure of a wellbore. As an example, a numerical solution of single-phase flow equations using finite difference methods is presented, for which these methods for transient (parabolic), stationary (elliptic), and wave (hyperbolic) problems are reviewed. For applications of other numerical methods such as finite volume and finite element methods to the numerical solution of the flow equations, the reader may refer to the books by Chen (2005) and Chen, Huan, and Ma (2006). Much attention in this chapter is paid to the treatment of practical issues in reservoir simulation, such as gridblock transmissibility, material balance, and the treatment of nonlinearity. The solution of linear systems of algebraic equations will not be discussed here. For more information on linear solvers, the reader should consult with reservoir simulation books by Aziz and Settari (1979), Ertekin, Abou-Kassem, and King (2001), and Chen, Huan, and Ma (2006). In particular, this last book contains all advanced solvers to date such as Krylov subspace linear solvers.

3.1 Basic Differential Equations

3.1.1 Mass Conservation

We briefly derive the governing differential equations for the flow and transport of a fluid in a porous medium in order to introduce the terminologies and notation used throughout this book.

The spatial and temporal variables will be represented by $\mathbf{x} = (x_1, x_2, x_3)$ and t, respectively. Denote by ϕ the *porosity* of the porous medium, by ρ the density of the fluid per unit volume, by $\mathbf{u} = (u_1, u_2, u_3)$ the superficial *Darcy velocity*, and by q the external sources and sinks. Consider a rectangular cube such that its faces are parallel to the coordinate axes (cf. Fig. 3.1). The centroid of this cube is denoted (x_1, x_2, x_3), and its

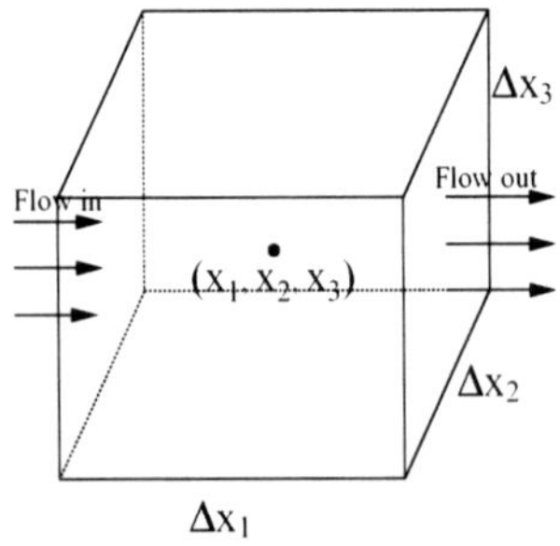

Figure 3.1. *A differential volume.*

length in the x_i-coordinate direction is Δx_i, $i = 1, 2, 3$. The x_i-component of the *mass flux* (mass flow per unit area per unit time) of the fluid is ρu_i. Referring to Fig. 3.1, the mass inflow across the surface at $x_1 - \frac{\Delta x_1}{2}$ per unit time is

$$(\rho u_1)_{x_1 - \frac{\Delta x_1}{2}, x_2, x_3} \Delta x_2 \Delta x_3,$$

and the mass outflow at $x_1 + \frac{\Delta x_1}{2}$ is

$$(\rho u_1)_{x_1 + \frac{\Delta x_1}{2}, x_2, x_3} \Delta x_2 \Delta x_3.$$

Similarly, in the x_2- and x_3-coordinate directions, the mass inflows and outflows across the surfaces are, respectively,

$$(\rho u_2)_{x_1, x_2 - \frac{\Delta x_2}{2}, x_3} \Delta x_1 \Delta x_3, \quad (\rho u_2)_{x_1, x_2 + \frac{\Delta x_2}{2}, x_3} \Delta x_1 \Delta x_3,$$

and

$$(\rho u_3)_{x_1, x_2, x_3 - \frac{\Delta x_3}{2}} \Delta x_1 \Delta x_2, \quad (\rho u_3)_{x_1, x_2, x_3 + \frac{\Delta x_3}{2}} \Delta x_1 \Delta x_2.$$

With $\partial/\partial t$ being the time differentiation, *mass accumulation* due to compressibility per unit time is

$$\frac{\partial(\phi\rho)}{\partial t} \Delta x_1 \Delta x_2 \Delta x_3,$$

and the removal of mass from the cube, i.e., the mass decrement (accumulation) due to a sink of strength q (mass per unit volume per unit time), is

$$-q \Delta x_1 \Delta x_2 \Delta x_3.$$

The difference between the mass inflow and outflow equals the sum of mass accumulation within this volume:

$$\begin{aligned}
&\left[(\rho u_1)_{x_1 - \frac{\Delta x_1}{2}, x_2, x_3} - (\rho u_1)_{x_1 + \frac{\Delta x_1}{2}, x_2, x_3}\right] \Delta x_2 \Delta x_3 \\
&+ \left[(\rho u_2)_{x_1, x_2 - \frac{\Delta x_2}{2}, x_3} - (\rho u_2)_{x_1, x_2 + \frac{\Delta x_2}{2}, x_3}\right] \Delta x_1 \Delta x_3 \\
&+ \left[(\rho u_3)_{x_1, x_2, x_3 - \frac{\Delta x_3}{2}} - (\rho u_3)_{x_1, x_2, x_3 + \frac{\Delta x_3}{2}}\right] \Delta x_1 \Delta x_2 \\
&= \left(\frac{\partial(\phi\rho)}{\partial t} - q\right) \Delta x_1 \Delta x_2 \Delta x_3.
\end{aligned}$$

Divide this equation by $\Delta x_1 \Delta x_2 \Delta x_3$ to see that

$$-\frac{(\rho u_1)_{x_1+\frac{\Delta x_1}{2},x_2,x_3} - (\rho u_1)_{x_1-\frac{\Delta x_1}{2},x_2,x_3}}{\Delta x_1}$$
$$-\frac{(\rho u_2)_{x_1,x_2+\frac{\Delta x_2}{2},x_3} - (\rho u_2)_{x_1,x_2-\frac{\Delta x_2}{2},x_3}}{\Delta x_2}$$
$$-\frac{(\rho u_3)_{x_1,x_2,x_3+\frac{\Delta x_3}{2}} - (\rho u_3)_{x_1,x_2,x_3-\frac{\Delta x_3}{2}}}{\Delta x_3} = \frac{\partial(\phi\rho)}{\partial t} - q.$$

Letting $\Delta x_i \to 0$, $i = 1, 2, 3$, we obtain the *mass conservation equation*

$$\frac{\partial(\phi\rho)}{\partial t} = -\nabla \cdot (\rho \mathbf{u}) + q, \tag{3.1}$$

where $\nabla\cdot$ is the *divergence* operator:

$$\nabla \cdot \mathbf{u} = \frac{\partial u_1}{\partial x_1} + \frac{\partial u_2}{\partial x_2} + \frac{\partial u_3}{\partial x_3}.$$

Note that q is negative for sinks and positive for sources.

The *formation volume factor* B of the fluid is

$$\rho = \frac{\rho_s}{B},$$

where ρ_s is the fluid density at standard conditions. Substituting ρ into (3.1), we have

$$\frac{\partial}{\partial t}\left(\frac{\phi}{B}\right) = -\nabla \cdot \left(\frac{1}{B}\mathbf{u}\right) + \frac{q}{\rho_s}. \tag{3.2}$$

Equations (3.1) and (3.2) are equivalent; they are the most general forms of the single-phase equation.

3.1.2 Darcy's Law

Darcy's law (Darcy, 1856) was originally a law for single-phase flow that relates the total volumetric flow rate of a fluid through a porous medium to the pressure gradient and the properties of the fluid (viscosity, μ) and the medium (permeability, k, and a cross-sectional area, A). It can be used to define the permeability in one of the flow directions, for example, in the x_1-direction (cf. Fig. 3.2):

$$q = -\frac{kA}{\mu}\frac{\partial p}{\partial x_1}.$$

Darcy's velocity is calculated by $u = q/A$, so

$$u = -\frac{k}{\mu}\frac{\partial p}{\partial x_1}. \tag{3.3}$$

Note that the pore velocity v is the fluid velocity: $v = u/\phi$.

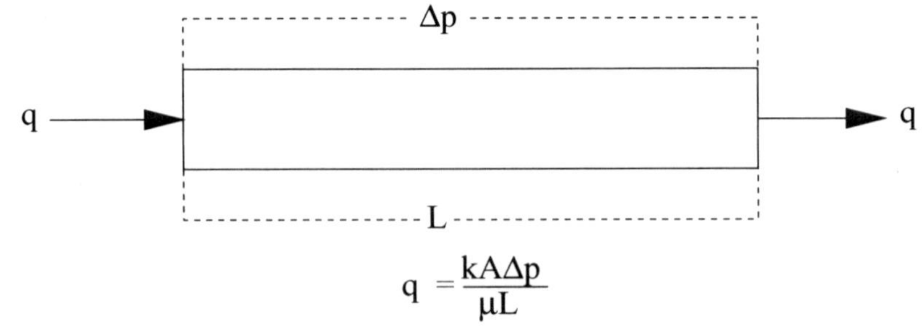

Figure 3.2. *Volumetric flow rate.*

For a 3D flow system with the gravitational force, the differential form of Darcy's law is

$$\mathbf{u} = -\frac{1}{\mu}\mathbf{k}\left(\nabla p - \rho\wp\nabla z\right), \tag{3.4}$$

where $\mathbf{k}$ is the *absolute permeability tensor* of the porous medium, $\wp$ is the magnitude of the gravitational acceleration, z is the depth, and ∇ is the *gradient* operator:

$$\nabla p = \left(\frac{\partial p}{\partial x_1}, \frac{\partial p}{\partial x_2}, \frac{\partial p}{\partial x_3}\right).$$

The x_3-coordinate in equation (3.4) is in the vertical downward direction.

3.1.3 Units

In this book the basic units are customary (English) and metric. Their conversion factors are stated in the preceding chapter. Here we describe these units for the variables used in single-phase flow; multiphase flow uses similar variables and units. One multiplies a customary unit by a conversion factor to obtain the corresponding metric unit. STB (standard barrel) and SCF (standard cubic feet) are measured at 60°F and 14.696 psia, while std m^3 (at standard conditions) is measured at 15°C and 100 kPa. In Table 3.1, FVF stands for formation volume factor.

3.1.4 Different Forms of Flow Equations

Substituting (3.4) into (3.1) yields

$$\frac{\partial(\phi\rho)}{\partial t} = \nabla\cdot\left(\frac{\rho}{\mu}\mathbf{k}\left(\nabla p - \rho\wp\nabla z\right)\right) + q. \tag{3.5}$$

An equation of state is expressed in terms of the *fluid compressibility* c_f:

$$c_f = -\frac{1}{V}\frac{\partial V}{\partial p}\bigg|_T = \frac{1}{\rho}\frac{\partial \rho}{\partial p}\bigg|_T \tag{3.6}$$

Table 3.1. *Customary and metric units.*

Quantity	Symbol	Customary	Metric	Conversion factor
Time	t	day	day	1.0
Angle	θ	rad	rad	1.0
Length	x_i, z	ft	m	0.3048
Area	A	ft^2	m^2	0.09290304
Volume	V	ft^3	m^2	0.02831685
Porosity	ϕ	fraction	fraction	1.0
Permeability	$\mathbf{k}$	darcy	μm^2	0.9869233
Density	ρ	lbm/ft^3	kg/m^3	16.01846
Gravitational acceleration	$\wp$	$32.174 ft/s^2$	$9.8066352\ m/s^2$	0.3048
Fluid gravity	γ	psi/ft	kPa/m	22.62059
Pressure	p	psia	kPa	6.894757
Velocity	$\mathbf{u}$	ft/D	m/d	0.3048
Viscosity	μ	cp	Pa·s	0.001
Compressibility	c	psi^{-1}	kPa^{-1}	0.1450377
Compressibility factor	Z	dimensionless	dimensionless	1.0
Flow rate	q	$lbm/(ft^3 \cdot d)$	$kg/(m^3 \cdot d)$	16.018653
Liquid FVF	B_w, B_o	RB/STB	$m^3/std\ m^3$	1.0
Gas FVF	B_g	RB/SCF	$m^3/std\ m^3$	5.5519314
Solution gas/oil ratio	R_{so}	SCF/STB	$std\ m^3/std\ m^3$	0.1801175
Gravity conversion factor	γ_c	0.21584E-3	1.0E-3	

at a fixed temperature T, where V stands for the volume occupied by the fluid at reservoir conditions. Combining equations (3.5) and (3.6) gives a closed system for the main unknown p or ρ. Simplified expressions such as a linear relationship between p and ρ for a *slightly compressible fluid* can be used.

It is sometimes convenient in mathematical analysis to write equation (3.5) in a form without the explicit appearance of gravity by the introduction of a *pseudopotential* (Hubbert, 1956):

$$\Phi' = \int_{p^o}^{p} \frac{1}{\rho(\xi)\wp} d\xi - z, \tag{3.7}$$

where p^o is a reference pressure. Using (3.7), equation (3.5) reduces to

$$\frac{\partial(\phi\rho)}{\partial t} = \nabla \cdot \left(\frac{\rho^2 \wp}{\mu} \mathbf{k} \nabla \Phi' \right) + q. \tag{3.8}$$

In numerical computations, we often use the usual *potential* (piezometric head)

$$\Phi = p - \rho\wp z,$$

which is related to Φ' (with, e.g., $p^o = 0$ and constant ρ) by

$$\Phi = \rho \wp \Phi'.$$

If we neglect the term $\wp z \nabla \rho$, in terms of Φ equation (3.5) becomes

$$\frac{\partial(\phi\rho)}{\partial t} = \nabla \cdot \left(\frac{\rho}{\mu} \mathbf{k} \nabla \Phi \right) + q. \tag{3.9}$$

Incompressible flow

When the rock and fluid are incompressible, the density ρ and porosity ϕ are assumed to be constant. In this case, equation (3.9) reduces to

$$\nabla \cdot \left(\frac{\rho}{\mu} \mathbf{k} \nabla \Phi \right) + q = 0, \tag{3.10}$$

which is an *elliptic equation* in Φ (cf. Section 5.1.3). For the flow of an incompressible fluid in a homogeneous and isotropic medium with a constant viscosity, equation (3.10) further becomes

$$\Delta \Phi = -\frac{\mu q}{\rho k}, \tag{3.11}$$

where the *Laplacian operator* Δ is defined by

$$\Delta \Phi = \frac{\partial^2 \Phi}{\partial x_1^2} + \frac{\partial^2 \Phi}{\partial x_2^2} + \frac{\partial^2 \Phi}{\partial x_3^2}.$$

Equation (3.11) is the *Poisson equation* in Φ. If there is no external source/sink term (well), it is called the *Laplace equation*.

Slightly compressible flow

It is sometimes possible to assume that the fluid compressibility c_f is constant over a certain range of pressures. Then, after integration, we write (3.6) as

$$\rho = \rho^o e^{c_f(p-p^o)}, \tag{3.12}$$

where ρ^o is the density at the reference pressure p^o. Using a Taylor series expansion, we see that

$$\rho = \rho^o \left\{ 1 + c_f(p - p^o) + \frac{1}{2!} c_f^2 (p - p^o)^2 + \cdots \right\},$$

so an approximation results:

$$\rho \approx \rho^o \big(1 + c_f(p - p^o)\big). \tag{3.13}$$

The *rock compressibility* is defined by

$$c_R = \frac{1}{\phi} \frac{d\phi}{dp}. \tag{3.14}$$

After integration, it is given by

$$\phi = \phi^o e^{c_R(p-p^o)}, \tag{3.15}$$

where ϕ^o is the porosity at p^o. Similarly, it is approximated by

$$\phi \approx \phi^o\big(1 + c_R(p - p^o)\big). \tag{3.16}$$

Then it follows that

$$\frac{d\phi}{dp} = \phi^o c_R. \tag{3.17}$$

After carrying out the time differentiation on the left-hand side of equation (3.5), this equation becomes

$$\left(\phi\frac{\partial\rho}{\partial p} + \rho\frac{d\phi}{dp}\right)\frac{\partial p}{\partial t} = \nabla\cdot\left(\frac{\rho}{\mu}\mathbf{k}\left(\nabla p - \rho\wp\nabla z\right)\right) + q. \tag{3.18}$$

Substituting (3.6) and (3.17) into equation (3.18) gives

$$\rho\left(\phi c_f + \phi^o c_R\right)\frac{\partial p}{\partial t} = \nabla\cdot\left(\frac{\rho}{\mu}\mathbf{k}\left(\nabla p - \rho\wp\nabla z\right)\right) + q.$$

Defining the *total compressibility*

$$c_t = c_f + \frac{\phi^o}{\phi}c_R, \tag{3.19}$$

we see that

$$\phi\rho c_t\frac{\partial p}{\partial t} = \nabla\cdot\left(\frac{\rho}{\mu}\mathbf{k}\left(\nabla p - \rho\wp\nabla z\right)\right) + q, \tag{3.20}$$

which is a *parabolic equation* in p (cf. Section 5.1.3), with ρ given by (3.12).

Compressible flow

For gas flow, the compressibility c_g of gas is usually not assumed to be constant. In such a case, the general equation (3.18) applies; i.e.,

$$c(p)\frac{\partial p}{\partial t} = \nabla\cdot\left(\frac{\rho}{\mu}\mathbf{k}\left(\nabla p - \rho\wp\nabla z\right)\right) + q, \tag{3.21}$$

where

$$c(p) = \phi\frac{\partial\rho}{\partial p} + \rho\frac{d\phi}{dp}. \tag{3.22}$$

A different form of equation (3.21) can be derived if we use the real *gas law* (the pressure-volume-temperature (PVT) relation)

$$\rho = \frac{pW}{ZRT}, \tag{3.23}$$

where W is the molecular weight, Z is the gas compressibility factor, and R is the *universal gas constant*. For a pure gas reservoir, the gravitational constant is usually small and neglected. We assume that the porous medium is isotropic; i.e., $\mathbf{k} = k\mathbf{I}$, where $\mathbf{I}$ is the identity tensor. Furthermore, we assume that ϕ and μ are constants. Then, substituting (3.23) into (3.5), we see that

$$\frac{\phi}{k}\frac{\partial}{\partial t}\left(\frac{p}{Z}\right) = \nabla\cdot\left(\frac{p}{\mu Z}\nabla p\right) + \frac{RT}{Wk}q. \tag{3.24}$$

Note that $2p\nabla p = \nabla p^2$, so (3.24) becomes

$$\frac{2\phi\mu Z}{k}\frac{\partial}{\partial t}\left(\frac{p}{Z}\right) = \Delta p^2 + 2pZ\frac{d}{dp}\left(\frac{1}{Z}\right)|\nabla p|^2 + \frac{2\mu ZRT}{Wk}q. \tag{3.25}$$

Because

$$c_g = \frac{1}{\rho}\frac{d\rho}{dp}\bigg|_T = \frac{1}{p} - \frac{1}{Z}\frac{dZ}{dp},$$

we have

$$\frac{\partial}{\partial t}\left(\frac{p}{Z}\right) = \frac{pc_g}{Z}\frac{\partial p}{\partial t}.$$

Inserting this equation into (3.25) and neglecting the term involving $|\nabla p|^2$ (often smaller than other terms in (3.25)), we obtain

$$\frac{\phi\mu c_g}{k}\frac{\partial p^2}{\partial t} = \Delta p^2 + \frac{2ZRT\mu}{Wk}q, \tag{3.26}$$

which is a parabolic equation in p^2.

There is another way to derive an equation similar to (3.26). Define a *pseudopressure* by

$$\psi = 2\int_{p^o}^{p}\frac{p}{Z\mu}\,dp.$$

Note that

$$\nabla\psi = \frac{2p}{Z\mu}\nabla p, \qquad \frac{\partial\psi}{\partial t} = \frac{2p}{Z\mu}\frac{\partial p}{\partial t}.$$

Equation (3.24) becomes

$$\frac{\phi\mu c_g}{k}\frac{\partial\psi}{\partial t} = \Delta\psi + \frac{2RT}{Wk}q. \tag{3.27}$$

The derivation of (3.27) does not require us to neglect the second term on the right-hand side of (3.25).

Boundary and initial conditions

The mathematical model described so far for single-phase flow is not complete unless necessary *boundary* and *initial conditions* are specified. Below we present boundary conditions of three kinds that are relevant to equation (3.5). We denote by Γ the external boundary or a boundary segment of the porous medium domain Ω under consideration.

Prescribed pressure. When the pressure is specified as a known function of position and time on Γ, the boundary condition is

$$p = g_1 \quad \text{on } \Gamma.$$

In the theory of partial differential equations, such a condition is termed a boundary condition of the *first kind*, or a *Dirichlet boundary condition.*

Prescribed mass flux. When the total mass flux is known on Γ, the boundary condition is

$$\rho \mathbf{u} \cdot \boldsymbol{\nu} = g_2 \quad \text{on } \Gamma,$$

where $\boldsymbol{\nu}$ indicates the outward unit normal to Γ. This condition is called a boundary condition of the *second kind*, or a *Neumann boundary condition.* For an *impervious boundary*, $g_2 = 0$ (i.e., a no-flow boundary condition).

Mixed boundary condition. A boundary condition of *mixed kind* (or *third kind*) takes the form

$$g_p p + g_u \rho \mathbf{u} \cdot \boldsymbol{\nu} = g_3 \quad \text{on } \Gamma,$$

where g_p, g_u, and g_3 are given functions. This condition is referred to as a *Robin* or *Dankwerts boundary condition.* Such a condition occurs when Γ is a semipervious boundary.

Finally, the initial condition can be defined in terms of p:

$$p(\mathbf{x}, 0) = p^0(\mathbf{x}), \qquad \mathbf{x} \in \Omega.$$

In general, in reservoir simulation, an initial pressure is given only at a *datum level depth.* The pressure at other locations is determined by the *gravity equilibrium condition* (cf. Section 3.4.1).

3.2 An Analytic Solution

In this section, we obtain an *analytic solution* for equation (3.20) that can be used to check the approximation accuracy for a numerical method for fluid flow in porous media and find the pressure near a wellbore. We assume that Ω is an isotropic medium (cf. Section 2.1), so $\mathbf{k} = k\mathbf{I}$, where $\mathbf{I}$ is the identity tensor. In cylindrical coordinates (r, θ, x_3), equation (3.20) takes the form

$$\begin{aligned} \phi \rho c_t \frac{\partial p}{\partial t} = & \frac{1}{r}\frac{\partial}{\partial r}\left[\frac{r\rho k}{\mu}\left(\frac{\partial p}{\partial r} - \rho \wp \frac{\partial z}{\partial r}\right)\right] \\ & + \frac{1}{r^2}\frac{\partial}{\partial \theta}\left[\frac{\rho k}{\mu}\left(\frac{\partial p}{\partial \theta} - \rho \wp \frac{\partial z}{\partial \theta}\right)\right] \\ & + \frac{\partial}{\partial x_3}\left[\frac{\rho k}{\mu}\left(\frac{\partial p}{\partial x_3} - \rho \wp \frac{\partial z}{\partial x_3}\right)\right]. \end{aligned} \tag{3.28}$$

We consider a reservoir Ω with an infinite extent in the horizontal direction. Assume that there is an isolated production well (located at $(0, 0, x_3)$) in this reservoir, all its properties are symmetric with respect to the axis of this well, and the reservoir is homogeneous in the vertical direction (cf. Fig. 3.3). In addition, if the gravity effect and density change are ignored, equation (3.28) reduces to

$$\frac{1}{\chi}\frac{\partial p}{\partial t} = \frac{\partial^2 p}{\partial r^2} + \frac{1}{r}\frac{\partial p}{\partial r}, \tag{3.29}$$

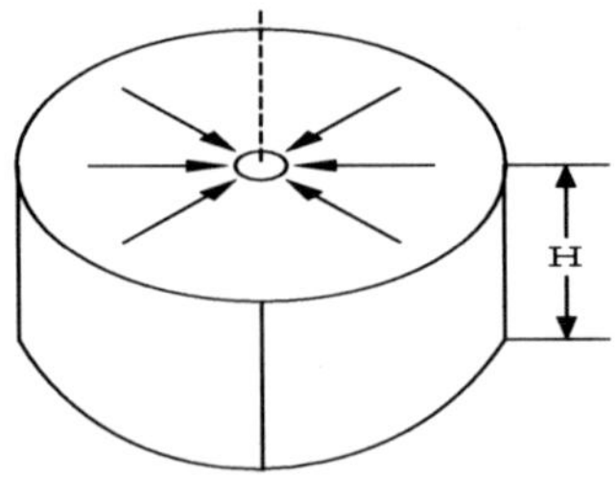

Figure 3.3. *1D radial flow.*

where

$$\chi = \frac{k}{\phi \mu c_t}.$$

Thus pressure p is a function of r and t only. That is, the flow is 1D in the radial direction. We find an analytic solution to this 1D equation. Initially, we assume that

$$p(r, 0) = p^0, \qquad 0 \leq r < \infty, \tag{3.30}$$

where p^0 is constant. The boundary conditions are given by

$$\begin{aligned} p(r, t) &= p^0 && \text{as } r \to \infty,\ t \geq 0, \\ r\frac{\partial p}{\partial r} &= \frac{Q\mu}{2\pi k H} && \text{as } r \to 0, \quad t > 0, \end{aligned} \tag{3.31}$$

where Q is a fixed production rate of the well and H is the thickness of the reservoir.

To solve (3.29), we introduce the *Boltzmann change of variable*

$$y = \frac{r^2}{4t\chi}, \qquad t > 0.$$

Then we see that

$$\begin{aligned} \frac{\partial p}{\partial r} &= \frac{dp}{dy}\frac{\partial y}{\partial r} = \frac{dp}{dy}\frac{r}{2t\chi}, \\ \frac{\partial^2 p}{\partial r^2} &= \frac{\partial}{\partial r}\left(\frac{dp}{dy}\frac{r}{2t\chi}\right) = \frac{d^2 p}{dy^2}\left(\frac{r}{2t\chi}\right)^2 + \frac{dp}{dy}\frac{1}{2t\chi}, \\ \frac{\partial p}{\partial t} &= \frac{dp}{dy}\frac{\partial y}{\partial t} = -\frac{dp}{dy}\frac{r^2}{4t^2\chi}. \end{aligned} \tag{3.32}$$

Substituting (3.32) into (3.29) yields

$$y\frac{d^2 p}{dy^2} + (1 + y)\frac{dp}{dy} = 0. \tag{3.33}$$

Using the *method of separation of variables*, from (3.33) we obtain

$$\frac{dp}{dy} = \frac{C}{y}e^{-y}, \tag{3.34}$$

where C is an arbitrary constant. Applying the boundary condition (3.31) to (3.34) gives

$$\frac{dp}{dy} = \frac{Q\mu}{4\pi kH}\frac{e^{-y}}{y}. \tag{3.35}$$

Note that

$$p = p^0 \qquad \text{when } y = \infty, \quad t = 0,$$

$$p = p(r, t) \qquad \text{when } y = \frac{r^2}{4t\chi}, \quad t > 0.$$

Integration of (3.35) from $t = 0$ to any t implies

$$p(r, t) = p^0 - \frac{Q\mu}{4\pi kH}\int_{r^2/(4t\chi)}^{\infty} \frac{e^{-y}}{y}\, dy. \tag{3.36}$$

The function $\int_{r^2/(4t\chi)}^{\infty} \frac{e^{-y}}{y}\, dy$ is the *exponential integral function* and is usually written as

$$\int_{r^2/(4t\chi)}^{\infty} \frac{e^{-y}}{y}\, dy = -Ei\left(-\frac{r^2}{4t\chi}\right) = -Ei(-y).$$

Consequently, it follows from (3.36) that pressure at any r is

$$p(r, t) = p^0 + \frac{Q\mu}{4\pi kH} Ei\left(-\frac{r^2}{4t\chi}\right), \qquad t > 0. \tag{3.37}$$

The graph of $-Ei(-y)$ in terms of y is displayed in Fig. 3.4, which shows that as y increases (r increases or t decreases), $-Ei(-y)$ decreases, so $p(r, t)$ increases and $p^0 - p$ decreases. That is, the farther we are from the well, the larger the pressure but the smaller the pressure drop. The same phenomenon can be observed as t decreases.

If the well starts to operate at $t = t_0$ instead of $t = 0$, the pressure becomes

$$p(r, t) = p^0 + \frac{Q\mu}{4\pi kH} Ei\left(-\frac{r^2}{4(t - t_0)\chi}\right), \qquad t > t_0. \tag{3.38}$$

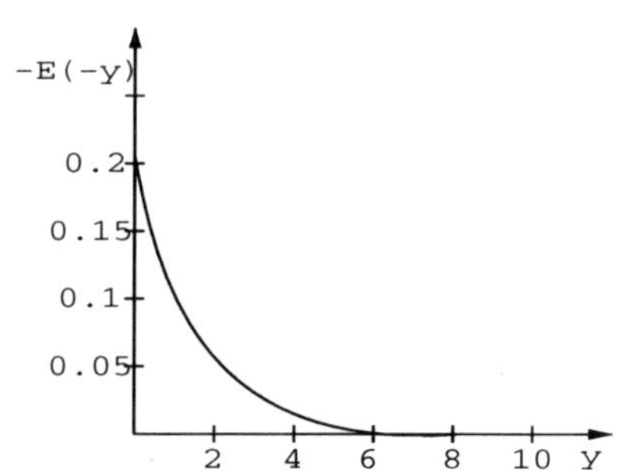

Figure 3.4. *The graph of* $-Ei(-y)$.

Similarly, if the well is located at $(x_{1,0}, x_{2,0})$ instead of $(0, 0)$, the pressure becomes

$$p(r,t) = p^0 + \frac{Q\mu}{4\pi kH} Ei\left(-\frac{(x_1 - x_{1,0})^2 + (x_2 - x_{2,0})^2}{4t\chi}\right), \qquad t > 0. \tag{3.39}$$

The exponential integral function can be expanded in the series

$$Ei\left(-\frac{r^2}{4t\chi}\right) = -\ln\left(\frac{4t\chi}{r^2}\right) + 0.5772 - \frac{r^2}{4t\chi} + \frac{1}{4}\left(\frac{r^2}{4t\chi}\right)^2 - \cdots, \qquad t > 0.$$

When $r^2/(4t\chi) < 0.01$, this function can be approximated by

$$Ei\left(-\frac{r^2}{4t\chi}\right) \approx -\ln\left(\frac{4t\chi}{r^2}\right) + 0.5772 = -\ln\left(\frac{2.25t\chi}{r^2}\right),$$

and the resulting approximation error is less than 0.25%. The corresponding simplified analytic solution from (3.37) is

$$p(r,t) \approx p^0 - \frac{Q\mu}{4\pi kH} \ln\left(\frac{2.25t\chi}{r^2}\right). \tag{3.40}$$

At $r = r_w$ (the radius of the well), $r^2/(4t\chi)$ is small because r_w is small. Then, in a few seconds $r^2/(4t\chi) < 0.01$. Hence equation (3.40) can be used to find the pressure of the wellbore:

$$p_w(t) = p^0 - \frac{Q\mu}{4\pi kH} \ln\left(\frac{2.25t\chi}{r_w^2}\right). \tag{3.41}$$

3.3 Finite Difference Methods

As shown above, an analytic solution can be obtained only for a simplified reservoir system. In general, a numerical solution must be sought. For this, we very briefly touch on the finite difference methods. The books by Peaceman (1977a), Aziz and Settari (1979), and Ertekin, Abou-Kassem, and King (2001) gave detailed information on the use of these methods in reservoir simulation.

3.3.1 First Difference Quotients

We describe *first* and *second difference quotients* for functions of two space variables, x_1 and x_2, and of time, t. Reduction to functions of one space variable and extension to functions of three space variables are straightforward.

Consider a function $p(x_1, x_2, t)$ of x_1, x_2, and t. The first partial derivative of p with respect to x_1 can be defined in one of the following ways:

$$\frac{\partial p}{\partial x_1}(x_1, x_2, t) = \lim_{h_1 \to 0} \frac{p(x_1 + h_1, x_2, t) - p(x_1, x_2, t)}{h_1},$$

$$\frac{\partial p}{\partial x_1}(x_1, x_2, t) = \lim_{h_1 \to 0} \frac{p(x_1, x_2, t) - p(x_1 - h_1, x_2, t)}{h_1},$$

$$\frac{\partial p}{\partial x_1}(x_1, x_2, t) = \lim_{h_1 \to 0} \frac{p(x_1 + h_1, x_2, t) - p(x_1 - h_1, x_2, t)}{2h_1}.$$

We replace this derivative by a difference quotient. For this, we utilize the *Taylor series expansion*

$$p(x_1 + h_1, x_2, t) = p(x_1, x_2, t) + \frac{\partial p}{\partial x_1}(x_1, x_2, t)h_1 + \frac{\partial^2 p}{\partial x_1^2}(x_1^\star, x_2, t)\frac{h_1^2}{2},$$

where $x_1 \le x_1^\star \le x_1 + h_1$ and $h_1 > 0$ is a fixed number. The last term in this equation is a *remainder* that involves a second partial derivative of p. Then $\partial p/\partial x_1$ can be obtained:

$$\frac{\partial p}{\partial x_1}(x_1, x_2, t) = \frac{p(x_1 + h_1, x_2, t) - p(x_1, x_2, t)}{h_1} - \frac{\partial^2 p}{\partial x_1^2}(x_1^\star, x_2, t)\frac{h_1}{2}. \tag{3.42}$$

The expression

$$\frac{p(x_1 + h_1, x_2, t) - p(x_1, x_2, t)}{h_1}$$

is referred to as a *forward difference quotient*, and it approximates the derivative $\partial p/\partial x_1$ with an *error* of the first order in h_1.

Similarly, we have

$$\frac{\partial p}{\partial x_1}(x_1, x_2, t) = \frac{p(x_1, x_2, t) - p(x_1 - h_1, x_2, t)}{h_1} - \frac{\partial^2 p}{\partial x_1^2}(x_1^{\star\star}, x_2, t)\frac{h_1}{2}, \tag{3.43}$$

where $x_1 - h_1 \le x_1^{\star\star} \le x_1$, and the quantity

$$\frac{p(x_1, x_2, t) - p(x_1 - h_1, x_2, t)}{h_1}$$

is called a *backward difference quotient*. This quantity also gives a first order approximation to $\partial p/\partial x_1$.

Next, we use the Taylor series expansions with remainders involving a third partial derivative of p:

$$\begin{aligned} p(x_1 + h_1, x_2, t) = {} & p(x_1, x_2, t) + \frac{\partial p}{\partial x_1}(x_1, x_2, t)h_1 \\ & + \frac{\partial^2 p}{\partial x_1^2}(x_1, x_2, t)\frac{h_1^2}{2!} + \frac{\partial^3 p}{\partial x_1^3}(x_1^\star, x_2, t)\frac{h_1^3}{3!}, \\ p(x_1 - h_1, x_2, t) = {} & p(x_1, x_2, t) - \frac{\partial p}{\partial x_1}(x_1, x_2, t)h_1 \\ & + \frac{\partial^2 p}{\partial x_1^2}(x_1, x_2, t)\frac{h_1^2}{2!} - \frac{\partial^3 p}{\partial x_1^3}(x_1^{\star\star}, x_2, t)\frac{h_1^3}{3!}, \end{aligned}$$

where $x_1 \le x_1^\star \le x_1 + h_1$ and $x_1 - h_1 \le x_1^{\star\star} \le x_1$. Subtracting these two equations and solving for $\partial p/\partial x_1$ yields

$$\begin{aligned} \frac{\partial p}{\partial x_1}(x_1, x_2, t) = {} & \frac{p(x_1 + h_1, x_2, t) - p(x_1 - h_1, x_2, t)}{2h_1} \\ & - \left(\frac{\partial^3 p}{\partial x_1^3}(x_1^\star, x_2, t) + \frac{\partial^3 p}{\partial x_1^3}(x_1^{\star\star}, x_2, t)\right)\frac{h_1^2}{12}. \end{aligned} \tag{3.44}$$

The quotient

$$\frac{p(x_1+h_1,x_2,t)-p(x_1-h_1,x_2,t)}{2h_1}$$

is termed a *centered difference quotient*, and it approximates $\partial p/\partial x_1$ with a higher order, i.e., second order in h_1.

From equations (3.42), (3.43), and (3.44), it would appear preferable to employ the centered difference approximation to $\partial p/\partial x_1$. This is not always the case. Which quotient is used depends on the particular problem (cf. Section 3.3.8).

It is sometimes necessary to use a difference quotient to approximate $\partial p/\partial x_1$ computed halfway between x_1 and x_1+h_1. Analogously to (3.44), we can obtain

$$\begin{aligned}\frac{\partial p}{\partial x_1}\left(x_1+\frac{h_1}{2},x_2,t\right) &= \frac{p(x_1+h_1,x_2,t)-p(x_1,x_2,t)}{h_1}\\ &\quad -\left(\frac{\partial^3 p}{\partial x_1^3}(x_1^\star,x_2,t)+\frac{\partial^3 p}{\partial x_1^3}(x_1^{\star\star},x_2,t)\right)\frac{h_1^2}{48},\end{aligned} \tag{3.45}$$

where $x_1 \le x_1^\star,\ x_1^{\star\star} \le x_1+h_1$. In summary, we have defined three first difference quotients in x_1. The same quotients can be introduced in x_2 and t.

3.3.2 Second Difference Quotients

We exploit the Taylor series expansions with remainders involving a fourth partial derivative of p:

$$\begin{aligned}p(x_1+h_1,x_2,t) &= p(x_1,x_2,t)+\frac{\partial p}{\partial x_1}(x_1,x_2,t)h_1\\ &\quad +\frac{\partial^2 p}{\partial x_1^2}(x_1,x_2,t)\frac{h_1^2}{2!}+\frac{\partial^3 p}{\partial x_1^3}(x_1,x_2,t)\frac{h_1^3}{3!}+\frac{\partial^4 p}{\partial x_1^4}(x_1^\star,x_2,t)\frac{h_1^4}{4!},\\ p(x_1-h_1,x_2,t) &= p(x_1,x_2,t)-\frac{\partial p}{\partial x_1}(x_1,x_2,t)h_1\\ &\quad +\frac{\partial^2 p}{\partial x_1^2}(x_1,x_2,t)\frac{h_1^2}{2!}-\frac{\partial^3 p}{\partial x_1^3}(x_1,x_2,t)\frac{h_1^3}{3!}+\frac{\partial^4 p}{\partial x_1^4}(x_1^{\star\star},x_2,t)\frac{h_1^4}{4!},\end{aligned}$$

where $x_1 \le x_1^\star \le x_1+h_1$ and $x_1-h_1 \le x_1^{\star\star} \le x_1$. Adding these two equations and solving for $\partial^2 p/\partial x_1^2$ yields

$$\begin{aligned}\frac{\partial^2 p}{\partial x_1^2}(x_1,x_2,t) &= \frac{p(x_1+h_1,x_2,t)-2p(x_1,x_2,t)+p(x_1-h_1,x_2,t)}{h_1^2}\\ &\quad -\left(\frac{\partial^4 p}{\partial x_1^4}(x_1^\star,x_2,t)+\frac{\partial^4 p}{\partial x_1^4}(x_1^{\star\star},x_2,t)\right)\frac{h_1^2}{24}.\end{aligned} \tag{3.46}$$

The expression

$$\Delta^2_{x_1} p(x_1, x_2, t) = \frac{p(x_1 + h_1, x_2, t) - 2p(x_1, x_2, t) + p(x_1 - h_1, x_2, t)}{h_1^2} \tag{3.47}$$

defines a *centered second difference quotient*, which approximates the partial derivative $\partial^2 p/\partial x_1^2$ with a second order accuracy in h_1.

Equation (3.46) is derived with the left and right intervals at x_1 of equal length. We now consider p on the intervals $(x_1 - h_1', x_1)$ and $(x_1, x_1 + h_1'')$, where h_1' and h_1'' are not necessarily the same, and introduce a difference quotient for the second derivative

$$\frac{\partial}{\partial x_1}\left(a(x_1, x_2, t)\frac{\partial p}{\partial x_1}\right),$$

where a is a given function. Using Taylor series expansions as above, the following approximations hold:

$$\begin{aligned}\left(a\frac{\partial p}{\partial x_1}\right)\left(x_1 - \frac{h_1'}{2}, x_2, t\right) &\approx a\left(x_1 - \frac{h_1'}{2}, x_2, t\right)\frac{p(x_1, x_2, t) - p(x_1 - h_1', x_2, t)}{h_1'},\\ \left(a\frac{\partial p}{\partial x_1}\right)\left(x_1 + \frac{h_1''}{2}, x_2, t\right) &\approx a\left(x_1 + \frac{h_1''}{2}, x_2, t\right)\frac{p(x_1 + h_1'', x_2, t) - p(x_1, x_2, t)}{h_1''}.\end{aligned} \tag{3.48}$$

Note that

$$\begin{aligned}\frac{\partial}{\partial x_1}\left(a\frac{\partial p}{\partial x_1}\right)(x_1, x_2, t) \approx &\left\{\left(a\frac{\partial p}{\partial x_1}\right)\left(x_1 + \frac{h_1''}{2}, x_2, t\right)\right.\\ &\left.- \left(a\frac{\partial p}{\partial x_1}\right)\left(x_1 - \frac{h_1'}{2}, x_2, t\right)\right\}\\ &\Big/\left(\left(x_1 + \frac{h_1''}{2}\right) - \left(x_1 - \frac{h_1'}{2}\right)\right).\end{aligned}$$

Consequently, using (3.48), we see that

$$\begin{aligned}&\frac{\partial}{\partial x_1}\left(a\frac{\partial p}{\partial x_1}\right)(x_1, x_2, t)\\ &\approx \left\{a\left(x_1 + \frac{h_1''}{2}, x_2, t\right)\frac{p(x_1 + h_1'', x_2, t) - p(x_1, x_2, t)}{h_1''}\right.\\ &\left.-a\left(x_1 - \frac{h_1'}{2}, x_2, t\right)\frac{p(x_1, x_2, t) - p(x_1 - h_1', x_2, t)}{h_1'}\right\}\Big/\frac{h_1' + h''}{2},\end{aligned}$$

which we write as

$$\Delta_{x_1}(a\Delta_{x_1} p). \tag{3.49}$$

This approximation to $\frac{\partial}{\partial x_1}(a\frac{\partial p}{\partial x_1})$ is of second order in h_1, where $h_1 = \max\{h_1', h_1''\}$. A similar definition can be given for $\Delta_{x_2}(a\Delta_{x_2} p)$.

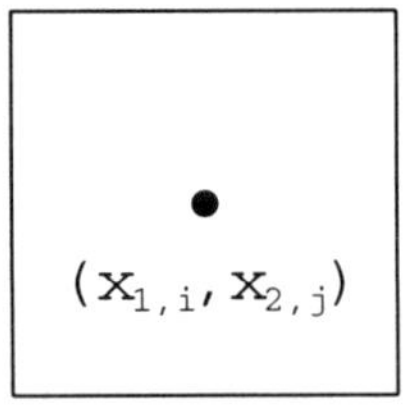

Figure 3.5. *A block-centered grid.*

3.3.3 Grid Systems

There are two types of *grid systems* commonly employed in reservoir simulation: *block-centered* and *point-distributed*. Let the integer i indicate the index in the x_1-direction, and the integer j denote the index in the x_2-direction. Furthermore, let $x_{1,i}$ and $x_{2,j}$ represent the ith and jth values of x_1 and x_2, respectively. Then we set

$$p_{ij} = p(x_{1,i}, x_{2,j}).$$

Block-centered grid

A rectangular solution domain Ω is divided into rectangles, and the point $(x_{1,i}, x_{2,j})$ is at the center of the rectangle (i, j), as in Fig. 3.5. The left side of the rectangle is at $x_{1,i-\frac{1}{2}}$, and the right side is at $x_{1,i+\frac{1}{2}}$. Similarly, $x_{2,j-\frac{1}{2}}$ and $x_{2,j+\frac{1}{2}}$ are the bottom and top sides of the rectangle (i, j). This type of grid is called a *block-centered grid*. It is specified by the sequences $0 = x_{1,\frac{1}{2}} < x_{1,\frac{3}{2}} < \cdots$ and $0 = x_{2,\frac{1}{2}} < x_{2,\frac{3}{2}} < \cdots$ if $\Omega = (0, 1)^2$ is the unit square, for example. Also, we see that

$$\begin{aligned} x_{1,i} &= \frac{1}{2}\left(x_{1,i-\frac{1}{2}} + x_{1,i+\frac{1}{2}}\right), \\ h_{1,i} &= x_{1,i+\frac{1}{2}} - x_{1,i-\frac{1}{2}}, \\ h_{1,i-\frac{1}{2}} &= x_{1,i} - x_{1,i-1}. \end{aligned}$$

Similar notation can be given for the x_2 variable.

Point-distributed grid

In the other type of grid, the point $(x_{1,i}, x_{2,j})$ is now a vertex of a rectangle, as in Fig. 3.6. This grid is referred to as a *point-distributed grid*. In this case, the grid is specified by the sequences $0 = x_{1,0} < x_{1,1} < \cdots$ and $0 = x_{2,0} < x_{2,1} < \cdots$ for $\Omega = (0, 1)^2$. Also, note that

$$\begin{aligned} x_{1,i-\frac{1}{2}} &= \frac{1}{2}\left(x_{1,i-1} + x_{1,i}\right), \\ h_{1,i} &= x_{1,i} - x_{1,i-1}. \end{aligned}$$

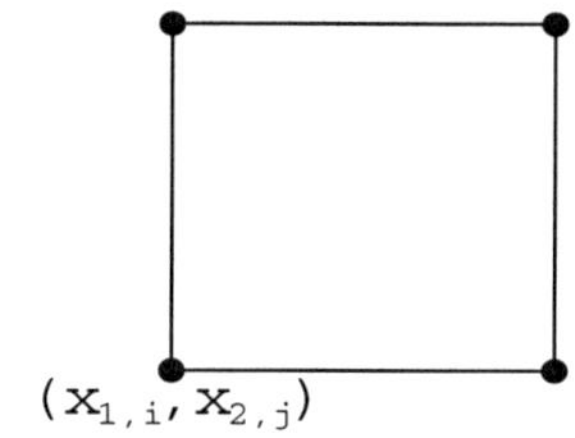

Figure 3.6. *A point-distributed grid.*

3.3.4 Treatment of Boundary Conditions

As we will see, the difference equations written for the two grid systems are the same in form. There are, however, significant differences between them. Specifically, when the grids are not uniform, the locations of points and block boundaries do not coincide. Also, the treatment of boundary conditions is different. Here we introduce difference equations to approximate the boundary conditions described in Section 3.1.4.

Boundary conditions of first kind

Suppose that we are given the boundary condition at $x_1 = 0$:

$$p(0, x_2, t) = g(x_2, t). \tag{3.50}$$

This is a boundary condition of the *first kind*, i.e., the Dirichlet kind. In reservoir simulation, Dirichlet boundary conditions arise when pressure on the reservoir boundary or at a well is specified. For a point-distributed grid (cf. Fig. 3.7), this boundary condition is given by

$$p_{0j} = g_j. \tag{3.51}$$

Equation (3.51) is utilized whenever p^n_{0j} is required in a difference equation.

For a block-centered grid, the closest point to the boundary is $(x_{1,1}, x_{2,j})$ (cf. Fig. 3.8). The value of p^n_{1j} must be *extrapolated* to this point. The simplest approach is

$$p_{1j} = g_j, \tag{3.52}$$

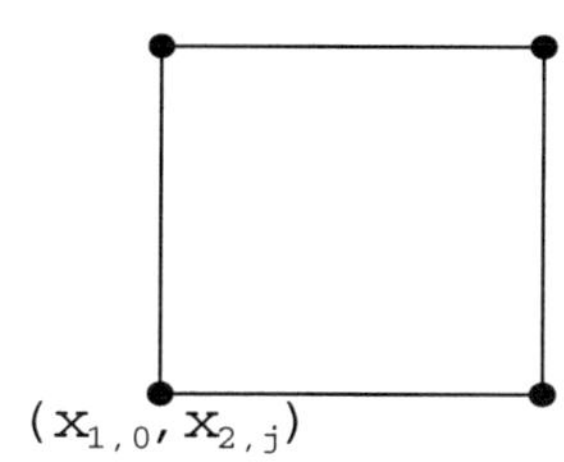

Figure 3.7. *The Dirichlet boundary condition for a point-distributed grid.*

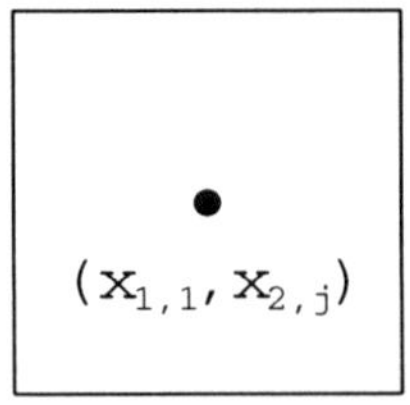

Figure 3.8. *The Dirichlet boundary condition for a block-centered grid.*

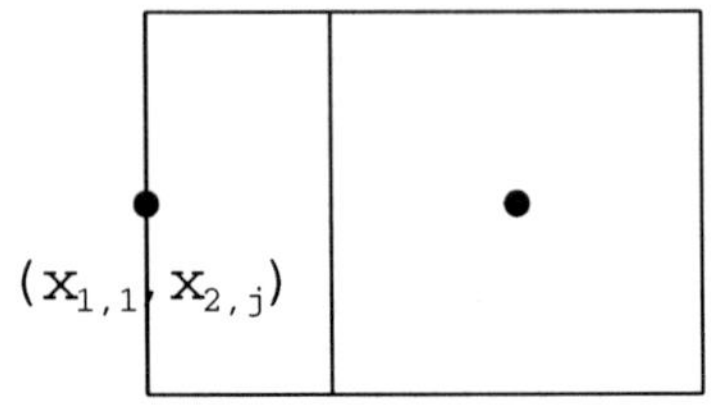

Figure 3.9. *The use of half blocks at the Dirichlet boundary.*

which is only of first order accuracy in space. A second order approximation uses

$$\frac{1}{2}\left(3p_{1j} - p_{2j}\right) = g_j. \tag{3.53}$$

Note that equation (3.53) must be included in the system of difference equations to be solved. For this reason, the block-centered grid is sometimes modified by use of half blocks at Dirichlet boundaries (cf. Fig. 3.9).

Boundary conditions of the second kind

Consider the following boundary condition at $x_1 = 0$:

$$\frac{\partial p}{\partial x_1}(0, x_2, t) = g(x_2, t). \tag{3.54}$$

This is a boundary condition of the *second kind*, i.e., the Neumann kind, and can be used to express a flow rate across a boundary or to specify an injection or production rate at a well. For a point-distributed grid, equation (3.54) can be approximated by

$$\frac{p_{1j} - p_{0j}}{h_{1,1}} = g_j, \tag{3.55}$$

which is a first order approximation. A second order accurate scheme uses a *reflection* (*ghost*) point; for each j, we introduce an auxiliary point $(x_{1,-1}, x_{2,j})$ (cf. Fig. 3.10). The boundary condition (3.54) is discretized using the centered difference at $x_1 = 0$:

$$\frac{p_{1j} - p_{-1j}}{2h_{1,1}} = g_j. \tag{3.56}$$

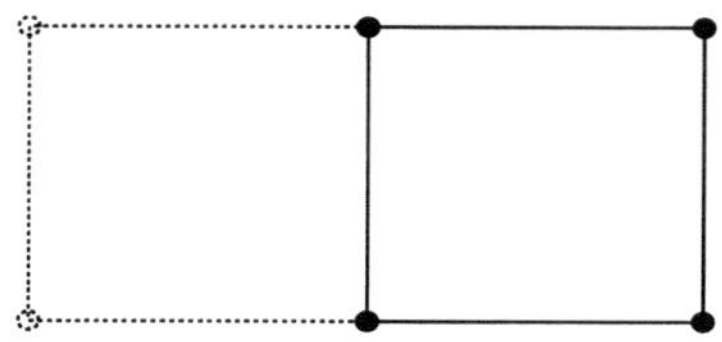

Figure 3.10. *A reflection point for a point-distributed grid.*

Equation (3.56) is exploited to eliminate p^n_{-1j} from the difference equation for the differential equation at $x_1 = 0$. The first and second order approximations for (3.54) can also be defined for a block-centered grid, using a modification similar to that for (3.50).

Boundary conditions of the third kind

A boundary condition of the *third kind* has the form

$$\left(a\frac{\partial p}{\partial x_1} + bp \right)(0, x_2, t) = g(x_2, t), \tag{3.57}$$

where the functions a and b are given. Such a condition occurs when part of the external boundary is semipervious. For a point-distributed grid, this equation can be approximated by

$$a_{0j}\frac{p_{1j} - p_{-1j}}{2h_{1,1}} + b_{0j}p_{0j} = g_j, \tag{3.58}$$

where we recall that $(x_{1,-1}, x_{2,j})$ is a reflection point. It is difficult to approximate (3.57) for a block-centered grid.

3.3.5 Finite Differences for Stationary Problems

We consider the *stationary problem* in two dimensions on a rectangular domain Ω,

$$-\nabla \cdot (a\nabla p) = f(x_1, x_2), \quad (x_1, x_2) \in \Omega, \tag{3.59}$$

where the functions a and f are given. Function a is assumed to be positive on Ω. A pressure equation for incompressible flow is stationary, for example. As pointed out earlier, there are two types of grids widely used in reservoir simulation; the difference equations are the same in form for both grids. Equation (3.59) at grid point (i, j) can be approximated by

$$\begin{aligned} &-\frac{a_{i+\frac{1}{2},j}\dfrac{p_{i+1,j} - p_{i,j}}{h_{1,i+\frac{1}{2}}} - a_{i-\frac{1}{2},j}\dfrac{p_{i,j} - p_{i-1,j}}{h_{1,i-\frac{1}{2}}}}{h_{1,i}} \\ &-\frac{a_{i,j+\frac{1}{2}}\dfrac{p_{i,j+1} - p_{i,j}}{h_{2,j+\frac{1}{2}}} - a_{i,j-\frac{1}{2}}\dfrac{p_{i,j} - p_{i,j-1}}{h_{2,j-\frac{1}{2}}}}{h_{2,j}} = f_{ij}, \end{aligned} \tag{3.60}$$

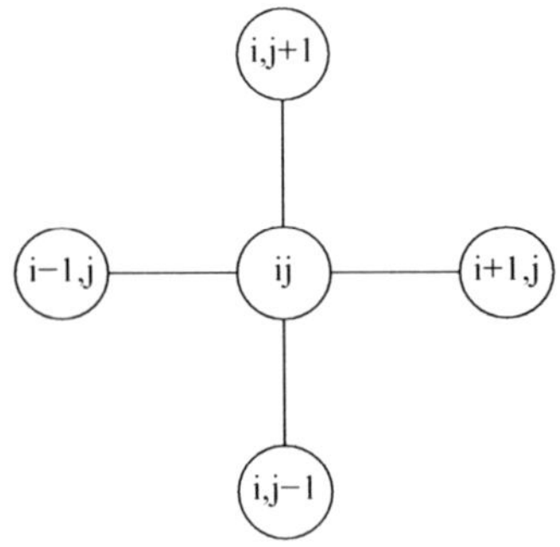

Figure 3.11. *A five-point stencil scheme.*

where $p_{ij} = p(x_{1,i}, x_{2,j})$ and $a_{i+\frac{1}{2},j} = a(x_{1,i+\frac{1}{2}}, x_{2,j})$. If we define

$$a_{1,i+\frac{1}{2},j} = \frac{a_{i+\frac{1}{2},j} h_{2,j}}{h_{1,i+\frac{1}{2}}},$$

$$a_{2,i,j+\frac{1}{2}} = \frac{a_{i,j+\frac{1}{2}} h_{1,i}}{h_{2,j+\frac{1}{2}}},$$

equation (3.60) can be then written as

$$\begin{aligned} &-a_{1,i+\frac{1}{2},j}(p_{i+1,j} - p_{i,j}) + a_{1,i-\frac{1}{2},j}(p_{i,j} - p_{i-1,j}) \\ &- a_{2,i,j+\frac{1}{2}}(p_{i,j+1} - p_{i,j}) + a_{2,i,j-\frac{1}{2}}(p_{i,j} - p_{i,j-1}) = F_{ij}, \end{aligned} \tag{3.61}$$

where $F_{ij} = f_{ij} h_{1,i} h_{2,j}$. F_{ij} may be interpreted as the integral of $f(x_1, x_2)$ over a rectangle with area $h_{1,i}h_{2,j}$. The *truncation error* is the error incurred by replacing a differential equation by a difference equation. From the discussion in Section 3.3.2, the truncation error in the approximation of the difference scheme (3.61) to (3.59) is of second order in both h_1 and h_2. This scheme is the commonly used *five-point stencil* scheme for 2D problems (cf. Fig. 3.11). For some points near or on the boundary of the solution domain, it involves one or two fictitious points outside the domain. The values of p at these points are eliminated, depending on which type of grid and boundary condition is employed. Equation (3.61) can be written in matrix form involving unknowns $\{p_{i,j}\}$ and must be solved via a direct or iterative algorithm (Chen, Huan, and Ma, 2006).

3.3.6 Finite Differences for Parabolic Problems

We turn to the *transient* (*parabolic*) *problem* in two dimensions on a rectangular domain Ω,

$$\phi \frac{\partial p}{\partial t} - \nabla \cdot (a \nabla p) = f(x_1, x_2, t), \quad (x_1, x_2) \in \Omega, \; t > 0, \tag{3.62}$$

where a, f, and ϕ are given functions of x_1, x_2, and t. Functions a and ϕ are assumed to be positive and nonnegative on Ω, respectively. A pressure equation for slightly compressible and compressible flow is parabolic. For a parabolic problem, in addition to a boundary condition, an initial condition is also needed:

$$p(x_1, x_2, 0) = p^0(x_1, x_2).$$

Let $\{t^n\}$ be a sequence of real numbers such that

$$0 = t^0 < t^1 < \cdots < t^n < t^{n+1} < \cdots.$$

For the transient problem, we proceed from the initial solution at t^0 to a solution at t^1; in general, we obtain a solution at t^{n+1} from solutions at the previous time levels. Thus the solution procedure advances through time. Set

$$\Delta t^n = t^{n+1} - t^n, \qquad n = 1, 2, \ldots,$$

and

$$p_{ij}^n = p(x_{1,i}, x_{2,j}, t^n).$$

Forward difference scheme

The simplest difference scheme for (3.62) is to replace the second partial derivatives in space by a second difference at t^n and $\partial p/\partial t$ by a forward difference. The resulting scheme is a centered second difference in space and a forward difference in time and is called the *forward difference* scheme (or *forward Euler* scheme):

$$\begin{aligned}\phi_{ij}^n \frac{p_{i,j}^{n+1} - p_{i,j}^n}{\Delta t^n} h_{1,i}h_{2,j} - a_{1,i+\frac{1}{2},j}^n (p_{i+1,j}^n - p_{i,j}^n) + a_{1,i-\frac{1}{2},j}^n (p_{i,j}^n - p_{i-1,j}^n) \\ - a_{2,i,j+\frac{1}{2}}^n (p_{i,j+1}^n - p_{i,j}^n) + a_{2,i,j-\frac{1}{2}}^n (p_{i,j}^n - p_{i,j-1}) = F_{ij}^n\end{aligned} \tag{3.63}$$

for $n = 0, 1, 2, \ldots$. Note that this equation can be solved *explicitly* for $p_{i,j}^{n+1}$. The use of an explicit scheme brings about a *stability* problem. For $a = \phi = 1$ and $f = 0$, for example, a stability analysis (cf. Section 3.3.7) shows that the time and space step sizes must satisfy the condition

$$\Delta t \left(\frac{1}{h_1^2} + \frac{1}{h_2^2} \right) \leq \frac{1}{2} \tag{3.64}$$

to obtain stability, where $\Delta t = \max\{\Delta t^n : n = 0, 1, \ldots\}$. Hence the forward difference scheme is *conditionally stable*.

Backward difference scheme

The stability condition (3.64) on the time steps is inherent in the forward difference scheme and can be removed by evaluating the second partial derivatives at t^{n+1}:

$$\begin{aligned}&\phi_{ij}^{n+1} \frac{p_{i,j}^{n+1} - p_{i,j}^n}{\Delta t^n} h_{1,i}h_{2,j} \\ &- a_{1,i+\frac{1}{2},j}^{n+1} (p_{i+1,j}^{n+1} - p_{i,j}^{n+1}) + a_{1,i-\frac{1}{2},j}^{n+1} (p_{i,j}^{n+1} - p_{i-1,j}^{n+1}) \\ &- a_{2,i,j+\frac{1}{2}}^{n+1} (p_{i,j+1}^{n+1} - p_{i,j}^{n+1}) + a_{2,i,j-\frac{1}{2}}^{n+1} (p_{i,j}^{n+1} - p_{i,j-1}^{n+1}) = F_{ij}^{n+1}.\end{aligned} \tag{3.65}$$

As we go from n to $n+1$, equation (3.65) defines $p_{i,j}^{n+1}$ *implicitly* and is termed the *backward difference* (or *backward Euler*) scheme. At each time level t^{n+1}, a linear system of algebraic equations must be solved. This system has the same form as that arising from the stationary problem. A stability analysis indicates that scheme (3.65) is *unconditionally stable*; that is, there is no restriction on the time step Δt that can be used (cf. Section 3.3.7).

The truncation errors for both the forward and backward difference schemes are of second order in h_1 and h_2 and of first order in Δt. To improve accuracy in time, the Crank–Nicolson difference scheme can be exploited, for example.

Crank–Nicolson difference scheme

Another implicit difference scheme for equation (3.62) is to replace the average $(\partial p(t^{n+1})/\partial t + \partial p(t^n)/\partial t)/2$ by the difference quotient $(p^{n+1} - p^n)/\Delta t^n$:

$$
\begin{aligned}
&\phi_{ij}^{n+1} \frac{p_{i,j}^{n+1} - p_{i,j}^{n}}{\Delta t^n} h_{1,i} h_{2,j} \\
&\quad - \frac{1}{2}\Big\{ a_{1,i+\frac{1}{2},j}^{n+1}(p_{i+1,j}^{n+1} - p_{i,j}^{n+1}) - a_{1,i-\frac{1}{2},j}^{n+1}(p_{i,j}^{n+1} - p_{i-1,j}^{n+1}) \\
&\quad + a_{2,i,j+\frac{1}{2}}^{n+1}(p_{i,j+1}^{n+1} - p_{i,j}^{n+1}) - a_{2,i,j-\frac{1}{2}}^{n+1}(p_{i,j}^{n+1} - p_{i,j-1}^{n+1}) \\
&\quad + a_{1,i+\frac{1}{2},j}^{n}(p_{i+1,j}^{n} - p_{i,j}^{n}) - a_{1,i-\frac{1}{2},j}^{n}(p_{i,j}^{n} - p_{i-1,j}^{n}) \\
&\quad + a_{2,i,j+\frac{1}{2}}^{n}(p_{i,j+1}^{n} - p_{i,j}^{n}) - a_{2,i,j-\frac{1}{2}}^{n}(p_{i,j}^{n} - p_{i,j-1}^{n}) \Big\} \\
&\quad = \frac{1}{2}\left(F_{ij}^{n+1} + F_{ij}^{n}\right).
\end{aligned}
\tag{3.66}
$$

The truncation error for this scheme is of second order in h_1, h_2, and Δt. This implicit scheme is also unconditionally stable. Moreover, it gives rise to a system of simultaneous equations that is of the same form as that arising from the backward difference scheme.

3.3.7 Consistency, Stability, and Convergence

We give the basic definitions of *consistency*, *stability*, and *convergence* of a finite difference scheme. We concentrate on pure initial value problems. When boundary conditions are included, the definitions must be extended to initial boundary value problems (Thomas, 1995). Furthermore, we focus on 1D transient problems, and the solution domain is the entire x_1-axis; i.e., $-\infty < x_1 < \infty$. Let $x_{1,i} = ih$, $i = 0, \pm 1, \pm 2, \ldots$, and $t^n = n\Delta t$, $n = 0, 1, 2, \ldots$

Consistency

For two real numbers ϵ and $h > 0$, we write

$$\epsilon = \mathcal{O}(h)$$

if there is a positive constant C such that

$$|\epsilon| \le Ch.$$

A finite difference scheme $L_i^n P_i^n = G_i^n$ is (pointwise) *consistent* with the partial differential equation $\mathcal{L}p = \mathcal{F}$ at point (x, t) if for any smooth function $v = v(x, t)$

$$R_i^n \equiv (\mathcal{L}v - \mathcal{F})\,|_i^n - \left\{L_i^n v(ih, n\Delta t) - G_i^n\right\} \to 0 \tag{3.67}$$

as h, $\Delta t \to 0$ and $(ih, n\Delta t) \to (x, t)$. Note that the truncation errors for the forward difference scheme (3.63) and the backward difference scheme (3.65) take the form

$$R_i^n = \mathcal{O}(h^2) + \mathcal{O}(\Delta t),$$

whereas the truncation error for the Crank–Nicolson scheme (3.66) has the form

$$R_i^n = \mathcal{O}(h^2) + \mathcal{O}\big((\Delta t)^2\big).$$

Hence these schemes are consistent with equation (3.62).

Stability

A finite difference scheme is *stable* if the effect of an error (or perturbation) made in any stage of computation is not propagated into larger errors in later stages of the computation; i.e., if local errors are not magnified by further computation. A difference scheme can be examined for stability by substituting into it *perturbed values* of the solution.

We consider the 1D version of equation (3.62) (with $x = x_1$):

$$\frac{\partial p}{\partial t} = \frac{\partial^2 p}{\partial x^2}. \tag{3.68}$$

Let P_i^n be a solution of the corresponding forward difference scheme, and let its perturbation $P_i^n + \epsilon_i^n$ satisfy the same scheme:

$$\begin{aligned}\frac{(P_i^{n+1} + \epsilon_i^{n+1}) - (P_i^n + \epsilon_i^n)}{\Delta t}&\\ = \frac{(P_{i+1}^n + \epsilon_{i+1}^n) - 2(P_i^n + \epsilon_i^n) + (P_{i-1}^n + \epsilon_{i-1}^n)}{h^2}&.\end{aligned}$$

Because of the definition of P_i^n, we see that

$$\frac{\epsilon_i^{n+1} - \epsilon_i^n}{\Delta t} = \frac{\epsilon_{i+1}^n - 2\epsilon_i^n + \epsilon_{i-1}^n}{h^2}. \tag{3.69}$$

We expand the error ϵ_i^n in a Fourier series of the form

$$\epsilon_i^n = \sum_k \bar{\gamma}_k^n \exp(\bar{i}kx_i),$$

where $\bar{i} = \sqrt{-1}$. The analysis can be simplified somewhat if we assume that a solution to the error equation (3.69) has one term (dropping the subscript k in $\bar{\gamma}_k^n$),

$$\epsilon_i^n = \bar{\gamma}^n \exp(\bar{i}kx_i). \tag{3.70}$$

We substitute (3.70) into (3.69) and solve for the *amplification factor*

$$\bar{\gamma} = \bar{\gamma}^{n+1}/\bar{\gamma}^n.$$

The *von Neumann criterion* for stability is that the modulus of this factor must not be greater than 1 (Thomas, 1995). Using (3.69) and (3.70), we see that

$$\frac{\bar{\gamma}^{n+1} - \bar{\gamma}^n}{\Delta t} = \frac{\bar{\gamma}^n \exp(\bar{i}kh) - 2\bar{\gamma}^n + \bar{\gamma}^n \exp(-\bar{i}kh)}{h^2}. \tag{3.71}$$

Since

$$\exp(\bar{i}kh) - 2 + \exp(-\bar{i}kh) = 2\cos(kh) - 2 = -4\sin^2(kh/2),$$

it follows from (3.71) that

$$\bar{\gamma}^{n+1} = \left(1 - \frac{4\Delta t}{h^2}\sin^2(kh/2)\right)\bar{\gamma}^n.$$

Dividing this equation by $\bar{\gamma}^n$, we obtain

$$\bar{\gamma} = 1 - \frac{4\Delta t}{h^2}\sin^2(kh/2).$$

Thus the von Neumann criterion for stability is satisfied if

$$\left|1 - \frac{4\Delta t}{h^2}\sin^2(kh/2)\right| \leq 1. \tag{3.72}$$

Inequality (3.72) is satisfied when the *stability condition*

$$\frac{\Delta t}{h^2} \leq \frac{1}{2} \tag{3.73}$$

holds. Therefore, the forward difference scheme for equation (3.68) is stable under condition (3.73); i.e., this scheme is *conditionally stable*, as noted earlier.

We perform a similar von Neumann stability analysis for the backward difference scheme (3.65) for (3.68). In this case, the error equation takes the form

$$\frac{\epsilon_i^{n+1} - \epsilon_i^n}{\Delta t} = \frac{\epsilon_{i+1}^{n+1} - 2\epsilon_i^{n+1} + \epsilon_{i-1}^{n+1}}{h^2}. \tag{3.74}$$

Substituting (3.70) into (3.74) and performing simple algebraic calculations yields the equation for the amplification factor $\bar{\gamma}$,

$$\bar{\gamma} = \frac{1}{1 + (4\Delta t/h^2)\sin^2(kh/2)},$$

which is always less than or equal to 1 for any choice of k, Δt, and h. Hence the backward difference scheme is *unconditionally stable*. An analogous analysis shows that the Crank–Nicolson scheme is also unconditionally stable.

Convergence

Finite difference schemes are used because their solutions approximate the solutions to certain partial differential equations. What we really need is that the solutions of difference schemes can be made to approximate the solutions of the differential equations to any desired accuracy. Namely, we need *convergence* of the finite difference solutions to those of the differential equations. Specifically, a finite difference scheme $L_i^n P_i^n = G_i^n$ approximating the partial differential equation $\mathcal{L}p = \mathcal{F}$ is (pointwise) *convergent* if, for any (x, t), P_i^n converges to $p(x, t)$ as h, $\Delta t \to 0$ and $(ih, n\Delta t) \to (x, t)$.

As an example, we consider the forward difference scheme (3.63) for equation (3.68):

$$\frac{P_i^{n+1} - P_i^n}{\Delta t} = \frac{P_{i+1}^n - 2P_i^n + P_{i-1}^n}{h^2}. \tag{3.75}$$

Using the analysis in Section 3.3.7, it follows from (3.68) that

$$\frac{p_i^{n+1} - p_i^n}{\Delta t} = \frac{p_{i+1}^n - 2p_i^n + p_{i-1}^n}{h^2} + \mathcal{O}(h^2) + \mathcal{O}(\Delta t). \tag{3.76}$$

Define the error

$$z_i^n = P_i^n - p_i^n,$$

and subtract (3.76) from (3.75) to yield

$$z_i^{n+1} = (1 - 2\mathcal{R})z_i^n + \mathcal{R}(z_{i+1}^n + z_{i-1}^n) + \mathcal{O}(h^2 \Delta t) + \mathcal{O}\left((\Delta t)^2\right),$$

where $\mathcal{R} = \Delta t / h^2$. If $0 < \mathcal{R} \leq 1/2$, the coefficients on the right-hand side of this equation are nonnegative. Thus we see that

$$\begin{aligned} \left|z_i^{n+1}\right| &\leq (1 - 2\mathcal{R}) \left|z_i^n\right| + \mathcal{R}\left(\left|z_{i+1}^n\right| + \left|z_{i-1}^n\right|\right) + C\Delta t \left(h^2 + \Delta t\right) \\ &\leq z^n + C\Delta t \left(h^2 + \Delta t\right), \end{aligned} \tag{3.77}$$

where $z^n = \sup_i\{\left|z_i^n\right|\}$ and the constant C is a uniform constant used to bound the "big $\mathcal{O}$" terms. Taking the supremum over i on the left-hand side of (3.77), we obtain

$$z^{n+1} \leq z^n + C\Delta t \left(h^2 + \Delta t\right). \tag{3.78}$$

Applying inequality (3.78) repeatedly implies

$$z^{n+1} \leq z^0 + C(n+1)\Delta t \left(h^2 + \Delta t\right).$$

Initially, let $z^0 = 0$. Then we have

$$\begin{aligned} \left|P_i^{n+1} - p(ih, (n+1)\Delta t)\right| &\leq z^{n+1} \\ &\leq C(n+1)\Delta t \left(h^2 + \Delta t\right) \\ &\to 0, \end{aligned}$$

as $(n+1)\Delta t \to t$ and h, $\Delta t \to 0$. Therefore, we have proven convergence of the forward difference scheme for (3.68) under condition (3.73). Convergence of the backward and Crank–Nicolson difference schemes can be shown without such a condition.

There is a connection between stability and convergence. In fact, a consistent, two-level difference scheme (i.e., it involves two time levels) for a well-posed linear initial value problem is stable if and only if it is convergent. This is the *Lax equivalence theorem* (Thomas, 1995).

3.3.8 Finite Differences for Hyperbolic Problems

For the introduction of finite differences for hyperbolic problems, we consider the model problem

$$\frac{\partial p}{\partial t} + b\frac{\partial p}{\partial x} = 0, \tag{3.79}$$

where b is a constant and $x = x_1$. This problem is a *one-way wave problem*. The 1D Buckley–Leverett equation is of this form (cf. Section 5.1.3). A boundary condition for (3.79) depends on the sign of b. If this problem is imposed on a bounded interval (l_1, l_2), for example, only an *inflow boundary condition* is needed. That is, p is given at l_1 if $b > 0$, and it is given at l_2 if $b < 0$. For brevity of presentation, we consider problem (3.79) over the entire real line $\mathbb{R}$. Of course, in any case, an initial condition must be given:

$$p(x, 0) = p^0(x).$$

Explicit schemes

We consider an *explicit scheme* for problem (3.79):

$$\frac{p_i^{n+1} - p_i^n}{\Delta t} + b\frac{p_{i+1}^n - p_i^n}{h} = 0, \tag{3.80}$$

which is consistent with (3.79). The amplification factor $\bar{\gamma}$ for (3.80) satisfies

$$\bar{\gamma} = 1 + \frac{b\Delta t}{h}(1 - \cos(kh)) - \bar{i}\frac{b\Delta t}{h}\sin(kh).$$

In the case $b > 0$, $|\bar{\gamma}| > 1$. Thus, by the von Neumann criterion for stability, the difference scheme (3.80) is always *unstable*. In the case $b < 0$, it can be checked that scheme (3.80) is stable, provided that

$$\frac{|b|\Delta t}{h} \leq 1. \tag{3.81}$$

This is the Courant–Friedrichs–Lewy (CFL) condition. That is, scheme (3.80) is conditionally stable if $b < 0$.

It is not surprising that scheme (3.80) is a good choice for problem (3.79) when $b < 0$, and a bad choice when $b > 0$. When $b < 0$, the characteristic for (3.80) through any point runs down to the right towards the x-axis (cf. Fig. 3.12). Scheme (3.80) must then follow back in the same direction. For this reason, when $b > 0$, a good choice for (3.79) is

$$\frac{p_i^{n+1} - p_i^n}{\Delta t} + b\frac{p_i^n - p_{i-1}^n}{h} = 0. \tag{3.82}$$

In fact, when $b > 0$, it can be seen that scheme (3.82) is stable under condition (3.81). (It is always unstable for $b < 0$.)

The explicit difference schemes (3.80) and (3.82) are *one-sided*. Based on the stability analysis above, only the *upwind* versions are conditionally stable.

There are other difference schemes for solving problem (3.79). The centered scheme in space is

$$\frac{p_i^{n+1} - p_i^n}{\Delta t} + b\frac{p_{i+1}^n - p_{i-1}^n}{2h} = 0. \tag{3.83}$$

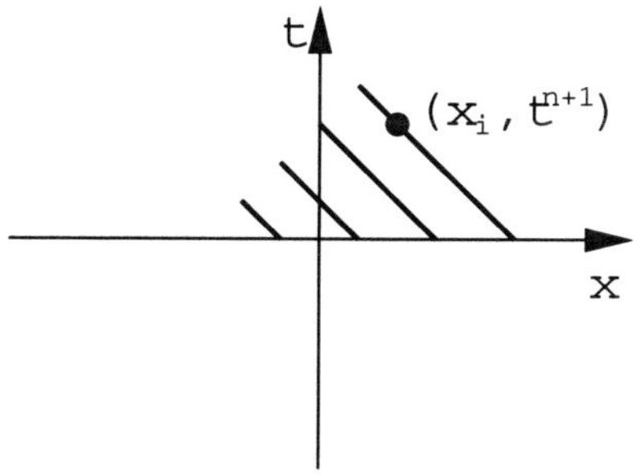

Figure 3.12. *Characteristics for problem* (3.79) *when* $b < 0$.

This scheme yields the amplification factor $\bar{\gamma}$:

$$\bar{\gamma} = 1 - \bar{i}\frac{b\Delta t}{h}\sin(kh).$$

Since $|\bar{\gamma}|^2 = 1 + b^2(\Delta t)^2 \sin^2(kh)/h^2 \geq 1$, we see that scheme (3.83) is always unstable.

Implicit schemes

A stability analysis analogous to that in the explicit case shows that one-sided, stable, fully implicit difference schemes must be upwind. When $b < 0$, the *upwind implicit scheme* is

$$\frac{p_i^{n+1} - p_i^n}{\Delta t} + b\frac{p_{i+1}^{n+1} - p_i^{n+1}}{h} = 0, \tag{3.84}$$

and when $b > 0$,

$$\frac{p_i^{n+1} - p_i^n}{\Delta t} + b\frac{p_i^{n+1} - p_{i-1}^{n+1}}{h} = 0. \tag{3.85}$$

Scheme (3.84) has the amplification factor $\bar{\gamma}$,

$$\bar{\gamma} = \left(1 - \frac{b\Delta t}{h}(1 - \cos(kh)) + \bar{i}\frac{b\Delta t}{h}\sin(kh)\right)^{-1},$$

so

$$|\bar{\gamma}|^2 = \left(1 - 4\frac{b\Delta t}{h}\sin^2\left(\frac{kh}{2}\right)\left(1 - \frac{b\Delta t}{h}\right)\right)^{-1} \leq 1 \quad \text{if } b < 0.$$

Hence scheme (3.84) is unconditionally stable when $b < 0$. A similar argument can be used to prove that scheme (3.85) has the same stability property when $b > 0$.

Now, we consider a fully implicit analogue to scheme (3.83):

$$\frac{p_i^{n+1} - p_i^n}{\Delta t} + b\frac{p_{i+1}^{n+1} - p_{i-1}^{n+1}}{2h} = 0. \tag{3.86}$$

The amplification factor $\bar{\gamma}$ of this scheme is

$$\bar{\gamma} = \left(1 + \bar{i}\frac{b\Delta t}{h}\sin(kh)\right)^{-1},$$

which satisfies $|\bar{\gamma}| \leq 1$. Therefore, scheme (3.86) is unconditionally stable, in contrast with the always unstable (3.83). A centered scheme in time (e.g., the Crank–Nicolson scheme) can be also defined for the solution of problem (3.79).

Numerical dispersion

The local *truncation error* associated with the upwind version of the difference scheme (3.80) for problem (3.79) with $b < 0$ is

$$R_i^n = -\frac{bh}{2}\frac{\partial^2 p}{\partial x^2}(x_i, t^n) - \frac{\Delta t}{2}\frac{\partial^2 p}{\partial t^2}(x_i, t^n) + \mathcal{O}(h^2) + \mathcal{O}\big((\Delta t)^2\big). \tag{3.87}$$

Differentiation of equation (3.79) with respect to t gives

$$\frac{\partial^2 p}{\partial t^2} = -b\frac{\partial^2 p}{\partial x \partial t},$$

and differentiation with respect to x yields

$$\frac{\partial^2 p}{\partial x \partial t} = -b\frac{\partial^2 p}{\partial x^2}.$$

Consequently,

$$\frac{\partial^2 p}{\partial t^2} = b^2\frac{\partial^2 p}{\partial x^2},$$

which is substituted into (3.87) to give

$$R_i^n = -\frac{bh}{2}\left(1 + \frac{b\Delta t}{h}\right)\frac{\partial^2 p}{\partial x^2}(x_i, t^n) + \mathcal{O}(h^2) + \mathcal{O}\big((\Delta t)^2\big). \tag{3.88}$$

This is the local truncation error associated with scheme (3.80).

By the definition (3.67) of the local truncation error, equation (3.88) can be written as

$$\begin{aligned}\frac{p_i^{n+1} - p_i^n}{\Delta t} + b\frac{p_{i+1}^n - p_i^n}{h} &= \left\{\frac{\partial p}{\partial t} + b\frac{\partial p}{\partial x} + a_{num}\frac{\partial^2 p}{\partial x^2}\right\}(x_i, t^n) \\ &\quad + \mathcal{O}(h^2) + \mathcal{O}\big((\Delta t)^2\big),\end{aligned} \tag{3.89}$$

where

$$a_{num} = \frac{bh}{2}\left(1 + \frac{b\Delta t}{h}\right). \tag{3.90}$$

Therefore, we are, in fact, solving the difference equation (3.80) for the diffusion-convection problem

$$\frac{\partial p}{\partial t} + b\frac{\partial p}{\partial x} + a_{num}\frac{\partial^2 p}{\partial x^2} = 0,$$

rather than for the pure hyperbolic problem (3.79). That is, the truncation error of (3.80) includes the *numerical dispersion* term a_{num}.

If we consider the diffusion-convection problem

$$\frac{\partial p}{\partial t} + b\frac{\partial p}{\partial x} - a\frac{\partial^2 p}{\partial x^2} = 0, \quad a > 0,$$

and develop a difference scheme similar to (3.80), then the above truncation error analysis indicates that the solution of the resulting difference equation will be associated with the problem

$$\frac{\partial p}{\partial t} + b\frac{\partial p}{\partial x} - (a - a_{num})\frac{\partial^2 p}{\partial x^2} = 0.$$

When the physical diffusion coefficient a is small, a serious problem arises. If numerical dispersion is severe (it is frequently so), a_{num} can easily dominate a. Consequently, the numerical dispersion swamps the physical dispersion, leading to a sharp front being severely smeared.

Grid orientation effects

Another drawback of finite difference methods is that the solution of a partial differential problem using these methods heavily depends on spatial orientations of a computational grid, known as *grid orientation effects*. In petroleum reservoir simulation, this means that drastically different predictions from simulators can be obtained from different grid orientations.

If an upwind technique is used as in (3.80) for a 2D counterpart, the resulting numerical dispersion is related to the quantity (cf. equation (3.90))

$$\frac{h_1}{2}\frac{\partial^2 p}{\partial x_1^2} + \frac{h_2}{2}\frac{\partial^2 p}{\partial x_2^2},$$

which is not rotationally invariant and is thus directionally dependent. When modeling multiphase flow with a high mobility ratio (mainly due to a large viscosity ratio), once a preferential flow pattern has been established, the greater mobility of the less viscous fluid causes this flow path to dominate the flow pattern. With the five-point in 2D or seven-point in 3D finite difference stencil schemes, preferred flow paths are established along the coordinate directions. Then the use of an upwind stabilizing technique greatly enhances flow in these preferred directions. This grid orientation effect is dramatic in cases with very high mobility ratios.

3.4 Numerical Solution of Single-Phase Flow

The development of finite difference methods will be carried out for the general single-phase flow equation (3.21), which applies to all the cases: incompressible, slightly compressible, and compressible. The discretization of these cases will be different, particularly in the treatment of transmissibility terms, and the difference will be pointed out as appropriate. Most noticeably, the flow equation in the incompressible case is linear and does not contain the accumulation term, while the equations in the other two cases are nonlinear and involve the accumulation term; nonlinearity in the slightly compressible case is weaker than that in the compressible case.

Assume that the permeability tensor $\mathbf{k}$ is diagonal: $\mathbf{k} = \text{diag}(k_{11}, k_{22}, k_{33})$. Then equation (3.21) is expanded as follows:

$$\begin{aligned} c(p)\frac{\partial p}{\partial t} = &\frac{\partial}{\partial x_1}\left(\frac{\rho k_{11}}{\mu}\left(\frac{\partial p}{\partial x_1} - \gamma\frac{\partial z}{\partial x_1}\right)\right) + \frac{\partial}{\partial x_2}\left(\frac{\rho k_{22}}{\mu}\left(\frac{\partial p}{\partial x_2} - \gamma\frac{\partial z}{\partial x_2}\right)\right) \\ &+\frac{\partial}{\partial x_3}\left(\frac{\rho k_{33}}{\mu}\left(\frac{\partial p}{\partial x_3} - \gamma\frac{\partial z}{\partial x_3}\right)\right) + q, \end{aligned} \tag{3.91}$$

where $\gamma = \rho\wp$ is the fluid density in terms of pressure per distance (usually called *fluid gravity*). The grid point $(x_{1,i}, x_{2,j}, x_{3,k})$ and spatial steps $h_{1,i}$, $h_{2,j}$, and $h_{3,k}$ will be simply written as (i, j, k) and h_1, h_2, h_3, respectively. In addition, Δt will simply indicate Δt^n. Because the difference equations are the same in form for both types of grid systems, the block-centered grid structure is used as an example.

3.4.1 Treatment of Initial Conditions

In reservoir simulation, in general, an initial pressure p_d^0 is given at a *datum level depth* z_d. The pressure at other grid points (i, j, k) is determined by the *gravity equilibrium break condition*:

$$p_{i,j,k}^0 = p_d^0 + \gamma\left(z_{i,j,k} - z_d\right). \tag{3.92}$$

For slightly compressible and compressible flow, the fluid density γ depends on p. Hence an iteration procedure may be required in the application of equation (3.92).

3.4.2 Time Discretization

The time discretization of equation (3.91) is similar to that given in (3.65) for a linear parabolic problem:

$$\left(Vc(p)\frac{\partial p}{\partial t}\right)_{i,j,k}^n \approx \left(Vc\left(p^{n+1}\right)\frac{p^{n+1} - p^n}{\Delta t}\right)_{i,j,k}, \tag{3.93}$$

where $V_{i,j,k}$ is the volume of the gridblock at (i, j, k). For incompressible flow, c is zero. In slightly compressible case, $c(p) = \phi\rho c_t$, with ρ, ϕ, and c_t given by (3.13), (3.16), and (3.19), respectively. For compressible flow, $c(p)$ is defined by (3.22); particularly, for gas flow, the real gas law (3.23) can be applied.

3.4.3 Spatial Discretization

An extension of the difference equation (3.61) to equation (3.91) in three dimensions gives

$$\begin{aligned}
&\left(\frac{A_1\rho k_{11}}{\mu h_1}\right)_{i+1/2,j,k}\left(p_{i+1,j,k}-p_{i,j,k}\right)-\left(\frac{A_1\rho k_{11}}{\mu h_1}\right)_{i-1/2,j,k}\left(p_{i,j,k}-p_{i-1,j,k}\right)\\
&+\left(\frac{A_2\rho k_{22}}{\mu h_2}\right)_{i,j+1/2,k}\left(p_{i,j+1,k}-p_{i,j,k}\right)-\left(\frac{A_2\rho k_{22}}{\mu h_2}\right)_{i,j-1/2,k}\left(p_{i,j,k}-p_{i,j-1,k}\right)\\
&+\left(\frac{A_3\rho k_{33}}{\mu h_3}\right)_{i,j,k+1/2}\left(p_{i,j,k+1}-p_{i,j,k}\right)-\left(\frac{A_3\rho k_{33}}{\mu h_3}\right)_{i,j,k-1/2}\left(p_{i,j,k}-p_{i,j,k-1}\right)\\
&-\left(\frac{A_1\rho k_{11}\gamma}{\mu h_1}\right)_{i+1/2,j,k}\left(z_{i+1,j,k}-z_{i,j,k}\right)+\left(\frac{A_1\rho k_{11}\gamma}{\mu h_1}\right)_{i-1/2,j,k}\left(z_{i,j,k}-z_{i-1,j,k}\right)\\
&-\left(\frac{A_2\rho k_{22}\gamma}{\mu h_2}\right)_{i,j+1/2,k}\left(z_{i,j+1,k}-z_{i,j,k}\right)+\left(\frac{A_2\rho k_{22}\gamma}{\mu h_2}\right)_{i,j-1/2,k}\left(z_{i,j,k}-z_{i,j-1,k}\right)\\
&-\left(\frac{A_3\rho k_{33}\gamma}{\mu h_3}\right)_{i,j,k+1/2}\left(z_{i,j,k+1}-z_{i,j,k}\right)+\left(\frac{A_3\rho k_{33}\gamma}{\mu h_3}\right)_{i,j,k-1/2}\left(z_{i,j,k}-z_{i,j,k-1}\right)\\
&+Q_{i,j,k},
\end{aligned} \tag{3.94}$$

where $Q_{i,j,k} = q_{i,j,k}V_{i,j,k}$ is the injection/production and A_i is the cross-sectional area normal to the x_i-direction, $i = 1, 2, 3$.

3.4.4 Treatment of Block Transmissibility

The transmissibilities

$$T_{1,i\pm1/2,j,k} = \left(\frac{A_1\rho k_{11}}{\mu h_1}\right)_{i\pm1/2,j,k} \tag{3.95}$$

(and analogously in the x_2- and x_3-directions) are evaluated at gridblock boundaries (as shown by the subscripts $i \pm 1/2$, $j \pm 1/2$, and $k \pm 1/2$). The gridblock dimensions and labeling are displayed in Fig. 3.13. The grid dimension in each direction and other properties, such as formation thickness, permeability, and porosity, can significantly differ from one block to another. The rock and fluid properties are often given only at the block centers, but the transmissibilities are computed at the gridblock boundaries. Hence some sort of averaging techniques must be utilized to estimate these properties between two adjacent gridblocks.

Several averaging techniques are available for a set of real numbers $\{a_1, a_2, \ldots, a_m\}$:

- Arithmetic averaging,

$$A_A = \frac{a_1 + a_2 + \cdots + a_m}{m}.$$

- Geometric averaging,

$$A_G = (a_1 a_2 \ldots a_m)^{1/m}.$$

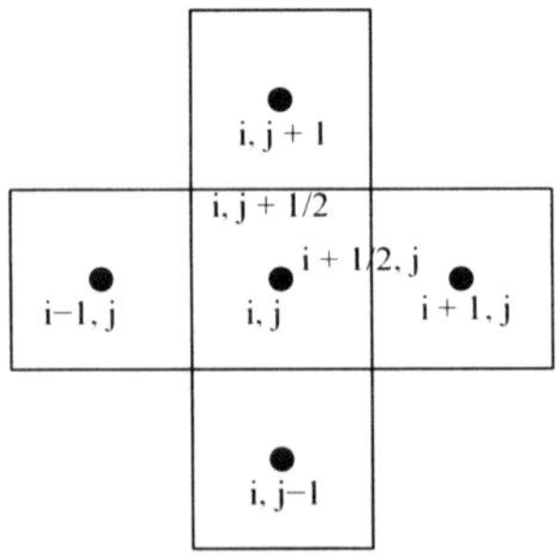

Figure 3.13. *Grid labeling.*

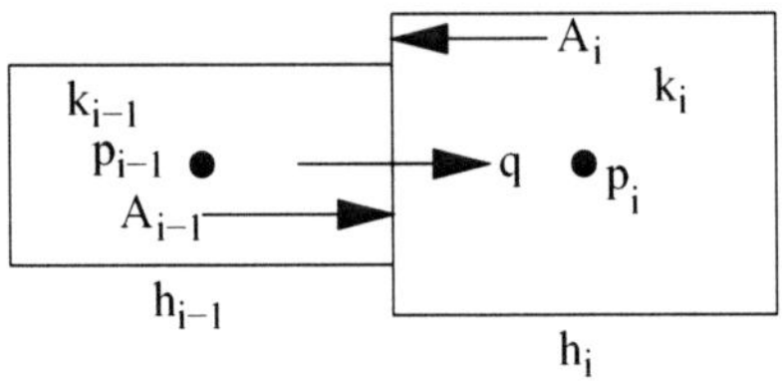

Figure 3.14. *Two cells.*

- Weighted averaging, with weights $w_1, w_2, \ldots, w_m$,

$$A_W = \frac{w_1 a_1 + w_2 a_2 + \cdots + w_m a_m}{w_1 + w_2 + \cdots + w_m}.$$

- Harmonic averaging,

$$\frac{1}{A_H} = \frac{1}{m}\left(\frac{1}{a_1} + \frac{1}{a_2} + \cdots + \frac{1}{a_m}\right).$$

Note that $A_H \leq A_G \leq A_A$. These averages are equal if $a_1 = a_2 = \cdots = a_m$.

The transmissibilities in equations (3.95) have two distinct quantities: $A_1 k_{11}/h_1$ represents the rock and grid properties and ρ/μ represents the fluid properties. What average should be used for these quantities?

Consider a 1D example with two cells (cf. Fig. 3.14) and a constant viscosity, without gravity. The flow rate q in the left and right cells is, respectively,

$$q = -\frac{k_{i-1}A_{i-1}}{\mu}\frac{p_{i-1/2} - p_{i-1}}{h_{i-1}/2}, \qquad q = -\frac{k_i A_i}{\mu}\frac{p_i - p_{i-1/2}}{h_i/2}.$$

Consequently,

$$p_{i-1/2} - p_{i-1} = -q\mu\frac{h_{i-1}}{2}\frac{1}{(kA)_{i-1}}, \qquad p_i - p_{i-1/2} = -q\mu\frac{h_i}{2}\frac{1}{(kA)_i}.$$

Adding these two equations yields

$$p_i - p_{i-1} = -\frac{q\mu}{2}\left(\frac{h_{i-1}}{(kA)_{i-1}} + \frac{h_i}{(kA)_i}\right). \tag{3.96}$$

On the other hand, the single-phase Darcy law is

$$q = -\frac{\overline{kA}}{\mu}\frac{p_i - p_{i-1}}{(h_{i-1}+h_i)/2}, \tag{3.97}$$

where $\overline{kA}$ is some average of kA at the two cell interface. Comparing equations (3.96) and (3.97), we see that

$$\overline{kA} = \frac{h_{i-1}+h_i}{\dfrac{h_{i-1}}{(kA)_{i-1}} + \dfrac{h_i}{(kA)_i}}, \tag{3.98}$$

which is the (weighted) harmonic average for kA. This average weights the effect of a lower permeability. To see this, letting $h_{i-1} = h_i$, $A_{i-1} = A_i$, $k_{i-1} = 200$ md, and $k_i = 2$ md, then we have

$$\frac{1}{k_H} = \frac{1}{2}\left(\frac{1}{k_{i-1}} + \frac{1}{k_i}\right),$$

i.e., $k_H = 3.9$ md. As a result, the flow between gridblocks is affected more by lower permeability blocks. The same argument carries over to the multiphase flow to be studied in the subsequent chapters.

From the above argument, the harmonic average is used for the rock and grid properties in the transmissibility $T_{1,i\pm1/2,j,k}$:

$$\left(\frac{A_1k_{11}}{h_1}\right)_{i\pm1/2,j,k} = \frac{2(A_1k_{11})_{i,j,k}(A_1k_{11})_{i\pm1,j,k}}{(A_1k_{11})_{i,j,k}(h_1)_{i\pm1,j,k} + (A_1k_{11})_{i\pm1,j,k}(h_1)_{i,j,k}}; \tag{3.99}$$

similar quantities can be defined in the x_2- and x_3-directions.

Since the fluid properties in the transmissibility term for the single-phase flow do not change much from block to block (they are slowly varying functions of pressure only), the usual arithmetic average can be used for them. Two of the common methods are used in petroleum industry practices: The first one averages the pressure before the properties are computed:

$$\begin{aligned} p_{i\pm1/2,j,k} &= \beta p_{i,j,k} + (1-\beta)p_{i\pm1,j,k}, \\ \left(\frac{\rho}{\mu}\right)_{i\pm1/2,j,k} &= \left(\frac{\rho}{\mu}\right)\left(p_{i\pm1/2,j,k}\right), \end{aligned} \tag{3.100}$$

where β is a weighting factor of either 0.5 for the standard arithmetic average or the fraction of the pore volumes for the pore-volume weighted average. The second method uses the block-centered pressures and then averages the properties:

$$\left(\frac{\rho}{\mu}\right)_{i\pm1/2,j,k} = \beta\left(\frac{\rho}{\mu}\right)\left(p_{i,j,k}\right) + (1-\beta)\left(\frac{\rho}{\mu}\right)\left(p_{i\pm1,j,k}\right). \tag{3.101}$$

We end with a remark that the transmissibilities are constant for incompressible flow. It is in the slightly compressible and compressible cases that they must be carefully evaluated. For multiphase flow, the relative permeabilities can change a great deal from block to block, which will require use of a different averaging (cf. Chapter 5).

3.4.5 Solution Approaches in Time

Note that the above definition of transmissibilities has not involved time. To obtain the pressure at a certain time level t^{n+1}, the pressure equation must be advanced from an initial datum. For the discretization of linear parabolic problems in Section 3.3.6, there are essentially three different solution approaches: forward, backward, and Crank–Nicolson. The flow equation under consideration raises another difficulty: nonlinearity; i.e., the coefficients in this equation depend on the pressure. Four approaches for linearizing the coefficients are discussed: explicit, linearization, extrapolation, and fully implicit.

Explicit approach

In the explicit solution approach, which is also called the forward difference scheme in Section 3.3.6, all the coefficients and pressure in spatial derivatives are evaluated at the previous time:

$$\begin{aligned}
&\left(Vc\left(p^{n+1}\right)\frac{p^{n+1}-p^{n}}{\Delta t}\right)_{i,j,k}\\
&= T^{n}_{1,i+1/2,j,k}\left(p^{n}_{i+1,j,k}-p^{n}_{i,j,k}\right)-T^{n}_{1,i-1/2,j,k}\left(p^{n}_{i,j,k}-p^{n}_{i-1,j,k}\right)\\
&\quad+T^{n}_{2,i,j+1/2,k}\left(p^{n}_{i,j+1,k}-p^{n}_{i,j,k}\right)-T^{n}_{2,i,j-1/2,k}\left(p^{n}_{i,j,k}-p^{n}_{i,j-1,k}\right)\\
&\quad+T^{n}_{3,i,j,k+1/2}\left(p^{n}_{i,j,k+1}-p^{n}_{i,j,k}\right)-T^{n}_{3,i,j,k-1/2}\left(p^{n}_{i,j,k}-p^{n}_{i,j,k-1}\right)\\
&\quad-(T\gamma)^{n}_{1,i+1/2,j,k}\left(z_{i+1,j,k}-z_{i,j,k}\right)+(T\gamma)^{n}_{1,i-1/2,j,k}\left(z_{i,j,k}-z_{i-1,j,k}\right)\\
&\quad-(T\gamma)^{n}_{2,i,j+1/2,k}\left(z_{i,j+1,k}-z_{i,j,k}\right)+(T\gamma)^{n}_{2,i,j-1/2,k}\left(z_{i,j,k}-z_{i,j-1,k}\right)\\
&\quad-(T\gamma)^{n}_{3,i,j,k+1/2}\left(z_{i,j,k+1}-z_{i,j,k}\right)+(T\gamma)^{n}_{3,i,j,k-1/2}\left(z_{i,j,k}-z_{i,j,k-1}\right)\\
&\quad+Q^{n}_{i,j,k}.
\end{aligned}\tag{3.102}$$

Each gridblock (i, j, k) contributes only one equation in one unknown $p^{n+1}_{i,j,k}$. Advancing the pressure from n to $n+1$ is achieved by moving through all blocks in a systematic fashion and solving equation (3.102) for $p^{n+1}_{i,j,k}$. To preserve material balance, some iteration may be needed in the evaluation of the compressibility coefficient c^{n+1}. For slightly compressible flow, an iteration may be not required in practice; this coefficient can be computed just at time level t^n, and equation (3.102) is solved in a noniterative manner. However, for compressible flow, this iteration is required.

As discussed in Section 3.3.6, the forward difference scheme is conditionally stable. That is, the time step must satisfy the stability condition for (3.102) to be stable:

$$\Delta t \le \min_{i,j,k}\left(\frac{\left(Vc(p)\right)_{(i,j,k)}}{D_{i,j,k}}\right),\tag{3.103}$$

where the minimum is taken over all gridblocks and

$$D_{i,j,k} = T_{1,i+1/2,j,k} + T_{1,i-1/2,j,k} + T_{2,i,j+1/2,k} + T_{2,i,j-1/2,k} + T_{3,i,j,k+1/2} + T_{3,i,j,k-1/2}.$$

Consequently, the explicit solution approach suffers from severe restrictions on time step sizes and is not used in numerical simulation of the single-phase flow.

Linearization approach

A way to avoid the stability problem in the explicit approach is to use the explicit treatment only in the transmissibilities:

$$
\begin{aligned}
&\left(Vc\left(p^{n+1}\right) \frac{p^{n+1} - p^n}{\Delta t} \right)_{i,j,k} \\
&= T^n_{1,i+1/2,j,k}\left(p^{n+1}_{i+1,j,k} - p^{n+1}_{i,j,k}\right) - T^n_{1,i-1/2,j,k}\left(p^{n+1}_{i,j,k} - p^{n+1}_{i-1,j,k}\right) \\
&\quad + T^n_{2,i,j+1/2,k}\left(p^{n+1}_{i,j+1,k} - p^{n+1}_{i,j,k}\right) - T^n_{2,i,j-1/2,k}\left(p^{n+1}_{i,j,k} - p^{n+1}_{i,j-1,k}\right) \\
&\quad + T^n_{3,i,j,k+1/2}\left(p^{n+1}_{i,j,k+1} - p^{n+1}_{i,j,k}\right) - T^n_{3,i,j,k-1/2}\left(p^{n+1}_{i,j,k} - p^{n+1}_{i,j,k-1}\right) \\
&\quad - (T\gamma)^n_{1,i+1/2,j,k}\left(z_{i+1,j,k} - z_{i,j,k}\right) + (T\gamma)^n_{1,i-1/2,j,k}\left(z_{i,j,k} - z_{i-1,j,k}\right) \\
&\quad - (T\gamma)^n_{2,i,j+1/2,k}\left(z_{i,j+1,k} - z_{i,j,k}\right) + (T\gamma)^n_{2,i,j-1/2,k}\left(z_{i,j,k} - z_{i,j-1,k}\right) \\
&\quad - (T\gamma)^n_{3,i,j,k+1/2}\left(z_{i,j,k+1} - z_{i,j,k}\right) + (T\gamma)^n_{3,i,j,k-1/2}\left(z_{i,j,k} - z_{i,j,k-1}\right) \\
&\quad + Q^{n+1}_{i,j,k}.
\end{aligned}
\tag{3.104}
$$

For slightly compressible flow, the pressure-dependent fluid properties, ρ (or B) and μ, are weakly nonlinear and can be evaluated at the previous time level. Fig. 3.15 shows the value of transmissibility used in the explicit treatment of the linearization approach. The source/sink term $Q^{n+1}_{i,j,k}$ needs special consideration, depending on the type of the constraint of the well (cf. Chapter 4).

Equation (3.104) can be rearranged in the *seven-point stencil scheme* (cf. Fig. 3.16):

$$
\begin{aligned}
&T^n_{3,i,j,k-1/2}p^{n+1}_{i,j,k-1} + T^n_{2,i,j-1/2,k}p^{n+1}_{i,j-1,k} + T^n_{1,i-1/2,j,k}p^{n+1}_{i-1,j,k} \\
&\quad -\left\{ \frac{1}{\Delta t}\left(Vc\left(p^{n+1}\right)\right)_{i,j,k} + T^n_{3,i,j,k-1/2} + T^n_{2,i,j-1/2,k} + T^n_{1,i-1/2,j,k} \right. \\
&\quad \left. + T^n_{1,i+1/2,j,k} + T^n_{2,i,j+1/2,k} + T^n_{3,i,j,k+1/2} \right\} p^{n+1}_{i,j,k} \\
&\quad + T^n_{1,i+1/2,j,k}p^{n+1}_{i+1,j,k} + T^n_{2,i,j+1/2,k}p^{n+1}_{i,j+1,k} + T^n_{3,i,j,k+1/2}p^{n+1}_{i,j,k+1} \\
&= -\frac{1}{\Delta t}\left(Vc\left(p^{n+1}\right)\right)_{i,j,k} p^n_{i,j,k} \\
&\quad + (T\gamma)^n_{1,i+1/2,j,k}\left(z_{i+1,j,k} - z_{i,j,k}\right) - (T\gamma)^n_{1,i-1/2,j,k}\left(z_{i,j,k} - z_{i-1,j,k}\right) \\
&\quad + (T\gamma)^n_{2,i,j+1/2,k}\left(z_{i,j+1,k} - z_{i,j,k}\right) - (T\gamma)^n_{2,i,j-1/2,k}\left(z_{i,j,k} - z_{i,j-1,k}\right) \\
&\quad + (T\gamma)^n_{3,i,j,k+1/2}\left(z_{i,j,k+1} - z_{i,j,k}\right) - (T\gamma)^n_{3,i,j,k-1/2}\left(z_{i,j,k} - z_{i,j,k-1}\right) \\
&\quad - Q^{n+1}_{i,j,k}.
\end{aligned}
\tag{3.105}
$$

Again, some iteration may be needed for the compressibility coefficient $c(p)$. Unlike the explicit solution approach, equation (3.105) is now a system of equations, which must be solved by either a direct algorithm or an iterative algorithm. When the entire grid system is written in *natural ordering* (i.e., when the unknowns are ordered by lines vertically or horizontally, in increasing orders of i, j, and k in a nested manner, where i is the innermost index, j is the intermediate index, and k is the outermost index) (Chen, Huan, and Ma, 2006),

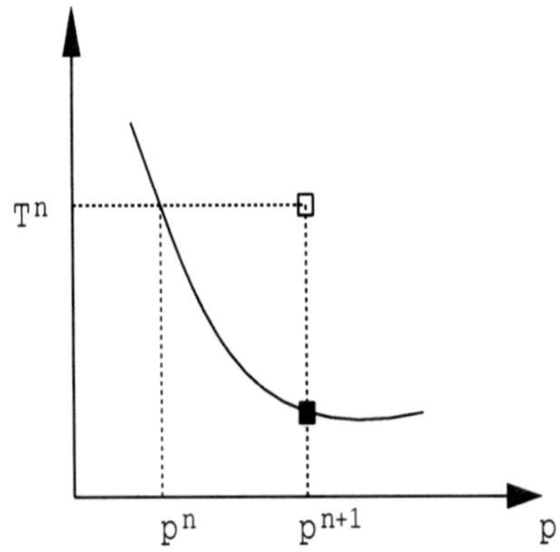

Figure 3.15. *Value of transmissibility in linearization.*

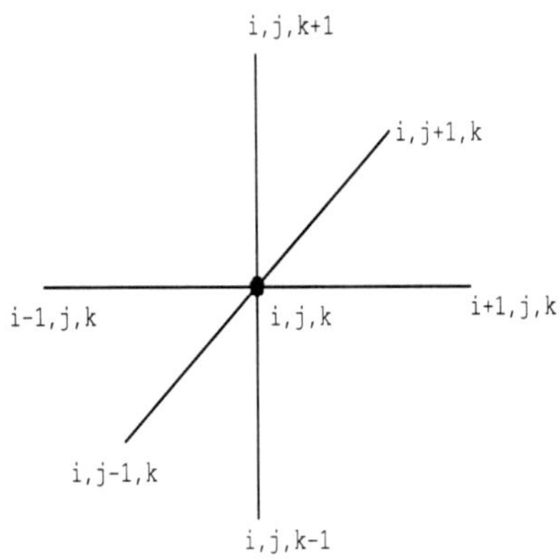

Figure 3.16. *Seven-point stencil.*

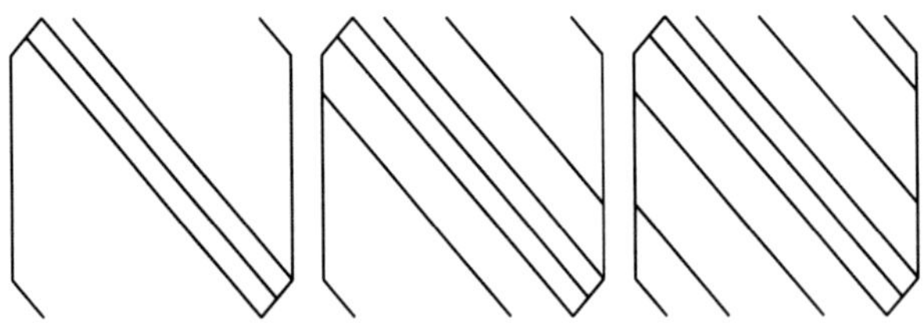

Figure 3.17. *Tridiagonal, pentadiagonal, and heptadiagonal matrices.*

the coefficient matrix associated with (3.105) is *heptadiagonal* (cf. Fig. 3.17). For 1D and 2D flow, the coefficient matrix will become *tridiagonal* and *pentadiagonal*, respectively (cf. Fig. 3.17). If there is no irregularity on the outer boundary of the reservoir, this matrix always exhibits a well-structured manner. However, this well-defined structure is usually spoiled by wells that perforate into many gridblocks and/or by irregular gridblocks.

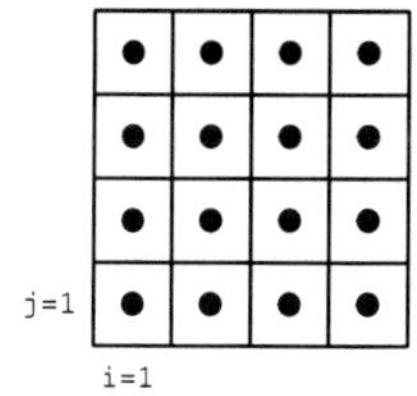

Figure 3.18. *Boundary of a domain.*

Treatment of boundary conditions

If a Dirichlet boundary condition is imposed on the external boundary of the reservoir domain, the pressures on the boundary are given. Thus, after the transmissibilities are properly evaluated on the boundary, the known pressures are moved to the right-hand side of equation (3.105).

When a Neumann boundary condition is given, for example, at the left boundary (cf. Fig. 3.18),

$$\frac{\partial p}{\partial x_1} = g,$$

the *reflection point* method (cf. Section 3.3.4) is used:

$$\frac{p_{1,j,k} - p_{0,j,k}}{h_1} = g_{j,k}. \tag{3.106}$$

A no-flow boundary condition (i.e., $g = 0$) is the most commonly used in reservoir simulation. This type of boundary condition can be alternatively modeled by assigning a zero value to the boundary transmissibility.

Extrapolation approach

The explicit handling of the transmissibility terms may result in a stability problem, especially for multiphase problems. Furthermore, the linearization will reduce the accuracy of time discretization if a more accurate discretization method in time, such as Crank–Nicolson (cf. Section 3.3.6), is used. That is true for any higher order time discretization method with the present linearization technique. This drawback can be overcome by using *extrapolation techniques* in the linearization of the transmissibility coefficients.

In the *extrapolation approach*, the pressure at the new time level is obtained from the previous two time levels,

$$p_{i,j,k}^{(n+1)^*} = p_{i,j,k}^n + \frac{\Delta t^{n+1}}{\Delta t^n}(p_{i,j,k}^n - p_{i,j,k}^{n-1}), \tag{3.107}$$

and the corresponding transmissibilities are evaluated at the extrapolated time level,

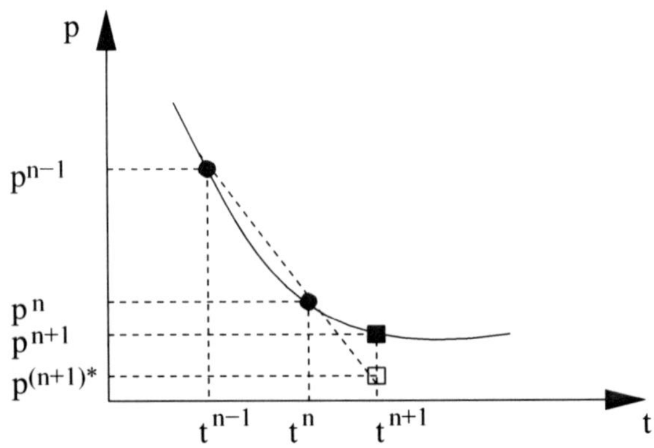

Figure 3.19. *Transmissibility in the extrapolation approach.*

$$
\begin{aligned}
&\left(Vc\left(p^{n+1}\right)\frac{p^{n+1}-p^{n}}{\Delta t}\right)_{i,j,k}\\
&= T^{(n+1)^*}_{1,i+1/2,j,k}\left(p^{n+1}_{i+1,j,k}-p^{n+1}_{i,j,k}\right)-T^{(n+1)^*}_{1,i-1/2,j,k}\left(p^{n+1}_{i,j,k}-p^{n+1}_{i-1,j,k}\right)\\
&\quad+T^{(n+1)^*}_{2,i,j+1/2,k}\left(p^{n+1}_{i,j+1,k}-p^{n+1}_{i,j,k}\right)-T^{(n+1)^*}_{2,i,j-1/2,k}\left(p^{n+1}_{i,j,k}-p^{n+1}_{i,j-1,k}\right)\\
&\quad+T^{(n+1)^*}_{3,i,j,k+1/2}\left(p^{n+1}_{i,j,k+1}-p^{n+1}_{i,j,k}\right)-T^{(n+1)^*}_{3,i,j,k-1/2}\left(p^{n+1}_{i,j,k}-p^{n+1}_{i,j,k-1}\right)\\
&\quad-(T\gamma)^{(n+1)^*}_{1,i+1/2,j,k}\left(z_{i+1,j,k}-z_{i,j,k}\right)+(T\gamma)^{(n+1)^*}_{1,i-1/2,j,k}\left(z_{i,j,k}-z_{i-1,j,k}\right)\\
&\quad-(T\gamma)^{(n+1)^*}_{2,i,j+1/2,k}\left(z_{i,j+1,k}-z_{i,j,k}\right)+(T\gamma)^{(n+1)^*}_{2,i,j-1/2,k}\left(z_{i,j,k}-z_{i,j-1,k}\right)\\
&\quad-(T\gamma)^{(n+1)^*}_{3,i,j,k+1/2}\left(z_{i,j,k+1}-z_{i,j,k}\right)+(T\gamma)^{(n+1)^*}_{3,i,j,k-1/2}\left(z_{i,j,k}-z_{i,j,k-1}\right)+Q^{n+1}_{i,j,k},
\end{aligned}
\tag{3.108}
$$

which can be also rewritten in system form as in (3.105). Fig. 3.19 indicates the pressure value used to compute the transmissibility coefficients in this approach. It shows that the closer the pressure depletion to a linear function, the more accurate the approximation $p^{(n+1)^*}$ to $p^{(n+1)}$. Note that p^{-1} is required at the first time level t^1, which is generally not given and needs special care in the first iteration.

Fully implicit approach

In the previous three approaches, all the transmissibilities have been linearized before an iterative procedure is used to obtain the pressure at the new time level. The resulting convergence of the iterative procedure is very slow because of the poor approximation of these nonlinear coefficients. Figs. 3.15 and 3.19 illustrate the approximations in these approaches. To speed up convergence, a *fully implicit approach* is used in the solution of the single-phase flow equation:

$$\begin{aligned}
&\left(Vc\left(p^{n+1}\right)\frac{p^{n+1}-p^{n}}{\Delta t}\right)_{i,j,k}\\
&= T^{n+1}_{1,i+1/2,j,k}\left(p^{n+1}_{i+1,j,k}-p^{n+1}_{i,j,k}\right)-T^{n+1}_{1,i-1/2,j,k}\left(p^{n+1}_{i,j,k}-p^{n+1}_{i-1,j,k}\right)\\
&+T^{n+1}_{2,i,j+1/2,k}\left(p^{n+1}_{i,j+1,k}-p^{n+1}_{i,j,k}\right)-T^{n+1}_{2,i,j-1/2,k}\left(p^{n+1}_{i,j,k}-p^{n+1}_{i,j-1,k}\right)\\
&+T^{n+1}_{3,i,j,k+1/2}\left(p^{n+1}_{i,j,k+1}-p^{n+1}_{i,j,k}\right)-T^{n+1}_{3,i,j,k-1/2}\left(p^{n+1}_{i,j,k}-p^{n+1}_{i,j,k-1}\right)\\
&-(T\gamma)^{n+1}_{1,i+1/2,j,k}\left(z_{i+1,j,k}-z_{i,j,k}\right)+(T\gamma)^{n+1}_{1,i-1/2,j,k}\left(z_{i,j,k}-z_{i-1,j,k}\right)\\
&-(T\gamma)^{n+1}_{2,i,j+1/2,k}\left(z_{i,j+1,k}-z_{i,j,k}\right)+(T\gamma)^{n+1}_{2,i,j-1/2,k}\left(z_{i,j,k}-z_{i,j-1,k}\right)\\
&-(T\gamma)^{n+1}_{3,i,j,k+1/2}\left(z_{i,j,k+1}-z_{i,j,k}\right)+(T\gamma)^{n+1}_{3,i,j,k-1/2}\left(z_{i,j,k}-z_{i,j,k-1}\right)\\
&+Q^{n+1}_{i,j,k}.
\end{aligned} \tag{3.109}$$

Now, system (3.109) is a system of nonlinear equations in p^{n+1}, which must be solved at each time step via an iteration method. Here the classic *Newton–Raphson method* is applied.

The Newton–Raphson method. Consider a general system of nonlinear differential equations:

$$\pounds_m\{F_m\left[\mathbf{p}(\mathbf{x})\right]\} = f_m(\mathbf{x}), \quad m = 1, 2, \ldots, M, \quad \mathbf{x} \in \Omega, \tag{3.110}$$

where $\pounds_m$ denotes a linear differential operator, $F_m(\cdot)$ is a nonlinear function, $\mathbf{p} = (p_1, p_2, \ldots, p_M)^T$ is the vector of dependent variables, $\mathbf{f} = (f_1, f_2, \ldots, f_M)^T$ is a given vector, M is the total number of equations, and the superscript T denotes the transpose. The *Newton–Raphson iteration* for solving (3.110) establishes an iterative equation system. Taylor's series expansion for $F_m(\mathbf{p} + \delta\mathbf{p})$ is

$$F_m(\mathbf{p} + \delta\mathbf{p}) = F_m(\mathbf{p}) + \nabla F_m(\mathbf{p}) \cdot \delta\mathbf{p} + \mathcal{O}(|\delta\mathbf{p}|^2), \tag{3.111}$$

where $|\delta\mathbf{p}|$ is the Euclidean norm of $\delta\mathbf{p}$. If the higher order term $\mathcal{O}(|\delta\mathbf{p}|^2)$ (relative to $|\delta\mathbf{p}|$) is truncated, $F_m(\mathbf{p} + \delta\mathbf{p})$ can be approximated:

$$F_m(\mathbf{p} + \delta\mathbf{p}) \approx F_m(\mathbf{p}) + \nabla F_m(\mathbf{p}) \cdot \delta\mathbf{p}. \tag{3.112}$$

If we substitute (3.112) into (3.110), we obtain the iterative equations

$$\begin{aligned}
\pounds_m\left[F_m(\mathbf{p}^l) + \nabla F_m(\mathbf{p}^l) \cdot \delta\mathbf{p}^{l+1}\right] &= f_m(\mathbf{x}),\\
m = 1, 2, \ldots, M, \quad &\mathbf{x} \in \Omega,
\end{aligned} \tag{3.113}$$

where $\mathbf{p}^l$ is the lth iterative solution of $\mathbf{p}$ and $\nabla F_m(\mathbf{p}^l)$ is $\nabla F_m(\mathbf{p})$ at $\mathbf{p} = \mathbf{p}^l$, with an initial solution $\mathbf{p}^0$. In the iterative equation system (3.113), the correction vector $\delta\mathbf{p}^{l+1}$ is the unknown. This system can be rewritten:

$$\pounds_m\left[\nabla F_m(\mathbf{p}^l) \cdot \delta\mathbf{p}^{l+1}\right] = g_m(\mathbf{x}), \quad m = 1, 2, \ldots, M, \quad \mathbf{x} \in \Omega, \tag{3.114}$$

where $g_m(\mathbf{x}) = f_m(\mathbf{x}) - \pounds_m\left[F_m(\mathbf{p}^l)\right]$, and $F_m(\mathbf{p}^l)$ and $\nabla F_m(\mathbf{p}^l)$ are treated as fixed. Now, (3.114) is a linear system for $\delta\mathbf{p}^{l+1}$. Note that $\nabla F_m(\mathbf{p}^l)$ is the Jacobian matrix of F_m and that g_m is the *residual* of equation (3.110) at $\mathbf{p}^l$.

A new solution vector $\mathbf{p}^{l+1}$ is obtained by adding the correction vector $\delta\mathbf{p}^{l+1}$ to the previous iterative solution vector $\mathbf{p}^l$; i.e.,

$$\mathbf{p}^{l+1} = \mathbf{p}^l + \delta\mathbf{p}^{l+1}.$$

This iteration proceeds until the Euclidean norm of $\delta\mathbf{p}^{l+1}$ is smaller than a prescribed value.

The Newton–Raphson method is now applied for linearization of equation (3.109). Below we write

$$p^{n+1,l+1} = p^{n+1,l} + \delta p^{n+1,l+1} \text{ simply by } p^{l+1} = p^l + \delta p^{l+1}.$$

The residual of equation (3.109) at the Newton–Raphson iteration level l is

$$\begin{aligned}
R^l_{i,j,k} = &\left(Vc\left(p^l\right)\frac{p^l - p^n}{\Delta t}\right)_{i,j,k} \\
&- T^l_{1,i+1/2,j,k}\left(p^l_{i+1,j,k} - p^l_{i,j,k}\right) + T^l_{1,i-1/2,j,k}\left(p^l_{i,j,k} - p^l_{i-1,j,k}\right) \\
&- T^l_{2,i,j+1/2,k}\left(p^l_{i,j+1,k} - p^l_{i,j,k}\right) + T^l_{2,i,j-1/2,k}\left(p^l_{i,j,k} - p^l_{i,j-1,k}\right) \\
&- T^l_{3,i,j,k+1/2}\left(p^l_{i,j,k+1} - p^l_{i,j,k}\right) + T^l_{3,i,j,k-1/2}\left(p^l_{i,j,k} - p^l_{i,j,k-1}\right) \\
&+ (T\gamma)^l_{1,i+1/2,j,k}\left(z_{i+1,j,k} - z_{i,j,k}\right) - (T\gamma)^l_{1,i-1/2,j,k}\left(z_{i,j,k} - z_{i-1,j,k}\right) \\
&+ (T\gamma)^l_{2,i,j+1/2,k}\left(z_{i,j+1,k} - z_{i,j,k}\right) - (T\gamma)^l_{2,i,j-1/2,k}\left(z_{i,j,k} - z_{i,j-1,k}\right) \\
&+ (T\gamma)^l_{3,i,j,k+1/2}\left(z_{i,j,k+1} - z_{i,j,k}\right) - (T\gamma)^l_{3,i,j,k-1/2}\left(z_{i,j,k} - z_{i,j,k-1}\right) \\
&- Q^l_{i,j,k},
\end{aligned}$$

where the source/sink term $Q^l_{i,j,k}$ requires special consideration (cf. Chapter 4). Then application of the Newton–Raphson iteration yields a linear system of equations in terms of δp^{l+1}:

$$\begin{aligned}
&\frac{\partial R^l_{i,j,k}}{\partial p_{i,j,k-1}}\delta p^{l+1}_{i,j,k-1} + \frac{\partial R^l_{i,j,k}}{\partial p_{i,j-1,k}}\delta p^{l+1}_{i,j-1,k} + \frac{\partial R^l_{i,j,k}}{\partial p_{i-1,j,k}}\delta p^{l+1}_{i-1,j,k} \\
&\quad + \frac{\partial R^l_{i,j,k}}{\partial p_{i,j,k}}\delta p^{l+1}_{i,j,k} + \frac{\partial R^l_{i,j,k}}{\partial p_{i+1,j,k}}\delta p^{l+1}_{i+1,j,k} + \frac{\partial R^l_{i,j,k}}{\partial p_{i,j+1,k}}\delta p^{l+1}_{i,j+1,k} \\
&\quad + \frac{\partial R^l_{i,j,k}}{\partial p_{i,j,k+1}}\delta p^{l+1}_{i,j,k+1} = -R^l_{i,j,k},
\end{aligned} \tag{3.115}$$

which is the seven-point stencil in the increment δp^{l+1}.

In system (3.115), the derivative of the residual R in pressure p can be computed either numerically or analytically. When the numerical approach is used, the following approximation can be adopted, for example:

$$\frac{\partial T_1}{\partial p} \approx \frac{T_1(p+\epsilon) - T_1(p)}{\epsilon}.$$

The parameter ϵ should be chosen in such a way that it is small enough to produce a reasonable approximation to the derivative but large enough that a machine roundoff error does

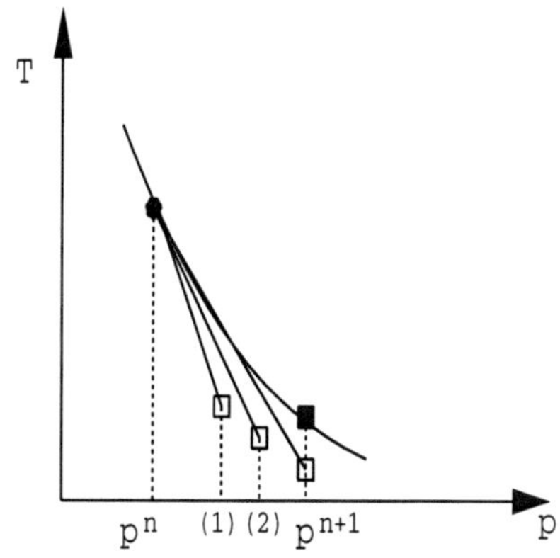

Figure 3.20. *Transmissibility in the fully implicit approach.*

not dominate the approximation. The convergence behavior of the fully implicit approach is indicated in Fig. 3.20.

In summary, we have developed explicit, linearization, and implicit time approximation approaches for numerically solving equation (3.21). In terms of computational efforts, the explicit approach is the simplest at each time step; however, it requires an impractical stability restriction. The linearization approach is more practical, but it reduces the order of accuracy in time for high order time discretization methods (unless extrapolations are exploited) and introduces stability problems. An efficient and accurate method is the fully implicit approach; the extra cost involved at each time step for this implicit method is usually more than compensated for by the fact that larger time steps may be taken. Modified implicit methods such as *semi-implicit methods* (Aziz and Settari, 1979) can be applied; for a given physical problem, the linearization approach should be applied for weak nonlinearity (e.g., the dependence of viscosity μ and density ρ on pressure p for slightly compressible flow), while the implicit one should be used for strong nonlinearity (e.g., the dependence of density ρ on p for compressible flow).

3.4.6 Material Balance Analysis

Material balance is engineering terminology for mass conservation over a fixed volume, which is the hydrocarbon reservoir. To ensure the accuracy of the numerical solution, the material balance must be checked over the entire reservoir: For a closed reservoir (no-flow boundary), the accumulation of mass must be equal to the net mass entering and leaving the boundary. That is, over each time step, it holds that

$$\sum_{i,j,k} \left(Vc\left(p^{n+1}\right) \frac{p^{n+1} - p^n}{\Delta t} \right)_{i,j,k} = \sum_{i,j,k} Q_{i,j,k}^{n+1}. \tag{3.116}$$

This material balance can also be checked over the whole time period:

$$\sum_{n} \sum_{i,j,k} \left(Vc\left(p^{n+1}\right) \left(p^{n+1} - p^n\right) \right)_{i,j,k} = \sum_{n} \sum_{i,j,k} Q_{i,j,k}^{n+1} \Delta t^n. \tag{3.117}$$

If the external boundary contains other types of boundary conditions (e.g., mass transport), equations (3.116) and (3.117) must be changed to account for these conditions. Because approximations occur in the solution process, a perfect material balance relation is seldom achieved.

Note that for an incompressible flow problem with a no-flow boundary, equation (3.116) reduces to

$$\sum_{i,j,k} Q_{i,j,k}^{n+1} = 0,$$

which is the incompressibility condition.

It follows from an addition of equation (3.109) over (i, j, k) that

$$\begin{aligned}
&\sum_{i,j,k} \left(Vc\left(p^{n+1}\right) \frac{p^{n+1} - p^n}{\Delta t} \right)_{i,j,k} \\
&\quad = \sum_{i,j,k} \Big\{ T_{1,i+1/2,j,k}^{n+1} \left(p_{i+1,j,k}^{n+1} - p_{i,j,k}^{n+1} \right) - T_{1,i-1/2,j,k}^{n+1} \left(p_{i,j,k}^{n+1} - p_{i-1,j,k}^{n+1} \right) \\
&\quad + T_{2,i,j+1/2,k}^{n+1} \left(p_{i,j+1,k}^{n+1} - p_{i,j,k}^{n+1} \right) - T_{2,i,j-1/2,k}^{n+1} \left(p_{i,j,k}^{n+1} - p_{i,j-1,k}^{n+1} \right) \\
&\quad + T_{3,i,j,k+1/2}^{n+1} \left(p_{i,j,k+1}^{n+1} - p_{i,j,k}^{n+1} \right) - T_{3,i,j,k-1/2}^{n+1} \left(p_{i,j,k}^{n+1} - p_{i,j,k-1}^{n+1} \right) \\
&\quad - (T\gamma)_{1,i+1/2,j,k}^{n+1} \left(z_{i+1,j,k} - z_{i,j,k} \right) + (T\gamma)_{1,i-1/2,j,k}^{n+1} \left(z_{i,j,k} - z_{i-1,j,k} \right) \\
&\quad - (T\gamma)_{2,i,j+1/2,k}^{n+1} \left(z_{i,j+1,k} - z_{i,j,k} \right) + (T\gamma)_{2,i,j-1/2,k}^{n+1} \left(z_{i,j,k} - z_{i,j-1,k} \right) \\
&\quad - (T\gamma)_{3,i,j,k+1/2}^{n+1} \left(z_{i,j,k+1} - z_{i,j,k} \right) + (T\gamma)_{3,i,j,k-1/2}^{n+1} \left(z_{i,j,k} - z_{i,j,k-1} \right) \Big\} \\
&\quad + \sum_{i,j,k} Q_{i,j,k}^{n+1}.
\end{aligned} \tag{3.118}$$

By the material balance equation (3.116), this equation becomes

$$\begin{aligned}
&\sum_{i,j,k} \Big\{ T_{1,i+1/2,j,k}^{n+1} \left(p_{i+1,j,k}^{n+1} - p_{i,j,k}^{n+1} \right) - T_{1,i-1/2,j,k}^{n+1} \left(p_{i,j,k}^{n+1} - p_{i-1,j,k}^{n+1} \right) \\
&\quad + T_{2,i,j+1/2,k}^{n+1} \left(p_{i,j+1,k}^{n+1} - p_{i,j,k}^{n+1} \right) - T_{2,i,j-1/2,k}^{n+1} \left(p_{i,j,k}^{n+1} - p_{i,j-1,k}^{n+1} \right) \\
&\quad + T_{3,i,j,k+1/2}^{n+1} \left(p_{i,j,k+1}^{n+1} - p_{i,j,k}^{n+1} \right) - T_{3,i,j,k-1/2}^{n+1} \left(p_{i,j,k}^{n+1} - p_{i,j,k-1}^{n+1} \right) \\
&\quad - (T\gamma)_{1,i+1/2,j,k}^{n+1} \left(z_{i+1,j,k}^{n+1} - z_{i,j,k}^{n+1} \right) + (T\gamma)_{1,i-1/2,j,k}^{n+1} \left(z_{i,j,k} - z_{i-1,j,k} \right) \\
&\quad - (T\gamma)_{2,i,j+1/2,k}^{n+1} \left(z_{i,j+1,k} - z_{i,j,k} \right) + (T\gamma)_{2,i,j-1/2,k}^{n+1} \left(z_{i,j,k} - z_{i,j-1,k} \right) \\
&\quad - (T\gamma)_{3,i,j,k+1/2}^{n+1} \left(z_{i,j,k+1} - z_{i,j,k} \right) + (T\gamma)_{3,i,j,k-1/2}^{n+1} \left(z_{i,j,k} - z_{i,j,k-1} \right) \Big\} \\
&\quad = 0.
\end{aligned} \tag{3.119}$$

At any interior gridblock boundary, the flow rate across the boundary is evaluated six times: once for each adjacent block that shares the boundary. Namely, these two terms,

$T_{1,i+1/2,j,k}\left(p_{i+1,j,k} - p_{i,j,k}\right)$ and $T_{1,i-1/2,j,k}\left(p_{i,j,k} - p_{i-1,j,k}\right)$, occur in the x_1-direction; analogous terms are present in the x_2- and x_3-directions. For a closed reservoir system, the transmissibilities on the boundary can be set to zero. Then, by changing the indices in the summation of equation (3.119), we obtain (with $n+1$ omitted)

$$
\begin{aligned}
\sum_{i,j,k}\Big\{ & \left(T^{+}_{1,i-1/2,j,k} - T^{-}_{1,i-1/2,j,k}\right)\left(p_{i,j,k} - p_{i-1,j,k}\right) \\
& + \left(T^{+}_{2,i,j-1/2,k} - T^{-}_{2,i,j-1/2,k}\right)\left(p_{i,j,k} - p_{i,j-1,k}\right) \\
& + \left(T^{+}_{3,i,j,k-1/2} - T^{-}_{3,i,j,k-1/2}\right)\left(p_{i,j,k} - p_{i,j,k-1}\right) \\
& - \left((T\gamma)^{+}_{1,i-1/2,j,k} - (T\gamma)^{-}_{1,i-1/2,j,k}\right)\left(z_{i,j,k} - z_{i-1,j,k}\right) \\
& - \left((T\gamma)^{+}_{2,i,j-1/2,k} - (T\gamma)^{-}_{2,i,j-1/2,k}\right)\left(z_{i,j,k} - z_{i,j-1,k}\right) \\
& - \left((T\gamma)^{+}_{3,i,j,k-1/2} - (T\gamma)^{-}_{3,i,j,k-1/2}\right)\left(z_{i,j,k} - z_{i,j,k-1}\right)\Big\} = 0,
\end{aligned}
\tag{3.120}
$$

where the summation is taken over all the interior gridblocks and $T^{+}_{1,i-1/2,j,k}$ and $T^{-}_{1,i-1/2,j,k}$ denote the transmissibility at an internal boundary evaluated from the positive and negative directions. In the presence of a pressure and depth gradient (i.e., $p_{i,j,k} \neq p_{i-1,j,k}$ and $z_{i,j,k} \neq z_{i-1,j,k}$), equation (3.120) holds only if

$$
T^{+}_{1,i-1/2,j,k} = T^{-}_{1,i-1/2,j,k} \quad \text{and} \quad (T\gamma)^{+}_{1,i-1/2,j,k} = (T\gamma)^{-}_{1,i-1/2,j,k}; \tag{3.121}
$$

similar equations hold in the other two coordinate directions. The physical implication of equation (3.121) is that the flow rates at any internal gridblock boundary evaluated from the positive and negative directions must be equal.

Chapter 4
Well Modeling

The numerical simulation of single-phase flow considered in the previous chapter must account for the treatment of wells. In fact, numerical simulation of fluid flow of any type in petroleum reservoirs must account for the presence of the wells. The pressure at a gridblock that contains a well is different from the average pressure in that block and different from the *flowing bottom hole pressure* for the well (Peaceman, 1977b). The difficulty in modeling wells in a field-scale numerical simulation is that the region where pressure gradients are the largest is closest to a well and is far smaller than the spatial size of gridblocks. Using local grid refinement around the well can alleviate this problem but can lead to an impractical restriction on time step sizes in the numerical simulation. The fundamental task in modeling wells is to model flows into the wellbore accurately and to develop accurate well equations that allow the computation of the bottom hole pressure when a production or injection rate is given, or the computation of the rate when this pressure is known. In this chapter, we develop well flow equations for numerical simulation of single-phase and multiphase flows in petroleum reservoirs using finite difference methods. Different numerical methods produce different forms of well representations (Chen, Huan, and Ma, 2006).

4.1 Introduction

In general, there is not a distributed mass source or sink in fluid flows in a 3D medium. However, as an approximation, we may consider the case where sources and sinks of a fluid are located at isolated points $\mathbf{x}^{(i)}$. Then these point sources and sinks can be surrounded by small spheres that are excluded from the medium. The surface of these spheres can be treated as part of the boundary of the medium, and the mass flow rate per unit volume of each source or sink specifies the total flux through its surface.

A more practical approach to handling *point sources* and *sinks* is to insert them into the mass conservation equation. That is, for point sinks for single-phase flow, for example, we define q in (3.5) by

$$q = -\sum_i \rho q^{(i)} \delta(\mathbf{x} - \mathbf{x}^{(i)}), \tag{4.1}$$

where ρ is the fluid density, $q^{(i)}$ indicates the volume of the fluid produced per unit time at $\mathbf{x}^{(i)}$, and δ is the Dirac delta function. For point sources, q is given by

$$q = \sum_i \rho^{(i)} q^{(i)} \delta(\mathbf{x} - \mathbf{x}^{(i)}), \tag{4.2}$$

where $q^{(i)}$ and $\rho^{(i)}$ denote the volume of the fluid injected per unit time and its density (which is known) at $\mathbf{x}^{(i)}$, respectively. In this chapter we discuss various well representations for q and their corresponding counterparts for multiphase flow.

4.2 Analytical Formulas

The derivation of well flow equations is based on a basic assumption that the flow is *radial* in a neighborhood of the well (cf. Section 3.2) and requires the use of *analytical formulas* for radial flow. These formulas are known only in simplified flow situations. Thus we consider single-phase incompressible flow in isotropic reservoirs. Furthermore, we focus on steady state flow; an unsteady state single-phase flow was described in Section 3.2. In the steady state case, the mass conservation equation is (cf. equations (3.1) and (4.1))

$$\nabla \cdot (\rho \mathbf{u}) = q\delta, \tag{4.3}$$

where $\mathbf{u}$ is the volumetric velocity of the fluid, δ is the Dirac delta function representing a well placed at the origin, for example, and q is the mass production/injection at this well. Darcy's law without the gravity term is (cf. equation (3.4))

$$\mathbf{u} = -\frac{1}{\mu}\mathbf{k}\nabla p, \tag{4.4}$$

where $\mathbf{k}$ is the absolute permeability tensor of the reservoir and p and μ are the fluid pressure and viscosity, respectively.

To obtain an *analytical solution* for equations (4.3) and (4.4), we assume the following:

- The flow is 2D in x_1 and x_2 (i.e., it is homogeneous in the x_3-direction, and gravity is neglected).
- The reservoir is homogeneous and isotropic; i.e., $\mathbf{k} = k\mathbf{I}$ and k is a constant (cf. Section 2.1).
- The viscosity μ and density ρ are constant.
- The flow is radial in a small neighborhood of the well.

With the last assumption, near the well the velocity $\mathbf{u}$ has the form

$$\mathbf{u}(r, \theta) = u(r)(\cos\theta, \sin\theta),$$

where (r, θ) is the polar coordinate system. Since the well is placed at the origin, substitution of this velocity into equation (4.3) gives

$$\frac{du}{dr} + \frac{1}{r}u = 0, \qquad r > 0, \tag{4.5}$$

whose solution is $u = C/r$. The constant C is proportional to q. Note that q represents the mass production/injection. Hence, when the well is an injector, for example, for any small neighborhood B of the origin (a small circle) q is the mass flux

$$q = h_3 \int_B \rho \mathbf{u} \cdot \boldsymbol{\nu}\, da(\mathbf{x}) = 2\pi\rho h_3 C; \quad \text{i.e.,} \quad C = \frac{q}{2\pi\rho h_3},$$

where $\boldsymbol{\nu}$ is the outward unit normal to B and h_3 is the reservoir thickness (or the height of the gridblock containing the well). Consequently, we obtain

$$\mathbf{u} = \frac{q}{2\pi\rho h_3 r}(\cos\theta, \sin\theta). \tag{4.6}$$

Substituting (4.6) into (4.4), taking a dot product of the resulting equation with $\boldsymbol{\nu} = (1, 0)$, and integrating from $(r^o, 0)$ to $(r, 0)$, we obtain

$$p(r) = p(r^o) - \frac{\mu q}{2\pi\rho k h_3} \ln\left(\frac{r}{r^o}\right), \tag{4.7}$$

where $(r^o, 0)$ is a reference point (e.g., r^o is the well radius r_w). Equation (4.7) is the *analytical flow model* near the well, on which the development of well equations for numerical simulation is based.

4.3 Single-Layer Well Models

Van Poolen, Breitenback, and Thurnau (1968) made one of the earliest attempts in the development of a reservoir simulation well model, which is not used today. It was Peaceman (1977b) who made the first comprehensive study of well equations for cell-centered finite difference methods on square grids for single-phase flow. Peaceman's study gave a proper interpretation of a well-block pressure, and indicated how it relates to the flowing bottom hole pressure. The importance of his study is that the computed block pressure is associated with the steady state pressure for the actual well at an *equivalent radius* r_e. For a square grid with a grid size h, Peaceman derived a formula for r_e by three different approaches: (1) analytically by assuming that the pressure in the blocks adjacent to the well block is computed exactly by the radial flow model, obtaining $r_e = 0.208h$; (2) numerically by solving the pressure equation on a sequence of grids, deriving $r_e = 0.2h$; and (3) by solving exactly the system of difference equations and using the equation for the pressure drop between the injector and producer in a repeated five-spot pattern problem, finding $r_e = 0.1987h$. From these approaches, he concluded that $r_e \approx 0.2h$. In this chapter, the first approach is adapted for finite difference methods.

4.3.1 Square Grids

For a square grid K_h, we solve equations (4.3) and (4.4) in the case where the well is located in the center of a grid cell. The adjacent cells are enumerated as in Fig. 4.1. Application of a five-point stencil scheme (cf. Section 3.3.5) to (4.3) and (4.4) gives

$$\frac{\rho k h_3}{\mu}(4p_0 - p_1 - p_2 - p_3 - p_4) = q. \tag{4.8}$$

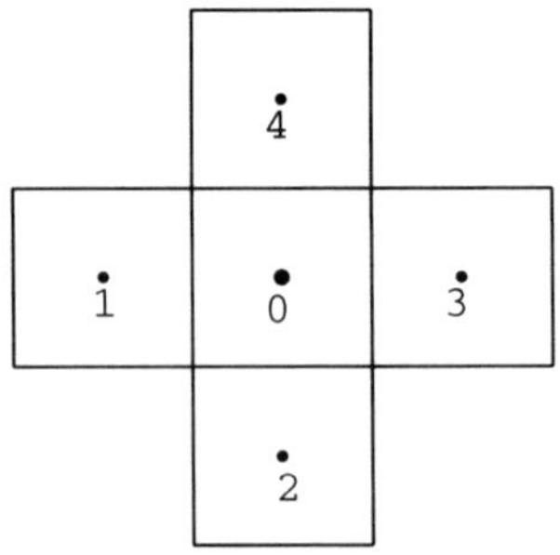

Figure 4.1. *A cell-centered finite difference on a square grid.*

Using the symmetry of the solution p, i.e., $p_1 = p_2 = p_3 = p_4$, we see that

$$\frac{\rho k h_3}{\mu}(p_0 - p_1) = \frac{q}{4}. \tag{4.9}$$

We assume that the pressure at the adjacent cells is computed accurately. In particular, this means that the analytical well model derived in the previous section can be an accurate approximation in cell 1. Thus, if a bottom hole pressure p_{bh} is given, then it follows from equation (4.7) that

$$p_1 = p_{bh} - \frac{\mu q}{2\pi \rho k h_3} \ln\left(\frac{r_1}{r_w}\right), \tag{4.10}$$

where we recall that r_w is the well radius and $r_1 = h$. Inserting equation (4.10) into (4.9) yields

$$\begin{aligned} p_0 &= p_{bh} - \frac{\mu q}{2\pi \rho k h_3} \ln\left(\frac{h}{r_w}\right) + \frac{q\mu}{4\rho k} \\ &= p_{bh} + \frac{\mu q}{2\pi \rho k h_3}\left(\ln\left(\frac{r_w}{h}\right) + \frac{\pi}{2}\right) \\ &= p_{bh} + \frac{\mu q}{2\pi \rho k h_3} \ln\left(\frac{r_w}{\alpha_1 h}\right), \end{aligned}$$

where $\alpha_1 = e^{-\pi/2} = 0.20788\ldots$. This is exactly Peaceman's well model:

$$q = \frac{2\pi \rho k h_3}{\mu \ln(r_e/r_w)}(p_{bh} - p), \tag{4.11}$$

where the *equivalent radius* equals $r_e = \alpha_1 h = 0.20788h$ and $p = p_0$ (cf. Fig. 4.2). The equivalent radius is the radius at which the steady state flowing pressure for the actual well equals the numerically computed pressure for the well cell. When the well is a producer, the flow term q is

$$q = \frac{2\pi \rho k h_3}{\mu \ln(r_e/r_w)}(p - p_{bh}). \tag{4.12}$$

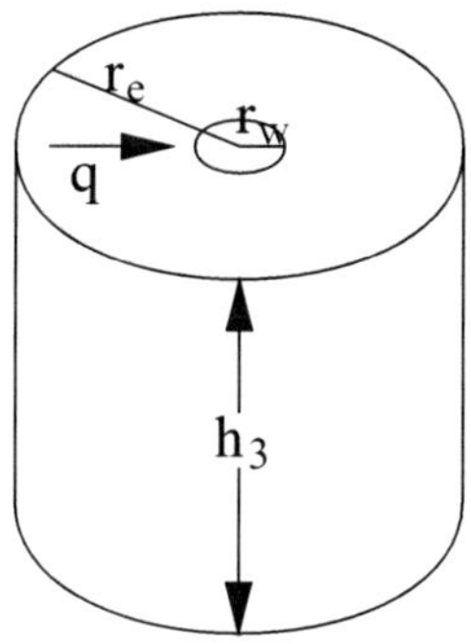

Figure 4.2. *Radial flow.*

4.3.2 Extensions

Extension to anisotropic media

The above well model needs be extended in various directions, including to rectangular grids and incorporating gravity force effects, anisotropic reservoirs, skin effects, horizontal wells, and multiphase flows. Here we consider an extension of the model in equation (4.11) to the first four effects. The gravitational effects must be treated on the same footing as pressure gradient effects. The *skin factor* s is a dimensionless number and accounts for the effect resulting from formation damage caused by drilling. With these effects for single-phase flow for an anisotropic permeability $\mathbf{k} = \text{diag}(k_{11}, k_{22}, k_{33})$, the well model is extended to

$$q = \frac{2\pi\rho h_3\sqrt{k_{11}k_{22}}}{\mu\left(\ln(r_e/r_w)+s\right)}\left(p_{bh} - p - \rho\wp(z_{bh} - z)\right), \tag{4.13}$$

where $\wp$ is the magnitude of the gravitational acceleration, z is the depth, and z_{bh} is the well datum level depth. The factor $\sqrt{k_{11}k_{22}}$ comes from the coordinate transformation: $x_1' = x_1/\sqrt{k_{11}}$ and $x_2' = x_2/\sqrt{k_{22}}$ (Chen, Huan, and Ma, 2006).

In the nonsquare grid and anisotropic medium case, the equivalent radius r_e is (Peaceman, 1983)

$$r_e = \frac{0.14\left((k_{22}/k_{11})^{1/2}h_1^2 + (k_{11}/k_{22})^{1/2}h_2^2\right)^{1/2}}{0.5\left((k_{22}/k_{11})^{1/4} + (k_{11}/k_{22})^{1/4}\right)}, \tag{4.14}$$

where h_1 and h_2 are the x_1- and x_2-grid sizes of the gridblock that contains the vertical well. The *well index* is defined by

$$WI = \frac{2\pi h_3\sqrt{k_{11}k_{22}}}{\ln(r_e/r_w)+s}. \tag{4.15}$$

The effect of near-wellbore factors on well productivity or injectivity can be incorporated by using the skin factor s. Effects that can be treated in this manner include well damage, perforation impacts, partial penetration, fracturing, acidizing, and inclined (deviated) wells. The well skin factors for these effects are additive:

$$s = \sum_i s_i,$$

where s_i indicates each individual skin factor.

Extension to horizontal wells

Horizontal wells in either the x_1- or the x_2-coordinate direction use the same well model equations as vertical ones. Only the parameters related to the direction of the wellbore need to be modified. The well index for a horizontal well parallel to the x_1-direction is calculated:

$$WI = \frac{2\pi h_1 \sqrt{k_{22}k_{33}}}{\ln(r_e/r_w) + s}; \tag{4.16}$$

if the well is parallel to the x_2-direction, it is

$$WI = \frac{2\pi h_2 \sqrt{k_{11}k_{33}}}{\ln(r_e/r_w) + s}. \tag{4.17}$$

Accordingly, in the x_1-direction the equivalent radius r_e is

$$r_e = \frac{0.14\left((k_{33}/k_{22})^{1/2} h_2^2 + (k_{22}/k_{33})^{1/2} h_3^2\right)^{1/2}}{0.5\left((k_{33}/k_{22})^{1/4} + (k_{22}/k_{33})^{1/4}\right)}, \tag{4.18}$$

and in the x_2-direction,

$$r_e = \frac{0.14\left((k_{33}/k_{11})^{1/2} h_1^2 + (k_{11}/k_{33})^{1/2} h_3^2\right)^{1/2}}{0.5\left((k_{33}/k_{11})^{1/4} + (k_{11}/k_{33})^{1/4}\right)}. \tag{4.19}$$

A well in an arbitrary direction (i.e., a *slanted well*) cannot be easily modeled via finite difference methods (Chen, Huan, and Ma, 2006).

The assumptions appropriate for successful implementation of the horizontal well equations (4.16)–(4.19) (Peaceman, 1991) are that the grid spacing and permeability are uniform, the well is isolated (not near any other well), and it is not located near any grid boundary. From Peaceman (1991), these assumptions can be satisfied if the following conditions hold:

- The distance between the well and any other well should be greater than ten times $h = \max\{h_1, h_3\}$ if it is parallel to the x_2-direction or $h = \max\{h_1, h_2\}$ if it is parallel to the x_3-direction.

- The well should not be closer than $5h_1 + h_3/(2h_1)$ (respectively, $5h_1 + h_2/(2h_1)$) from a vertical grid boundary.

- The well should not be closer than $5h_3 + h_1/(2h_3)$ (respectively, $5h_2 + h_1/(2h_2)$) from a horizontal grid boundary.

There are other horizontal well models, such as Babu and Odeh's (1989) pseudosteady state productivity model that is used to evaluate the productivity of a horizontal well. All the existing horizontal well model equations are similar to those of vertical wells in form. Finally, we mention that hybrid grids near a vertical or horizontal well are often used to improve the coupling of the wellbore to the reservoir and the accuracy of solution near the well (cf. Fig. 4.3).

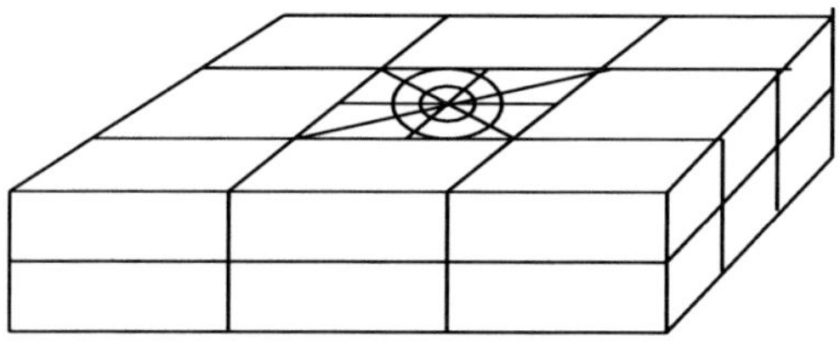

Figure 4.3. *Hybrid grid near a well.*

Extension to multiphase flow

The vertical well equations derived for single-phase flow can be extended to multiphase flow, e.g., to a flow system of water, oil, and gas:

$$q_\alpha = \frac{2\pi h_3 \sqrt{k_{11}k_{22}}}{\ln(r_e/r_w)+s} \frac{\rho_\alpha k_{r\alpha}}{\mu_\alpha} \left(p_{bh} - p_\alpha - \rho_\alpha \wp (z_{bh} - z) \right), \tag{4.20}$$

where ρ_α, $k_{r\alpha}$, μ_α, and p_α are the density, relative permeability, viscosity, and pressure of phase α, respectively, $\alpha = w, o, g$. Note that the definitions of the well index WI and equivalent radius r_e remain the same. The horizontal wells in either the x_1- or the x_2-coordinate direction for the single phase flow can also be extended to multiphase flow.

Off-centered wells

In general, in reservoir simulation, it is desirable to allow no more than one well to penetrate a gridblock. Furthermore, it is often necessary to have at least one or two empty blocks between the wells to accurately simulate pressure interference effects. In addition, in a block-centered grid, the well should be located in the center of a block; in a point-distributed grid, the location of the well should coincide with a grid node. There are cases where placing all the wells in the block centers results in a grid with a large number of blocks that are not computationally effective. In these cases, it is necessary to place some wells in off-center locations.

Abou-Kassem and Aziz (1985) derived an equivalent well radius equation that is applicable to wells located in off-center locations in a square or rectangular gridblock with the aspect ratio h_2/h_1 in the range of 1/2 to 2. The well equation is given by (4.13), with a complicated equivalent radius r_e that accounts for the interface transmissibility between the well block and its surrounding gridblocks.

The boundary of a reservoir also affects the magnitude of the equivalent radius. A well block is termed an interior block if all the reservoir boundary is outside the well block boundary; it is a boundary block, otherwise. Fig. 4.4 shows some examples of the well location and its corresponding equivalent radius (Kuniansky and Hillstad, 1980).

It is sometimes impossible to construct a grid that ensures the existence of only a single well in a gridblock; that is, a block may contain multiple wells. The simplest approach to handling this problem is to represent the multiple wells in the block with a single lumped source/sink term. Combining all these wells into a single hypothetical well using the principle of superposition results in this lumped term. If the combined wells have

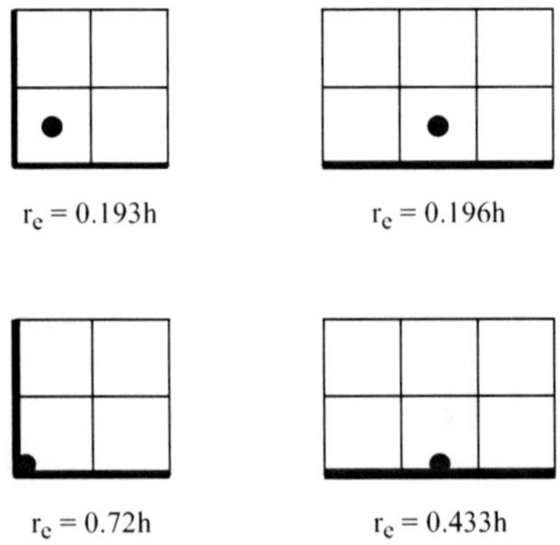

Figure 4.4. *Boundary elements and corresponding equivalent radii.*

analogous characteristics and histories, this lumping approach is physically reasonable. If a more accurate representation is required, pressure interference effects between the wells must be incorporated (Lingen, 1974; Williamson and Chappelear, 1981).

4.4 Multilayer Well Models

If a well penetrates multiple layers (cf. Fig. 4.5), then the mass production/injection at well block m is defined as in equation (4.13):

$$q_m = \frac{2\pi\rho_m h_{m,3}\sqrt{k_{m,11}k_{m,22}}}{\mu_m\left(\ln(r_{m,e}/r_w)+s_m\right)}\left(p_{bh}-p_m-\rho_m\wp(z_{bh}-z_m)\right). \tag{4.21}$$

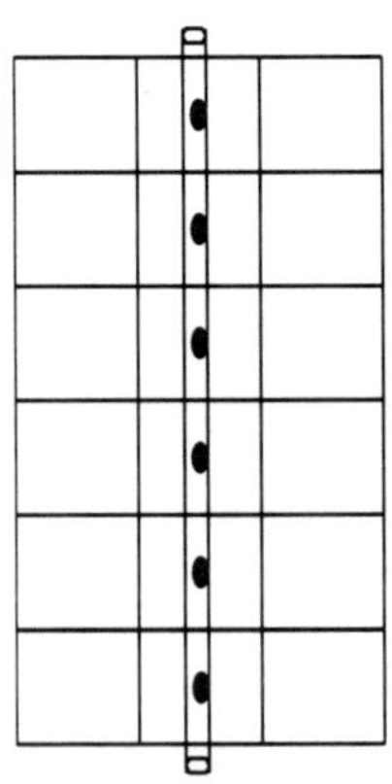

Figure 4.5. *Well penetration of multiple layers.*

Note that all the rock and fluid properties depend on the well-block number m. The total flow for the well in the reservoir is the sum of the rates for all perforated intervals:

$$q = \sum_m \frac{2\pi \rho_m h_{m,3}\sqrt{k_{m,11}k_{m,22}}}{\mu_m \left(\ln(r_{m,e}/r_w) + s_m\right)} \left(p_{bh} - p_m - \rho_m \wp (z_{bh} - z_m)\right). \tag{4.22}$$

For multiphase (e.g., water, oil, and gas) flow, following equation (4.20), this flow equation becomes

$$q_\alpha = \sum_m \frac{2\pi h_{m,3}\sqrt{k_{m,11}k_{m,22}}}{\ln(r_{m,e}/r_w) + s_m} \frac{\rho_{\alpha,m} k_{r\alpha,m}}{\mu_{\alpha,m}} \left(p_{bh} - p_{\alpha,m} - \rho_{\alpha,m} \wp (z_{bh} - z_m)\right) \tag{4.23}$$

for $\alpha = w, o, g$. Multilayer wells in the x_1- and x_2-directions can be similarly modeled.

The pressures at the gridblocks in the column containing the well are coupled. If desired, a specified well rate q can be allocated to each of the blocks in this column by

$$\begin{aligned} q_\nu = q &\frac{2\pi \rho_\nu h_{\nu,3}\sqrt{k_{\nu,11}k_{\nu,22}}}{\mu_\nu \left(\ln(r_{\nu,e}/r_w) + s_\nu\right)} \left(p_{bh} - p_\nu - \rho_\nu \wp (z_{bh} - z_\nu)\right) \\ &\Big/ \sum_m \frac{2\pi \rho_m h_{m,3}\sqrt{k_{m,11}k_{m,22}}}{\mu_m \left(\ln(r_{m,e}/r_w) + s_m\right)} \left(p_{bh} - p_m - \rho_m \wp (z_{bh} - z_m)\right). \end{aligned} \tag{4.24}$$

To simplify the computation, it is sometimes assumed that the pressure head drawdown $p_{bh} - p_m - \rho_m \wp (z_{bh} - z_m)$ and the skin factor s are the same at all well blocks in the column. Under these assumptions, equation (4.24) can be replaced by

$$q_\nu = q \frac{\rho_\nu h_{\nu,3}\sqrt{k_{\nu,11}k_{\nu,22}}/\mu_\nu}{\sum_m \rho_m h_{m,3}\sqrt{k_{m,11}k_{m,22}}/\mu_m}. \tag{4.25}$$

Equations (4.24) and (4.25) can also be generalized to multiphase flow.

4.5 Coupling of Flow and Well Equations

Well equations are coupled to the flow equations presented in the previous and subsequent chapters and need to be solved in a coupled or decoupled fashion, depending on the type of well constraints. In this section we focus on the solution of the single-phase flow and well equations; the subsequent chapters will deal with the multiphase flow and well coupled simulation.

We recall the single-phase flow equation from the previous chapter (cf. (3.21) or (3.91)),

$$c(p)\frac{\partial p}{\partial t} = \nabla \cdot \left(\frac{\rho}{\mu}\mathbf{k}\left(\nabla p - \rho \wp \nabla z\right)\right) + q, \tag{4.26}$$

where the source/sink term q is

$$q = \sum_{\nu=1}^{N_w} \sum_{m=1}^{M_{w\nu}} \frac{2\pi \rho \sqrt{k_{11}k_{22}} h_3}{\mu \left(\ln(r_e/r_w) + s\right)}\Bigg|_m^{(\nu)} \left(p_{bh}^{(\nu)} - p - \rho \wp (z_{bh}^{(\nu)} - z)\right) \delta\left(\mathbf{x} - \mathbf{x}_m^{(\nu)}\right), \tag{4.27}$$

where $\delta(\mathbf{x})$ is the Dirac delta function, N_w is the total number of wells, $M_{w\nu}$ is the total number of perforated zones of the νth well, $h_{m,3}^{(\nu)}$ and $\mathbf{x}_m^{(\nu)}$ are the segment length and central location of the mth perforated zone of the νth well, $p_{bh}^{(\nu)}$ is the *bottom hole pressure* at the datum level depth $z_{bh}^{(\nu)}$, $r_{m,e}^{(\nu)}$ is the equivalent well radius, and $r_w^{(\nu)}$ is the radius of the νth well. Introducing the *well index*

$$WI_m^{(\nu)} = \left.\frac{2\pi\sqrt{k_{11}k_{22}}h_3}{\ln(r_e/r_w)+s}\right|_m^{(\nu)},$$

the mass production/injection at the wells can be written as

$$q = \sum_{\nu=1}^{N_w}\sum_{m=1}^{M_{w\nu}} \left.\frac{WI\rho}{\mu}\right|_m^{(\nu)} \left(p_{bh}^{(\nu)} - p - \gamma(z_{bh}^{(\nu)} - z)\right)\delta\left(\mathbf{x} - \mathbf{x}_m^{(\nu)}\right), \tag{4.28}$$

where $\gamma = \rho\wp$ (the fluid gravity).

Two types of *well constraints* need to be taken into account for single-phase flow: Either the well bottom hole pressure p_{bh} is given (*pressure-specified wells*) or a flow (production or injection) rate is fixed (*rate-specified wells*). In the former case,

$$p_{bh}^{(\nu)} = P_{bh}^{(\nu)}, \tag{4.29}$$

where ν is the number of the well with this type of well control and $P_{bh}^{(\nu)}$ is the given bottom hole pressure at this well. In this case, the flow rate is unknown. In the latter case,

$$q_{spe}^{(\nu)} = \sum_{m=1}^{M_{w\nu}} \left.\frac{WI\rho}{\mu}\right|_m^{(\nu)} \left(p_{bh}^{(\nu)} - p - \gamma(z_{bh}^{(\nu)} - z)\right)\delta\left(\mathbf{x} - \mathbf{x}_m^{(\nu)}\right), \tag{4.30}$$

where $q_{spe}^{(\nu)}$ is specified at the νth well. In this case, the rate is specified for this entire well, and the bottom hole pressure of the νth well, $p_{bh}^{(\nu)}$, needs to be found.

The solution of either $q^{(\nu)}$ or $p_{bh}^{(\nu)}$ depends on the solution approach for the flow equation (4.26): explicit, linearization, extrapolation, or fully implicit (cf. Section 3.4.5).

Explicit approach

In the explicit solution approach, the source/sink term is handled explicitly (i.e., it is evaluated at the old time level t^n):

$$q^n = \sum_{\nu=1}^{N_w}\sum_{m=1}^{M_{w\nu}} \left.\frac{WI\rho^n}{\mu^n}\right|_m^{(\nu)} \left(\left(p_{bh}^{(\nu)}\right)^n - p^n - \gamma^n(z_{bh}^{(\nu)} - z)\right)\delta\left(\mathbf{x} - \mathbf{x}_m^{(\nu)}\right). \tag{4.31}$$

Thus both the source/sink term q and bottom hole pressure $p_{bh}^{(\nu)}$ are evaluated explicitly, and one can be found from the other using this well equation. Then $Q_{i,j,k}^n = (Vq^n)_{i,j,k}$ supplements the flow equation (3.102).

Linearization approach

The linearization solution approach results in the following well equation:

$$q^{n+1} = \sum_{\nu=1}^{N_w} \sum_{m=1}^{M_{w\nu}} \left. \frac{WI\rho^n}{\mu^n} \right|_m^{(\nu)} \left(\left(p_{bh}^{(\nu)} \right)^{n+1} - p^{n+1} - \gamma^n (z_{bh}^{(\nu)} - z) \right) \delta \left(\mathbf{x} - \mathbf{x}_m^{(\nu)} \right). \tag{4.32}$$

Equation (4.32) must be coupled to the flow equation (3.104) for the solution of the primary unknowns p and $p_{bh}^{(\nu)}$ (or $q^{(\nu)}$). For example, for a rate-specified well, in addition to the pressures of the neighboring gridblocks caused by interblock flow ($p_{i,j,k-1}$, $p_{i,j-1,k}$, $p_{i-1,j,k}$, $p_{i,j,k}$, $p_{i+1,j,k}$, $p_{i,j+1,k}$, and $p_{i,j,k+1}$) from equation (3.104), the pressures of the column of the gridblocks that contain the well and the bottom hole pressure also appear in the system of finite difference equations.

Extrapolation approach

Similarly, in the extrapolation approach, the source/sink term is

$$\begin{aligned} q^{n+1} = \sum_{\nu=1}^{N_w} \sum_{m=1}^{M_{w\nu}} \left. \frac{WI\rho^{(n+1)^*}}{\mu^{(n+1)^*}} \right|_m^{(\nu)} & \left(\left(p_{bh}^{(\nu)} \right)^{n+1} - p^{n+1} - \gamma^{(n+1)^*} (z_{bh}^{(\nu)} - z) \right) \\ & \times \delta \left(\mathbf{x} - \mathbf{x}_m^{(\nu)} \right), \end{aligned} \tag{4.33}$$

where $(n+1)^*$ is defined as in (3.107). The solution for the fluid pressure p and the bottom hole pressure $p_{bh}^{(\nu)}$ (or the flow rate $q^{(\nu)}$) can be carried out as in the linearization approach.

Fully implicit approach

Finally, the fully implicit solution approach requires an implicit treatment of the source/sink term:

$$\begin{aligned} q^{n+1} = \sum_{\nu=1}^{N_w} \sum_{m=1}^{M_{w\nu}} \left. \frac{WI\rho^{(n+1)}}{\mu^{(n+1)}} \right|_m^{(\nu)} & \left(\left(p_{bh}^{(\nu)} \right)^{n+1} - p^{n+1} - \gamma^{(n+1)} (z_{bh}^{(\nu)} - z) \right) \\ & \times \delta(\mathbf{x} - \mathbf{x}_m^{(\nu)}), \end{aligned} \tag{4.34}$$

which can be linearized via the Newton–Raphson method, for example. For the constraint of type (4.29), the well bottom hole pressure is given, so its increment at the Newton–Raphson iteration level $l+1$ is zero (cf. the residual equation (3.115)),

$$\delta \left(p_{bh}^{(\nu)} \right)^{n+1,l+1} = 0, \tag{4.35}$$

and system (3.115) involves only the computation of pressure after substitution of equation (4.34) into the fully implicit system (3.109).

For the constraint of type (4.30), $\delta(p_{bh}^{(\nu)})^{n+1,l+1}$ needs to be found through the equation (with the superscript $n+1$ omitted)

$$q_{spe}^{(\nu)} = \sum_{m=1}^{M_{w\nu}} \left. \frac{WI\rho^{(l+1)}}{\mu^{(l+1)}} \right|_m^{(\nu)} \left(\left(p_{bh}^{(\nu)} \right)^{l+1} - p^{l+1} - \gamma^{(l+1)} (z_{bh}^{(\nu)} - z) \right) \delta(\mathbf{x} - \mathbf{x}_m^{(\nu)}), \tag{4.36}$$

which is coupled to the flow system (3.115). In this case, the primary unknowns are δp^{l+1} and $\delta(p_{bh}^{(\nu)})^{l+1}$.

4.6 Coupling of Wellbore-Hydraulics and Reservoir Models

4.6.1 Single-Phase Flow

It is sometimes necessary to specify a rate or pressure at the wellhead or some other point in the surface facilities. This may be true when several wells are producing into a common manifold, flowline, or separator, for example. When the surface pressure p_{sp} is specified, it must be converted into the well bottom hole pressure p_{bh} for inclusion in the well model (4.13) or (4.20). To do so, a *hydraulics model* is used from either an energy or pressure balance, or from experiments; i.e., the total pressure drop in a length of tubing is

$$p_{bh} - p_{sp} = \Delta p_{\mathrm{g}} + \Delta p_{\mathrm{f}} + \Delta p_{\mathrm{a}} + \Delta p_{\mathrm{W}}, \tag{4.37}$$

where the four terms on the right-hand side of this equation, respectively, correspond to the pressure drops caused by the hydrostatic head or specific weight loss, friction loss, acceleration loss or kinetic energy change, and external work loss (e.g., driving a turbine; a pump would introduce a $-\Delta p_{\mathrm{W}}$), and the well direction l is positive in the downward direction. The gravitational pressure drop is given by changes in the hydrostatic head:

$$\Delta p_{\mathrm{g}} = \rho \wp \sin\theta \Delta l,$$

where $\theta = \pi/2$ for a true vertical well. For horizontal or deviated wells, $\Delta l > \Delta z$. The frictional pressure drop is related to the friction factor f_m, velocity u_3 in direction l, and the inner diameter D of tubing:

$$\Delta p_{\mathrm{f}} = \frac{f_m \rho u_3^2}{2D} \Delta l.$$

The acceleration pressure drop is

$$\Delta p_{\mathrm{a}} = \rho u_3^2 \frac{du_3}{dl} \Delta l,$$

which, in most practical reservoir applications, is ignored. The acceleration contribution is significant when flow is through critical flow provers or two-phase chokes (Nind, 1981). Finally, the pressure drop caused by external work normally does not appear in reservoir simulation. Common practice is to set boundary conditions at the sandface, at the pump suction in the case of pumping wells, or at the wellhead or separators in the case of flowing or gas lift wells.

Water injection

For a water injection well, equation (4.37) reduces to the relation for the wellhead pressure p_{wh} and the bottom hole pressure p_{bp},

$$p_{bh} = p_{wh} + \Delta p_{\mathrm{g}} + \Delta p_{\mathrm{f}}, \tag{4.38}$$

which can be used to find p_{bh} when p_{wh} is given. For single-phase flow, it is often adequate to express the frictional pressure drop in terms of the Reynolds number,

$$\Delta p_{\mathrm{f}} = \mathcal{K} q^a, \tag{4.39}$$

where the coefficient $\mathcal{K}$ is determined experimentally or derived from an analytic expression for the friction pressure gradient and the exponent a is experimentally determined (normally in the range of 1.75 to 1.85 for water). For injection, the injection rate q must be a negative quantity.

Gas production and injection

Because gas is highly compressible, an equation of state relating pressure to density is required for gas before integration over length. In this case, a simple hydraulics model is used (Smith, 1983):

$$p_{bh}^2 = \mathcal{A}p_{wh}^2 + \mathcal{B}(\mathcal{A} - 1)q^2, \tag{4.40}$$

where $\mathcal{A}$ is a function of gas gravity, compressibility factor Z, length, and temperature and $\mathcal{B}$ is a function of compressibility factor, diameter, temperature, and friction loss. Equation (4.40) holds for vertical flow through a circular conduit under the conditions of a linear temperature distribution, constant compressibility factor, and negligible acceleration effect.

For horizontal flowlines with a constant compressibility factor and temperature and negligible acceleration, equation (4.40) becomes

$$p_{bh}^2 = p_{sp}^2 + \mathcal{C}q^2, \tag{4.41}$$

where $\mathcal{C}$ is a function of gas gravity, compressibility factor, diameter, temperature, and friction loss.

4.6.2 Multiphase Flow

Solution for a multiphase hydraulics model similar to (4.37) is beyond the scope of this book. In general, multiphase flow data for the surface pressure p_{sp}, the water cut f_w, the gas/liquid ratio R_{gl} (equivalently, gas/oil ratio R_{go}), and the well rate q are generated separately for a number of cases that cover the circumstances likely to occur during simulation. This information is included in a tabular form into the simulator, and the flow from the bottom hole datum to the surface is determined by interpolation among the tabulated data (Brown, 1977; Nind, 1981). Fig. 4.6 shows a family of curves for R_{gl} with p_{sp} and f_w fixed, where we recall that q is positive for injection and negative for production (i.e., in the latter case, the horizontal axis uses $-q$).

Here we briefly discuss an algorithm for converting a specified surface pressure p_{sp} to a flowing bottom hole pressure p_{bh} and a well rate q for the black oil system of water, oil, and gas to be studied in Chapter 6. The capillary pressures are defined as in equation (2.19):

$$p_{cow}(S_w) = p - p_w, \quad p_{cgo}(S_g) = p_g - p, \tag{4.42}$$

where $p = p_o$. It follows from the multiphase well model (4.23) that

$$\begin{aligned} q_w &= \sum_k WI_k \lambda_{w,k} \left(p_{bh} - p_k + p_{cow,k} - \gamma_{w,k}(z_{bh} - z_k)\right) / B_{w,k}, \\ q_o &= \sum_k WI_k \lambda_{o,k} \left(p_{bh} - p_k - \gamma_{o,k}(z_{bh} - z_k)\right) / B_{o,k}, \\ q_g &= \sum_k WI_k \lambda_{g,k} \left(p_{bh} - p_k - p_{cgo,k} - \gamma_{g,k}(z_{bh} - z_k)\right) / B_{g,k}, \end{aligned} \tag{4.43}$$

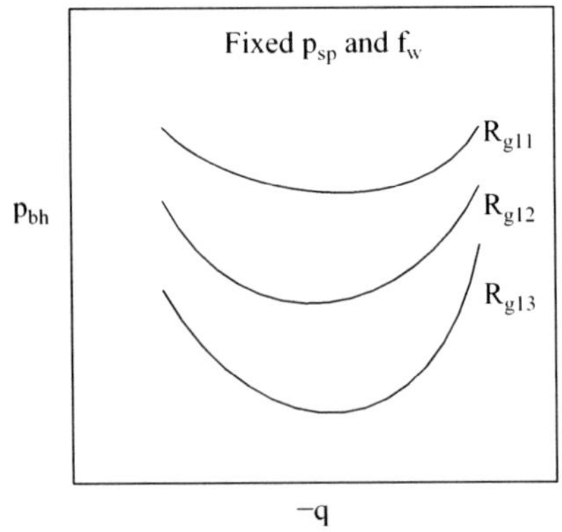

Figure 4.6. *A family of curves for R_{gl}.*

where the well index WI is defined in (4.15), B_α is the formation volume factor of the α-phase, and the mobility λ_α and fluid gravity γ_α are

$$\lambda_\alpha = \frac{k_{r\alpha}}{\mu_\alpha}, \quad \gamma_\alpha = \rho_\alpha \wp, \quad \alpha = w, o, g.$$

For the black oil model, the total well rate is (cf. Chapter 6)

$$q = q_w + (1 + R_{so})\, q_o + q_g, \tag{4.44}$$

where R_{so} is the gas solubility factor. Substituting (4.43) into equation (4.44) yields

$$\begin{aligned} q = {} & \sum_k WI_k \left(\frac{\lambda_{w,k}}{B_{w,k}} + (1 + R_{so}) \frac{\lambda_{o,k}}{B_{o,k}} + \frac{\lambda_{g,k}}{B_{g,k}} \right) (p_{bh} - p_k) \\ & + \sum_k WI_k \left(\frac{\lambda_{w,k}}{B_{w,k}} p_{cow,k} - \frac{\lambda_{g,k}}{B_{g,k}} p_{cgo,k} \right) \\ & - \sum_k WI_k \left(\frac{\lambda_{w,k}}{B_{w,k}} \gamma_{w,k} + (1 + R_{so}) \frac{\lambda_{o,k}}{B_{o,k}} \gamma_{o,k} + \frac{\lambda_{g,k}}{B_{g,k}} \gamma_{g,k} \right) (z_{bh} - z_k). \end{aligned} \tag{4.45}$$

Introduce the notation

$$\begin{aligned} W_t = {} & \sum_k WI_k \left(\frac{\lambda_{w,k}}{B_{w,k}} + (1 + R_{so}) \frac{\lambda_{o,k}}{B_{o,k}} + \frac{\lambda_{g,k}}{B_{g,k}} \right), \\ \bar{p} = {} & \frac{1}{W_t} \Bigg\{ \sum_k WI_k \left(\frac{\lambda_{w,k}}{B_{w,k}} + (1 + R_{so}) \frac{\lambda_{o,k}}{B_{o,k}} + \frac{\lambda_{g,k}}{B_{g,k}} \right) p_k \\ & - \sum_k WI_k \left(\frac{\lambda_{w,k}}{B_{w,k}} p_{cow,k} - \frac{\lambda_{g,k}}{B_{g,k}} p_{cgo,k} \right) \\ & + \sum_k WI_k \left(\frac{\lambda_{w,k}}{B_{w,k}} \gamma_{w,k} + (1 + R_{so}) \frac{\lambda_{o,k}}{B_{o,k}} \gamma_{o,k} + \frac{\lambda_{g,k}}{B_{g,k}} \gamma_{g,k} \right) (z_{bh} - z_k) \Bigg\}. \end{aligned} \tag{4.46}$$

Then equation (4.45) can be written as

$$p_{bh} = \bar{p} + \frac{q}{W_t}, \tag{4.47}$$

which shows a linear relationship between p_{bh} and q (cf. Fig. 4.7).

Because the well must obey both the R_{gl} (tubing performance) and $p_{bh} - q$ (inflow performance) relationships, it must operate at the intersection of these two curves. In general, however, there exist two intersections (cf. Fig. 4.7), so a further study is required before a proper operating point is selected. As an example, a production well is examined, and a perturbation is introduced in the production rate:

$$q' = q + \epsilon. \tag{4.48}$$

For the intersection with the higher value of q (i.e., the right intersection in Fig. 4.7), a positive ϵ gives a value of p_{bh} larger for the tubing performance curve than that for the inflow performance curve. That is, the tubing exerts additional backpressure against the sandface. As a result, the production rate drifts back to q. On the other hand, a negative ϵ generates a slightly underbalanced well, and again the production rate drifts back to q. Hence the intersection with the higher value of q is a *stable operating point.*

For the intersection with the lower value of q (i.e., the left intersection), a positive ϵ generates an underbalanced pressure, causing q to increase (ultimately terminating at the stable operating point). On the other hand, when ϵ is negative, the tubing exerts a larger backpressure and further chokes back the well. Ultimately, the well loads up and stops flowing. Therefore, this intersection is an *unstable operating point.*

We now state an algorithm for converting the specified surface pressure p_{sp} to the flowing bottom hole pressure p_{bh} and the well rate q:

1. Evaluate q_w, q_o, and q_g at the old time level t^n.
2. Compute the total well productivity W_t according to the definition given in (4.46).
3. Calculate the averaged pressure $\bar{p}$ according to (4.46).

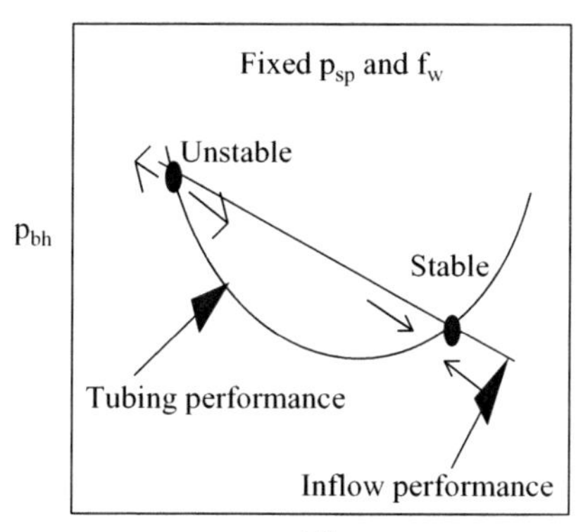

Figure 4.7. *Intersection of inflow and tubing performances.*

4. Evaluate the water cut f_w and the gas/liquid ratio R_{gl}:

$$f_w = \frac{B_w q_w}{B_w q_w + B_o q_o}, \qquad R_{gl} = \frac{B_g (q_g - R_{so} q_o)}{B_w q_w + B_o q_o}.$$

5. Use the specified p_{sp} and evaluated f_w and R_{gl} to choose a proper tubing performance curve from the tables that are read for the well.

6. Find the interactions of the tubing and inflow performance curves, and obtain p_{bh} and q from the stable operating point (one with the higher value of q).

7. Check the new values of q_w, q_o, and q_g. If they are sufficiently close to the old values, the solution converges. Otherwise, continue from step 2.

If the flowing bottom hole pressure p_{bh} and the well rate q are given, this algorithm can be employed to compute the surface pressure p_{sp} in a reverse manner. An analogous algorithm can be also defined for an injection well.

Chapter 5

Two-Phase Flow and Numerical Solution

In reservoir simulation, we are often interested in the *simultaneous flow* of two or more fluid phases within a porous medium. We now develop basic equations for multiphase flow in a porous medium and their numerical solution. In this chapter, we consider two-phase flow where the fluids are *immiscible* and there is no mass transfer between the phases. One phase (e.g., water) wets the porous medium more than the other (e.g., oil) and is called the *wetting phase*, indicated by a subscript w. The other phase is termed the *nonwetting phase* and indicated by o. In general, water is the wetting fluid relative to oil and gas, while oil is the wetting fluid relative to gas.

In addition to the basic differential equations for two-phase immiscible flow, alternative differential formulations for these differential equations are also discussed. A 1D case where an *analytic solution* can be obtained is studied. A solution approach, *IMPES* (implicit pressure–explicit saturation), for solving the two-phase differential equations is presented and compared with a recently introduced approach, an *improved IMPES*. Finally, numerical solution of the two-phase differential equations is described, and the treatment of transmissibilities at gridblock boundaries is stressed.

5.1 Basic Differential Equations

5.1.1 Mass Conservation

Several new quantities peculiar to multiphase flow, such as *saturation*, *capillary pressure*, and *relative permeability*, must be introduced (cf. Chapter 2). The saturation of a fluid phase is defined as the fraction of the void volume of a porous medium filled by this phase. The fact that the two fluids jointly fill the voids implies the relation

$$S_w + S_o = 1, \tag{5.1}$$

where S_w and S_o are the saturations of the wetting and nonwetting phases, respectively. Also, due to the *curvature* and *surface tension* of the interface between the two phases, the pressure in the wetting fluid is less than that in the nonwetting fluid. The pressure difference is given by the capillary pressure

$$p_c = p_o - p_w, \tag{5.2}$$

where p_w and p_o are the respective pressures of the wetting and nonwetting phases. Empirically, the capillary pressure is a function of saturation S_w (cf. Chapter 2).

Except for the accumulation term, the same derivation that led to equation (3.1) also applies to the mass conservation equation for each fluid phase. Mass accumulation in a differential volume per unit time is

$$\frac{\partial(\phi \rho_\alpha S_\alpha)}{\partial t} \Delta x_1 \Delta x_2 \Delta x_3,$$

where ϕ is the porosity of the porous medium, each phase has its own density ρ_α, and Δx_i is the length of the cube in the x_i-direction, $i = 1, 2, 3$, in Fig. 3.1. Taking into account this and the assumption that there is no mass transfer between phases in the immiscible flow, mass is conserved within each phase,

$$\frac{\partial(\phi \rho_\alpha S_\alpha)}{\partial t} = -\nabla \cdot (\rho_\alpha \mathbf{u}_\alpha) + q_\alpha, \qquad \alpha = w, o, \tag{5.3}$$

where each phase also has its own Darcy's velocity $\mathbf{u}_\alpha$ and mass flow rate q_α.

5.1.2 Darcy's Law

Darcy's law for single-phase flow can be directly extended to multiphase flow. In the present case it relates the total volumetric flow rate of each fluid phase through a porous medium to its pressure gradient and the properties of the fluid (viscosity, μ_w or μ_o) and the medium (*effective permeability*, k_w or k_o, and a cross-sectional area, A). For example, in the x_1-direction (cf. Fig. 5.1), it is written as

$$q_w = -\frac{k_w A}{\mu_w} \frac{\partial p_w}{\partial x_1}, \qquad q_o = -\frac{k_o A}{\mu_o} \frac{\partial p_o}{\partial x_1}. \tag{5.4}$$

Darcy's velocities for both phases are $u_w = q_w / A$ and $u_o = q_o / A$; as a result, we see that

$$u_w = -\frac{k_w}{\mu_w} \frac{\partial p_w}{\partial x_1}, \qquad u_o = -\frac{k_o}{\mu_o} \frac{\partial p_o}{\partial x_1}. \tag{5.5}$$

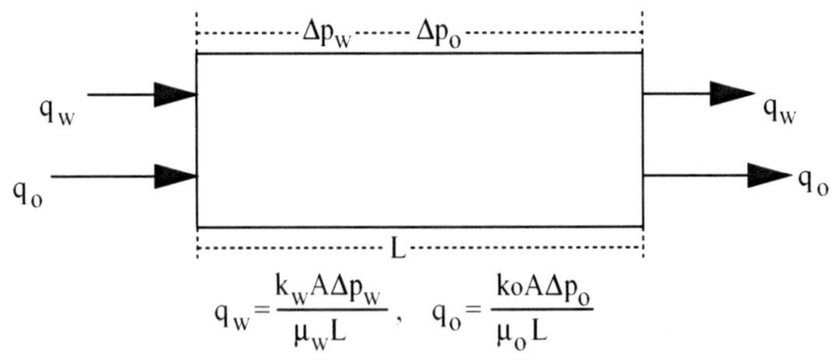

Figure 5.1. *Volumetric flow rates.*

For a 3D two-phase flow system with gravitational force, the differential form of Darcy's law is

$$\mathbf{u}_\alpha = -\frac{1}{\mu_\alpha}\mathbf{k}_\alpha\left(\nabla p_\alpha - \rho_\alpha \wp \nabla z\right), \qquad \alpha = w, o, \tag{5.6}$$

where $\wp$ is the magnitude of the gravitational acceleration and z is the depth.

Since the simultaneous flow of two fluids causes each to interfere with the other, the effective permeabilities are not greater than the absolute permeability $\mathbf{k}$ of the porous medium. The *relative permeabilities* $k_{r\alpha}$ are widely used in reservoir simulation:

$$\mathbf{k}_\alpha = k_{r\alpha}\mathbf{k}, \qquad \alpha = w, o. \tag{5.7}$$

The function $k_{r\alpha}$ indicates the tendency of phase α to wet the porous medium. Typical functions of $p_c(S_w)$ and $k_{r\alpha}(S_w)$ were described in Chapter 2.

5.1.3 Alternative Differential Equations

In this section, we derive several alternative formulations for the two-phase flow differential equations.

Formulation in phase pressures

Assume that the capillary pressure p_c has a unique inverse function:

$$S_w = p_c^{-1}(p_o - p_w).$$

We use p_w and p_o as the main unknowns. Then it follows from (5.1)–(5.3) and (5.6) that

$$\begin{aligned} \nabla \cdot \left(\frac{\rho_w}{\mu_w}\mathbf{k}_w\left(\nabla p_w - \rho_w \wp \nabla z\right)\right) &= \frac{\partial(\phi \rho_w p_c^{-1})}{\partial t} - q_w, \\ \nabla \cdot \left(\frac{\rho_o}{\mu_o}\mathbf{k}_o\left(\nabla p_o - \rho_o \wp \nabla z\right)\right) &= \frac{\partial\big(\phi \rho_o (1 - p_c^{-1})\big)}{\partial t} - q_o. \end{aligned} \tag{5.8}$$

This system was employed in the *simultaneous solution* (SS) scheme in petroleum reservoirs (Douglas, Peaceman, and Rachford, 1959). The equations in this system are strongly nonlinear and coupled.

Formulation in phase pressure and saturation

We use p_o and S_w as the main variables. Applying (5.1), (5.2), and (5.6), equation (5.3) can be rewritten as

$$\begin{aligned} \nabla \cdot \left(\frac{\rho_w}{\mu_w}\mathbf{k}_w\left(\nabla p_o - \frac{dp_c}{dS_w}\nabla S_w - \rho_w \wp \nabla z\right)\right) &= \frac{\partial(\phi \rho_w S_w)}{\partial t} - q_w, \\ \nabla \cdot \left(\frac{\rho_o}{\mu_o}\mathbf{k}_o\left(\nabla p_o - \rho_o \wp \nabla z\right)\right) &= \frac{\partial\big(\phi \rho_o (1 - S_w)\big)}{\partial t} - q_o. \end{aligned} \tag{5.9}$$

Carrying out the time differentiation in (5.9), dividing the first and second equations by ρ_w and ρ_o, respectively, and adding the resulting equations, we obtain

$$\frac{1}{\rho_w}\nabla\cdot\left(\frac{\rho_w}{\mu_w}\mathbf{k}_w\left(\nabla p_o-\frac{dp_c}{dS_w}\nabla S_w-\rho_w\wp\nabla z\right)\right)+\frac{1}{\rho_o}\nabla\cdot\left(\frac{\rho_o}{\mu_o}\mathbf{k}_o\left(\nabla p_o-\rho_o\wp\nabla z\right)\right)$$
$$=\frac{S_w}{\rho_w}\frac{\partial(\phi\rho_w)}{\partial t}+\frac{1-S_w}{\rho_o}\frac{\partial(\phi\rho_o)}{\partial t}-\frac{q_w}{\rho_w}-\frac{q_o}{\rho_o}. \tag{5.10}$$

Note that if the saturation S_w in (5.10) is explicitly evaluated, we can use this equation to solve for p_o. After computing this pressure, the second equation in (5.9) can be used to calculate S_w. This is the *implicit pressure–explicit saturation* (IMPES) scheme and has been widely exploited for two-phase flow in petroleum reservoirs (cf. Section 5.3).

Simplifications for incompressible fluids

We now develop three alternative formulations under the assumption that the two fluids are incompressible, which is physically reasonable for water and oil. The following three formulations also have similar counterparts for compressible fluids (Chen and Ewing, 1997a).

Phase formulation. Introduce the *phase mobilities*

$$\lambda_\alpha=\frac{k_{r\alpha}}{\mu_\alpha},\qquad \alpha=w,o,$$

and the *total mobility*

$$\lambda=\lambda_w+\lambda_o.$$

Also, define the *fractional flow* functions

$$f_\alpha=\frac{\lambda_\alpha}{\lambda},\qquad \alpha=w,o.$$

We use the oil pressure and water saturation as the *primary variables*

$$p=p_o,\qquad S=S_w. \tag{5.11}$$

Define the total velocity

$$\mathbf{u}=\mathbf{u}_w+\mathbf{u}_o. \tag{5.12}$$

Under the assumption that the fluids are incompressible, we apply (5.1) and (5.12) to (5.3) to see that

$$\nabla\cdot\mathbf{u}=\tilde{q}(p,S)\equiv\tilde{q}_w(p,S)+\tilde{q}_o(p,S), \tag{5.13}$$

and (5.2) and (5.12) to (5.6) to obtain

$$\mathbf{u}=-\mathbf{k}\left[\lambda(S)\nabla p-\lambda_w(S)\nabla p_c-\left(\lambda_w\rho_w+\lambda_o\rho_o\right)\wp\nabla z\right], \tag{5.14}$$

where $\tilde{q}_w=q_w/\rho_w$ and $\tilde{q}_o=q_o/\rho_o$. Substituting (5.14) into (5.13) yields the *pressure equation*

$$-\nabla\cdot\left(\mathbf{k}\lambda\nabla p\right)=\tilde{q}-\nabla\cdot\left(\mathbf{k}(\lambda_w\nabla p_c+(\lambda_w\rho_w+\lambda_o\rho_o)\wp\nabla z)\right). \tag{5.15}$$

The phase velocities $\mathbf{u}_w$ and $\mathbf{u}_o$ are related to the total velocity $\mathbf{u}$ by

$$\begin{aligned}\mathbf{u}_w &= f_w\mathbf{u} + \mathbf{k}\lambda_o f_w \nabla p_c + \mathbf{k}\lambda_o f_w(\rho_w - \rho_o)\wp\nabla z,\\ \mathbf{u}_o &= f_o\mathbf{u} - \mathbf{k}\lambda_w f_o \nabla p_c + \mathbf{k}\lambda_w f_o(\rho_o - \rho_w)\wp\nabla z.\end{aligned}$$

Similarly, we apply (5.2), (5.12), and (5.14) to (5.3) and (5.6) with $\alpha = w$ to obtain the *saturation equation*

$$\begin{aligned}\phi\frac{\partial S}{\partial t} + \nabla\cdot\bigg\{\mathbf{k} f_w(S)\lambda_o(S)\left(\frac{dp_c}{dS}\nabla S - (\rho_o - \rho_w)\wp\nabla z\right)\\ + f_w(S)\mathbf{u}\bigg\} = \tilde{q}_w(p, S),\end{aligned} \tag{5.16}$$

where, for notational convenience, we assume that $\phi = \phi(\mathbf{x})$.

Weighted formulation. We introduce a pressure that is smoother than the phase pressure:

$$p = S_w p_w + S_o p_o. \tag{5.17}$$

Even if a phase disappears (i.e., either S_w or S_o is zero), there is still a nonzero smooth variable p. Applying the same algebraic manipulations as in deriving the phase formulation, we obtain

$$\begin{aligned}\mathbf{u} = -\mathbf{k}\big\{&\lambda(S)\nabla p + \big(S\lambda(S) - \lambda_w(S)\big)\nabla p_c + \lambda(S)p_c\nabla S\\ &- \big(\lambda_w\rho_w + \lambda_o\rho_o\big)\wp\nabla z\big\}.\end{aligned} \tag{5.18}$$

Equations (5.13) and (5.16) remain the same.

Global formulation. Note that p_c appears in both (5.14) and (5.18). To remove it, we define a global pressure (Antontsev, 1972; Chavent and Jaffré, 1978):

$$p = p_o - \int^S \left(f_w\frac{dp_c}{dS}\right)(\xi)\, d\xi. \tag{5.19}$$

Using this pressure, the total velocity becomes

$$\mathbf{u} = -\mathbf{k}\left(\lambda(S)\nabla p - \big(\lambda_w\rho_w + \lambda_o\rho_o\big)\wp\nabla z\right). \tag{5.20}$$

It follows from (5.2) and (5.19) that

$$\lambda\nabla p = \lambda_w\nabla p_w + \lambda_o\nabla p_o,$$

which implies that the global pressure is the pressure that would produce flow of a fluid (with mobility λ) equal to the sum of the flows of fluids w and o. Again, equations (5.13) and (5.16) remain the same.

The coupling between the pressure and saturation equations in the global formulation is less than that in the phase and weighted formulations, and the nonlinearity is weakened as well. This formulation is most suitable for a mathematical analysis for two-phase flow (Antontsev, 1972; Chavent and Jaffré, 1978; Chen, 2001, 2002). When the capillary effect is neglected, the three formulations are the same. In this case, the saturation equation becomes the well-known *Buckley–Leverett equation*.

Classification of differential equations

There are basically three types of second order partial differential equations: elliptic, parabolic, and hyperbolic. We must be able to distinguish among these types when devising numerical methods for their solution.

If two independent variables (either (x_1, x_2) or (x_1, t)) are considered, then second order partial differential equations have the form, with $x = x_1$,

$$a\frac{\partial^2 p}{\partial x^2} + b\frac{\partial^2 p}{\partial t^2} = f\left(\frac{\partial p}{\partial x}, \frac{\partial p}{\partial t}, p\right).$$

This equation is (1) elliptic if $ab > 0$; (2) parabolic if $ab = 0$; (3) hyperbolic if $ab < 0$.

The simplest elliptic equation is the *Poisson equation*

$$\frac{\partial^2 p}{\partial x_1^2} + \frac{\partial^2 p}{\partial x_2^2} = f(x_1, x_2).$$

A typical parabolic equation is the *heat conduction equation*

$$\phi\frac{\partial p}{\partial t} = \frac{\partial^2 p}{\partial x_1^2} + \frac{\partial^2 p}{\partial x_2^2}.$$

Finally, the prototype hyperbolic equation is the *wave equation*

$$\frac{1}{v^2}\frac{\partial^2 p}{\partial t^2} = \frac{\partial^2 p}{\partial x_1^2} + \frac{\partial^2 p}{\partial x_2^2},$$

where v is a wave speed. In the 1D case, this equation can be "factorized" into two first order parts:

$$\left(\frac{1}{v}\frac{\partial}{\partial t} - \frac{\partial}{\partial x}\right)\left(\frac{1}{v}\frac{\partial}{\partial t} + \frac{\partial}{\partial x}\right)p = 0.$$

The second part gives the first order hyperbolic equation

$$\frac{\partial p}{\partial t} + v\frac{\partial p}{\partial x} = 0.$$

We now turn to the two-phase flow equations. While the phase mobilities λ_α can be zero (cf. Chapter 2), the total mobility λ is always positive, so the pressure equation (5.15) is elliptic in p. If one of the densities varies, this equation becomes parabolic. In general, $-\mathbf{k}\lambda_o f_w dp_c/dS$ is semipositive definite, so the saturation equation (5.16) is a parabolic equation in S, which is *degenerate* in the sense that the diffusion can be zero. This equation becomes hyperbolic if the capillary pressure is ignored. The total velocity is used in the global pressure formulation. This velocity is smoother than the phase velocities. It can be also used in the formulations (5.8) and (5.9) (Chen and Ewing, 1997b). Finally, with $p_c = 0$, equation (5.16) becomes the known *Buckley–Leverett equation* whose flux function f_w is generally nonconvex over the range of saturation values where this function is nonzero, as illustrated in Fig. 5.2; see the following for the formulation in hyperbolic form.

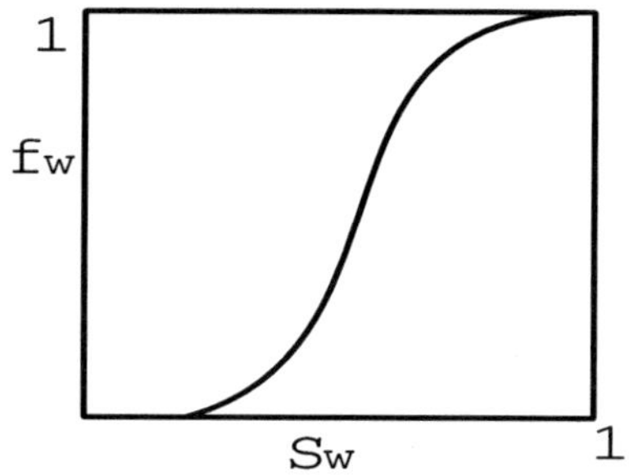

Figure 5.2. *A flux function* f_w.

Formulation in hyperbolic form

Assume that $p_c = 0$ and rock compressibility is neglected. Then equation (5.16) becomes

$$\phi \frac{\partial S}{\partial t} + \nabla \cdot (f_w \mathbf{u} - \lambda_o f_w (\rho_o - \rho_w) \wp \mathbf{k} \nabla z) = \frac{q_w}{\rho_w}. \tag{5.21}$$

Using (5.13) and the fact that $f_w + f_o = 1$, this equation can be manipulated into

$$\phi \frac{\partial S}{\partial t} + \left(\frac{df_w}{dS} \mathbf{u} - \frac{d(\lambda_o f_w)}{dS} (\rho_o - \rho_w) \wp \mathbf{k} \nabla z \right) \cdot \nabla S = \frac{f_o q_w}{\rho_w} - \frac{f_w q_o}{\rho_o}, \tag{5.22}$$

which is a hyperbolic equation in S. Finally, if we neglect the gravitational term, we obtain

$$\phi \frac{\partial S}{\partial t} + \frac{df_w}{dS} \mathbf{u} \cdot \nabla S = \frac{f_o q_w}{\rho_w} - \frac{f_w q_o}{\rho_o}, \tag{5.23}$$

which is the familiar form of waterflooding equation, i.e., the *Buckley–Leverett equation*. The source term in (5.23) is zero for production since

$$\frac{q_w}{\rho_w} = f_w \left(\frac{q_w}{\rho_w} + \frac{q_o}{\rho_o} \right)$$

by Darcy's law. For injection, this term may not be zero since it equals $(1 - f_w) q_w / \rho_w \neq 0$ in this case.

5.1.4 Boundary Conditions

As for single-phase flow in Chapter 3, the mathematical model described so far for two-phase flow is not complete unless necessary boundary and initial conditions are specified. Below we present boundary conditions of three kinds that are relevant to systems (5.8), (5.9), and (5.15) ((5.18) or (5.20)) and (5.16). We denote by Γ the external boundary or a boundary segment of the porous medium domain Ω under consideration.

Boundary conditions for system (5.8)

The symbol α, as a subscript, with $\alpha = w, o$, is used to indicate a considered phase. When a phase pressure is specified as a known function of position and time on Γ, the boundary condition reads

$$p_\alpha = g_{\alpha,1} \quad \text{on } \Gamma. \tag{5.24}$$

When the mass flux of phase α is known on Γ, the boundary condition is

$$\rho_\alpha \mathbf{u}_\alpha \cdot \boldsymbol{\nu} = g_{\alpha,2} \quad \text{on } \Gamma, \tag{5.25}$$

where $\boldsymbol{\nu}$ indicates the outward unit normal to Γ and $g_{\alpha,2}$ is given. For an impervious boundary for the α phase, $g_{\alpha,2} = 0$ (no flow for this phase).

When Γ is a semipervious boundary for the α phase, a boundary condition of mixed kind occurs:

$$g_{\alpha,p} p_\alpha + g_{\alpha,u} \rho_\alpha \mathbf{u}_\alpha \cdot \boldsymbol{\nu} = g_{\alpha,3} \quad \text{on } \Gamma, \tag{5.26}$$

where $g_{\alpha,p}$, $g_{\alpha,u}$, and $g_{\alpha,3}$ are given functions.

Initial conditions specify the values of the main unknowns p_w and p_o over the entire domain at some initial time, usually taken at $t = 0$:

$$p_\alpha(\mathbf{x}, 0) = p_\alpha^0(\mathbf{x}), \quad \alpha = w, o,$$

where $p_\alpha^0(\mathbf{x})$ are known functions. More details will be given about the choice of an appropriate set of initial conditions in Sections 5.3.1 and 6.2.1.

Boundary conditions for system (5.9)

Boundary conditions for system (5.9) can be imposed as for system (5.8); i.e., equations (5.24)–(5.26) are applicable to system (5.9). The only difference between the boundary conditions for these two systems is that a prescribed saturation is sometimes given on Γ for system (5.9):

$$S_w = g_4 \quad \text{on } \Gamma.$$

In practice, this prescribed saturation boundary condition seldom occurs. However, a condition $g_4 = 1$ does take place when a medium is in contact with a body of this wetting phase. The condition $S_w = 1$ can be exploited on the bottom of a water pond on the ground surface, for example. An initial saturation is also specified:

$$S_w(\mathbf{x}, 0) = S_w^0(\mathbf{x}),$$

where $S_w^0(\mathbf{x})$ is given (cf. Sections 5.3.1 and 6.2.1).

Boundary conditions for (5.15) ((5.18) or (5.20)) and (5.16)

Boundary conditions are usually specified in terms of phase quantities like those in (5.24)–(5.26). These conditions can be transformed into those in terms of the global quantities introduced in (5.19) and (5.20). For the prescribed pressure boundary condition in (5.24), for example, the corresponding boundary condition is given by

$$p = g_1 \quad \text{on } \Gamma,$$

where p is defined by (5.19) and g_1 is determined by

$$g_1 = g_{o,1} - \int^{g_{o,1}-g_{w,1}} f_w\big(p_c^{-1}(\xi)\big)d\xi.$$

Also, when the total mass flux is known on Γ, it follows from (5.25) that

$$\mathbf{u} \cdot \boldsymbol{\nu} = g_2 \quad \text{on } \Gamma,$$

where

$$g_2 = \frac{g_{o,2}}{\rho_o} + \frac{g_{w,2}}{\rho_w}.$$

For an impervious boundary for the total flow, $g_2 = 0$.

5.2 An Analytic Solution

As in the treatment of single-phase flow in Chapter 3, an analytic solution for a simple two-phase flow system is obtained.

5.2.1 Analytic Solution Before Water Breakthrough

The *breakthrough time* t_B is an important event in the water-oil displacement; as $t > t_B$, we are producing some of the water being injected. Assume that Ω is an isotropic medium and is homogeneous in the x_2- and x_3-directions (cf. Section 2.1). All its properties depend only on x_1. That is, we consider 1D flow in the x-direction ($x = x_1$). In addition, if the gravity and capillary effects are ignored, the mass conservation equations (5.3) become

$$\begin{aligned} \phi\frac{\partial S_w}{\partial t} + \frac{\partial u_w}{\partial x} &= 0, \\ \phi\frac{\partial S_o}{\partial t} + \frac{\partial u_o}{\partial x} &= 0, \end{aligned} \tag{5.27}$$

and Darcy's law (5.6) simplifies to

$$\begin{aligned} u_w &= -k\frac{k_{rw}(S_w)}{\mu_w}\frac{\partial p}{\partial x}, \\ u_o &= -k\frac{k_{ro}(S_o)}{\mu_o}\frac{\partial p}{\partial x}. \end{aligned} \tag{5.28}$$

We define the total velocity

$$u = u_w + u_o. \tag{5.29}$$

Using (5.1) and (5.27), we see that

$$\frac{\partial u}{\partial x} = 0, \tag{5.30}$$

so u is independent of x. Because $u_w = f_w(S_w)u$, it follows that

$$\frac{\partial u_w}{\partial x} = f_w\frac{\partial u}{\partial x} + u\frac{df_w(S_w)}{dS_w}\frac{\partial S_w}{\partial x} = uF_w(S_w)\frac{\partial S_w}{\partial x}, \tag{5.31}$$

where the *distribution function* F_w of saturation is defined by

$$F_w(S_w) = \frac{df_w(S_w)}{dS_w}.$$

Now, we substitute (5.31) into the first equation of (5.27) to see that

$$\phi\frac{\partial S_w}{\partial t} + uF_w(S_w)\frac{\partial S_w}{\partial x} = 0. \tag{5.32}$$

This equation defines a *characteristic* $x(t)$ along the *interstitial velocity* v by

$$\frac{dx}{dt} = v(x,t) \equiv \frac{uF_w(S_w)}{\phi}. \tag{5.33}$$

Along this characteristic, it follows from (5.32) that S_w is constant; i.e.,

$$\frac{dS_w(x(t),t)}{dt} = \frac{\partial S_w}{\partial x}\frac{dx}{dt} + \frac{\partial S_w}{\partial t} = 0. \tag{5.34}$$

Let A be the cross-sectional area (in the x_2x_3-plane) of Ω, and define the *cumulative liquid production*

$$V(t) = A\int_0^t u\,dt. \tag{5.35}$$

From (5.33), along the characteristic $x(t)$ we see that

$$\int_0^t dx = \frac{F_w(S_w)}{\phi}\int_0^t u\,dt,$$

so, by (5.35),

$$x(S_w,t) = \frac{F_w(S_w)}{\phi A}V(t), \tag{5.36}$$

from which we can find the saturation S_w before water breaks through.

5.2.2 Analytic Solution at the Water Front

Let S_{wf} be the water saturation at the water front, and let S_{wc} be the critical saturation (cf. Section 2.5). From the *material balance equation*

$$u_w\big|_{\text{at water front}} = \phi(S_{wf} - S_{wc})\frac{dx}{dt},$$

we have

$$\phi(S_{wf} - S_{wc})\frac{dx}{dt} = f_w u, \tag{5.37}$$

since $u_w = f_w(S_w)u$. Applying (5.33) to (5.37) gives

$$(S_{wf} - S_{wc})F_w = f_w;$$

i.e.,

$$\frac{df_w}{dS_w}(S_{wf}) = \frac{f_w(S_{wf})}{S_{wf} - S_{wc}}. \tag{5.38}$$

Equation (5.38) indicates that the slope of the tangent to the curve of f_w at S_{wf} equals the slope of the secant line through the points $(S_{wf}, f_w(S_{wf}))$ and $(S_{wc}, f_w(S_{wc}))$ (note that $f_w(S_{wc}) = 0$; cf. Section 2.5). Thus a graphical method based on this feature can be used to find the water saturation at the water front from equation (5.38).

5.2.3 Analytic Solution After Water Breakthrough

Let L be the length of Ω in the x-direction, and let S_{we} be the value of the saturation at $x = L$. At $x = L$, it follows from (5.36) that

$$V(t) = \frac{\phi AL}{F_w(S_{we})}. \tag{5.39}$$

We define the nondimensional cumulative liquid production

$$\bar{V}(t) = \frac{V(t)}{\phi AL}.$$

Then we see that

$$\bar{V}(t) = \frac{1}{F_w(S_{we})}. \tag{5.40}$$

Also, we introduce the *cumulative water production*

$$V_w(t) = \int_{t_B}^{t} f_w \, dV(t) = A \int_{t_B}^{t} u_w \, dt, \tag{5.41}$$

where we recall that t_B is the water breakthrough time (i.e., S_w equals the critical value S_{wc} at $t = t_B$), and we used the fact that $f_w dV = Au_w dt$ by equation (5.35). The nondimensional cumulative water production is

$$\bar{V}_w = \frac{V_w}{\phi AL}.$$

It follows from (5.41) and integration by parts that

$$\bar{V}_w = \frac{1}{\phi AL} \int_{t_B}^{t} f_w \, dV(t) = \frac{1}{\phi AL} \left(f_w V - \int_{t_B}^{t} V \, df_w \right),$$

since $f_w(S_{wc}) = 0$. Consequently, by the fact that $df_w = F_w \, dS_w$, we see that

$$\bar{V}_w = \frac{1}{\phi AL} \left(f_w V - \int_{t_B}^{t} VF_w \, dS_w \right).$$

Finally, applying (5.39), we obtain

$$\bar{V}_w = \frac{f_w(S_{we})}{F_w(S_{we})} - (S_{we} - S_{wc}), \tag{5.42}$$

which defines the value of S_{we}.

We also define the cumulative oil production

$$V_o(t) = \int_{t_B}^{t} f_o \, dV(t) = A \int_{t_B}^{t} u_o \, dt,$$

and the corresponding nondimensional value

$$\bar{V}_o = \frac{V_o}{\phi A L}.$$

Then we can derive

$$\bar{V}_o = \frac{1 - f_w(S_{we})}{F_w(S_{we})} + (S_{we} - S_{wc}) \tag{5.43}$$

and

$$\bar{V} = \bar{V}_w + \bar{V}_o.$$

Either of equations (5.42) and (5.43) can be utilized to find S_{we}.

5.3 Numerical Solution of Two-Phase Flow

Note that the differential equations (5.1)–(5.3) and (5.6) are nonlinear and coupled. There exist a variety of approaches for solving these equations, such as the IMPES, SS, *sequential*, and *adaptive implicit methods* (cf. Chapter 6). In light of the fact that the IMPES method is still popular in the petroleum industry and a very powerful method for solving two-phase flow (particularly for incompressible or slightly compressible fluids), we only discuss this solution approach for this type of flow. Other approaches will be discussed in the next chapter for the black oil model.

An IMPES method was originally developed by Sheldon, Zondek, and Cardwell (1959) and Stone and Garder (1961). The basic idea of this classical method for solving (5.1)–(5.3) and (5.6) is to separate the computation of pressure from that of saturation. Namely, the coupled system is split into a pressure equation and a saturation equation, and the pressure and saturation equations are solved using implicit and explicit time approximation approaches, respectively. This method is simple to set up and efficient to implement, and requires less computer memory than other methods such as the SS method (Douglas, Peaceman, and Rachford, 1959). However, for it to be stable, this classical method requires very small time steps for the saturation. This requirement is expensive and prohibitive, particularly for long time integration problems and for small gridblock problems such as coning problems. In this section, we first review classical IMPES and then introduce an *improved IMPES method*. We focus on incompressible flow; compressible flow will be treated in the next chapter.

As an example, the IMPES method is examined for the phase formulation. Assume that the permeability tensor $\mathbf{k}$ is diagonal: $\mathbf{k} = \text{diag}(k_{11}, k_{22}, k_{33})$. Introduce the fluid gravities

$$\gamma_w = \rho_w \wp, \qquad \gamma_o = \rho_o \wp.$$

We recall the *pressure equation* (5.15):

$$-\nabla \cdot \big(\mathbf{k}\lambda \nabla p\big) = \tilde{q} - \nabla \cdot \big(\mathbf{k}(\lambda_w \nabla p_c + (\lambda_w \gamma_w + \lambda_o \gamma_o)\nabla z)\big). \tag{5.44}$$

The saturation equation in S is

$$\phi \frac{\partial S}{\partial t} = \nabla \cdot \{\lambda_w(S)\mathbf{k}\,(\nabla p_w - \gamma_w \nabla z)\} = \tilde{q}_w, \tag{5.45}$$

where, for notational convenience, we assume that $\phi = \phi(\mathbf{x})$. Note that $p_w = p - p_c(S)$. Define the weighted fluid gravity

$$\bar{\gamma} = \lambda_w \gamma_w + \lambda_o \gamma_o.$$

Then, in terms of each coordinate direction, equations (5.44) and (5.45) are rewritten as

$$\begin{aligned} &-\frac{\partial}{\partial x_1}\left(k_{11}\lambda \frac{\partial p}{\partial x_1}\right) - \frac{\partial}{\partial x_2}\left(k_{22}\lambda \frac{\partial p}{\partial x_2}\right) - \frac{\partial}{\partial x_3}\left(k_{33}\lambda \frac{\partial p}{\partial x_3}\right) \\ &= \tilde{q} - \frac{\partial}{\partial x_1}\left(k_{11}\left(\lambda_w \frac{\partial p_c}{\partial x_1} + \bar{\gamma}\frac{\partial z}{\partial x_1}\right)\right) \\ &\quad - \frac{\partial}{\partial x_2}\left(k_{22}\left(\lambda_w \frac{\partial p_c}{\partial x_2} + \bar{\gamma}\frac{\partial z}{\partial x_2}\right)\right) - \frac{\partial}{\partial x_3}\left(k_{33}\left(\lambda_w \frac{\partial p_c}{\partial x_3} + \bar{\gamma}\frac{\partial z}{\partial x_3}\right)\right) \end{aligned} \tag{5.46}$$

and

$$\begin{aligned} \phi \frac{\partial S}{\partial t} &= \frac{\partial}{\partial x_1}\left(\lambda_w k_{11}\left(\frac{\partial p_w}{\partial x_1} - \gamma_w \frac{\partial z}{\partial x_1}\right)\right) \\ &\quad + \frac{\partial}{\partial x_2}\left(\lambda_w k_{22}\left(\frac{\partial p_w}{\partial x_2} - \gamma_w \frac{\partial z}{\partial x_2}\right)\right) + \frac{\partial}{\partial x_3}\left(\lambda_w k_{33}\left(\frac{\partial p_w}{\partial x_3} - \gamma_w \frac{\partial z}{\partial x_3}\right)\right) + \tilde{q}_w. \end{aligned} \tag{5.47}$$

5.3.1 Treatment of Initial Conditions

Initial conditions for two-phase flow involve the specification of phase pressures and/or saturations for each gridblock in the simulation model at the start of simulation. Differences in phase gravities and capillary pressures cause fluids to segregate until the reservoir system reaches gravity/capillary equilibrium. In a real reservoir there exist up to five different fluid zones vertically, and distinct initial data are specified in each of these five zones. The specification of these initial conditions will be discussed in the next chapter, in conjunction with the black oil model. In this chapter, for simplicity, we assume that the initial oil pressure and water saturation are known throughout the entire reservoir. For incompressible flow, only an initial saturation suffices.

5.3.2 Source/Sink Terms

Recall that $\tilde{q}_w = q_w/\rho_w$ and $\tilde{q}_o = q_o/\rho_o$, where the source/sink terms q_w and q_o are given by

$$q_\alpha = \sum_{\nu,m} q_{\alpha,m}^{(\nu)} \delta\left(\mathbf{x} - \mathbf{x}_m^{(\nu)}\right), \qquad \alpha = w, o, \tag{5.48}$$

where $q_{\alpha,m}^{(\nu)}$ indicates the volume of phase α produced or injected per unit time at the νth well and the mth perforated zone, $\mathbf{x}_m^{(\nu)}$, and δ is the Dirac delta function. From Section 4.4, $q_{\alpha,m}^{(\nu)}$ can be defined by

$$q_{\alpha,m}^{(\nu)} = \frac{2\pi\rho_\alpha \bar{k} k_{r\alpha} \Delta h}{\mu_\alpha \ln\left(r_e/r_w\right)}\bigg|_m^{(\nu)} \left(p_{bh}^{(\nu)} - p_\alpha - \gamma_\alpha\left(z_{bh}^{(\nu)} - z\right)\right), \tag{5.49}$$

where $\Delta h_m^{(\nu)}$ is the length (in the flow direction) of a gridblock (containing the νth well) at the mth perforated zone, $p_{bh}^{(\nu)}$ is the *bottom hole pressure* at the datum level depth $z_{bh}^{(\nu)}$, $r_e^{(\nu)}$ is the equivalent well radius, and $r_w^{(\nu)}$ is the radius of the νth well. The quantity $\bar{k}$ is some average of $\mathbf{k}$ at the wells; for the diagonal tensor $\mathbf{k}$, for example, $\bar{k} = \sqrt{k_{11}k_{22}}$ for a vertical well (cf. Section 4.4).

5.3.3 Spatial Discretization

As an example, the block-centered grid system is considered. The seven-point stencil scheme for the pressure equation (5.46) is

$$\begin{aligned}
&-\left(\frac{A_1\lambda k_{11}}{h_1}\right)_{i+1/2,j,k}\left(p_{i+1,j,k}-p_{i,j,k}\right)+\left(\frac{A_1\lambda k_{11}}{h_1}\right)_{i-1/2,j,k}\left(p_{i,j,k}-p_{i-1,j,k}\right)\\
&-\left(\frac{A_2\lambda k_{22}}{h_2}\right)_{i,j+1/2,k}\left(p_{i,j+1,k}-p_{i,j,k}\right)+\left(\frac{A_2\lambda k_{22}}{h_2}\right)_{i,j-1/2,k}\left(p_{i,j,k}-p_{i,j-1,k}\right)\\
&-\left(\frac{A_3\lambda k_{33}}{h_3}\right)_{i,j,k+1/2}\left(p_{i,j,k+1}-p_{i,j,k}\right)+\left(\frac{A_3\lambda k_{33}}{h_3}\right)_{i,j,k-1/2}\left(p_{i,j,k}-p_{i,j,k-1}\right)\\
&=-\left(\frac{A_1\lambda_w k_{11}}{h_1}\right)_{i+1/2,j,k}\left(p_{c,i+1,j,k}-p_{c,i,j,k}\right)+\left(\frac{A_1\lambda_w k_{11}}{h_1}\right)_{i-1/2,j,k}\left(p_{c,i,j,k}-p_{c,i-1,j,k}\right)\\
&-\left(\frac{A_2\lambda_w k_{22}}{h_2}\right)_{i,j+1/2,k}\left(p_{c,i,j+1,k}-p_{c,i,j,k}\right)+\left(\frac{A_2\lambda_w k_{22}}{h_2}\right)_{i,j-1/2,k}\left(p_{c,i,j,k}-p_{c,i,j-1,k}\right)\\
&-\left(\frac{A_3\lambda_w k_{33}}{h_3}\right)_{i,j,k+1/2}\left(p_{c,i,j,k+1}-p_{c,i,j,k}\right)+\left(\frac{A_3\lambda_w k_{33}}{h_3}\right)_{i,j,k-1/2}\left(p_{c,i,j,k}-p_{c,i,j,k-1}\right)\\
&-\left(\frac{A_1k_{11}\bar{\gamma}}{h_1}\right)_{i+1/2,j,k}\left(z_{i+1,j,k}-z_{i,j,k}\right)+\left(\frac{A_1k_{11}\bar{\gamma}}{h_1}\right)_{i-1/2,j,k}\left(z_{i,j,k}-z_{i-1,j,k}\right)\\
&-\left(\frac{A_2k_{22}\bar{\gamma}}{h_2}\right)_{i,j+1/2,k}\left(z_{i,j+1,k}-z_{i,j,k}\right)+\left(\frac{A_2k_{22}\bar{\gamma}}{h_2}\right)_{i,j-1/2,k}\left(z_{i,j,k}-z_{i,j-1,k}\right)\\
&-\left(\frac{A_3k_{33}\bar{\gamma}}{h_3}\right)_{i,j,k+1/2}\left(z_{i,j,k+1}-z_{i,j,k}\right)+\left(\frac{A_3k_{33}\bar{\gamma}}{h_3}\right)_{i,j,k-1/2}\left(z_{i,j,k}-z_{i,j,k-1}\right)\\
&+\tilde{Q}_{i,j,k},
\end{aligned} \tag{5.50}$$

where A_i is the cross-sectional area normal to the x_i-direction, $i = 1, 2, 3$, $\tilde{Q}_{i,j,k} = (\tilde{q}V)_{i,j,k}$, and $p_{c,i,j,k} = p_c(S_{i,j,k})$. The spatial discretization for the saturation equation (5.47) can be similarly defined:

$$\begin{aligned}
&\left(\frac{A_1\lambda_w k_{11}}{h_1}\right)_{i+1/2,j,k}(p_{w,i+1,j,k}-p_{w,i,j,k})-\left(\frac{A_1\lambda_w k_{11}}{h_1}\right)_{i-1/2,j,k}(p_{w,i,j,k}-p_{w,i-1,j,k})\\
&+\left(\frac{A_2\lambda_w k_{22}}{h_2}\right)_{i,j+1/2,k}(p_{w,i,j+1,k}-p_{w,i,j,k})-\left(\frac{A_2\lambda_w k_{22}}{h_2}\right)_{i,j-1/2,k}(p_{w,i,j,k}-p_{w,i,j-1,k})\\
&+\left(\frac{A_3\lambda_w k_{33}}{h_3}\right)_{i,j,k+1/2}(p_{w,i,j,k+1}-p_{w,i,j,k})-\left(\frac{A_3\lambda_w k_{33}}{h_3}\right)_{i,j,k-1/2}(p_{w,i,j,k}-p_{w,i,j,k-1})\\
&-\left(\frac{A_1\lambda_w k_{11}\gamma_w}{h_1}\right)_{i+1/2,j,k}(z_{i+1,j,k}-z_{i,j,k})+\left(\frac{A_1\lambda_w k_{11}\gamma_w}{h_1}\right)_{i-1/2,j,k}(z_{i,j,k}-z_{i-1,j,k})\\
&-\left(\frac{A_2\lambda_w k_{22}\gamma_w}{h_2}\right)_{i,j+1/2,k}(z_{i,j+1,k}-z_{i,j,k})+\left(\frac{A_2\lambda_w k_{22}\gamma_w}{h_2}\right)_{i,j-1/2,k}(z_{i,j,k}-z_{i,j-1,k})\\
&-\left(\frac{A_3\lambda_w k_{33}\gamma_w}{h_3}\right)_{i,j,k+1/2}(z_{i,j,k+1}-z_{i,j,k})+\left(\frac{A_3\lambda_w k_{33}\gamma_w}{h_3}\right)_{i,j,k-1/2}(z_{i,j,k}-z_{i,j,k-1})\\
&+\tilde{Q}_{w,i,j,k},
\end{aligned} \tag{5.51}$$

where $\tilde{Q}_{w,i,j,k}=(\tilde{q}_w V)_{i,j,k}$ and $p_{w,i,j,k}=p_{i,j,k}-p_{c,i,j,k}$.

5.3.4 Treatment of Block Transmissibility

As in single-phase flow the transmissibilities at the gridblock boundaries must be carefully calculated. For example, the transmissibility of water in the x_1-direction,

$$T_{w1,i\pm1/2,j,k}=\left(\frac{A_1\lambda_w k_{11}}{h_1}\right)_{i\pm1/2,j,k}=\left(\frac{A_1 k_{rw} k_{11}}{\mu_w h_1}\right)_{i\pm1/2,j,k}, \tag{5.52}$$

contains the rock and grid properties A_1k_{11}/h_1, the fluid property μ_w, and the rock/fluid property k_{rw}. For the first two quantities, as shown for single-phase flow (cf. Section 3.4.4), the (weighted) harmonic and arithmetic averages are appropriate, respectively. What average should be employed for the rock/fluid property?

Consider Fig. 5.3, where $k_{rw}=1$ and 0 in the $(i-1)$th and ith cells, respectively. Two possible averages for k_{rw} are

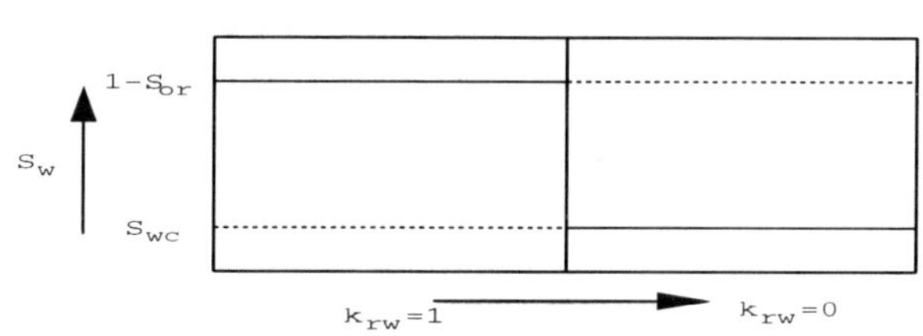

Figure 5.3. *Two cells.*

- harmonic average, which yields $k_{rw}|_{i-1/2} = 0$, but it is not true;
- arithmetic average, which gives $k_{rw}|_{i-1/2} = 0.5$; it is possible, but then $k_{ro}|_{i-1/2} = 0.5$, which is not possible because the oil phase does not flow from the $(i-1)$th cell to the ith cell.

The correct average is the *upstream weighting* value of k_{rw} for flow from cell $i-1$ to cell i: $k_{rw}|_{i-1/2} = k_{rw}|_i$.

Midpoint weighting

The midpoint weighting value of $k_{rw,i\pm1/2,j,k}$ is simply defined by

$$k_{rw,i\pm1/2,j,k} = \frac{1}{2}\left(k_{rw}(S_i) + k_{rw}(S_{i\pm1})\right),$$

or

$$k_{rw,i\pm1/2,j,k} = k_{rw}\left(S_{i\pm1/2}\right), \quad S_{i\pm1/2} = \frac{S_i V_{w,i} + S_{i\pm1} V_{w,i\pm1}}{V_{w,i} + V_{w,i\pm1}},$$

where V_w is the water volume. While it is second order accurate, the midpoint weighting of strong nonlinearity is seldom used since it may produce a physically incorrect solution, as outlined above.

Single-point upstream weighting

The water phase potential difference between gridblocks $(i-1,\ j,\ k)$ and $(i,\ j,\ k)$ is given by

$$\Delta\Phi_{w,i-1/2,j,k} = (p_{w,i,j,k} - p_{w,i-1,j,k}) - \gamma_w(z_{i,j,k} - z_{i-1,j,k}).$$

The *single-point upstream weighting* value of $k_{rw,i-1/2,j,k}$ is

$$k_{rw,i-1/2,j,k} = \begin{cases} k_{rw,i-1,j,k} & \text{if } \Delta\Phi_{w,i-1/2,j,k} < 0, \\ k_{rw,i,j,k} & \text{if } \Delta\Phi_{w,i-1/2,j,k} > 0. \end{cases} \tag{5.53}$$

If $\Delta\Phi_{w,i-1/2,j,k} < 0$, the flow of water is from block $(i-1,\ j,\ k)$ to block $(i,\ j,\ k)$, where block $(i-1,\ j,\ k)$ is the upstream block and block $(i,\ j,\ k)$ is the downstream block for water. Similarly, as $\Delta\Phi_{w,i-1/2,j,k} > 0$, the flow of water is from the upstream block $(i,\ j,\ k)$ to the downstream block $(i-1,\ j,\ k)$.

Two-point upstream weighting

Two-point upstream weighting is an extrapolation technique where $k_{rw,i-1/2,j,k}$ is evaluated using the values of k_{rw} at the two upstream points:

$$k_{rw,i-1/2,j,k} = \begin{cases} (1+\beta_{i-1})k_{rw,i-1,j,k} - \beta_{i-1}k_{rw,i-2,j,k} & \text{if } \Delta\Phi_{w,i-1/2,j,k} < 0, \\ (1+\beta_i')k_{rw,i,j,k} - \beta_i' k_{rw,i+1,j,k} & \text{if } \Delta\Phi_{w,i-1/2,j,k} > 0, \end{cases} \tag{5.54}$$

where $\beta_{i-1} = h_{i-1}/(2h_{i-3/2})$ and $\beta_i' = h_i/(2h_{i+1/2})$. For a point-distributed grid system, $\beta_{i-1} = h_{i-1/2}/(2h_{i-3/2})$ and $\beta_i' = h_{i-1/2}/(2h_{i+1/2})$. For uniform grids, $\beta_{i-1} = \beta_i' = 0.5$.

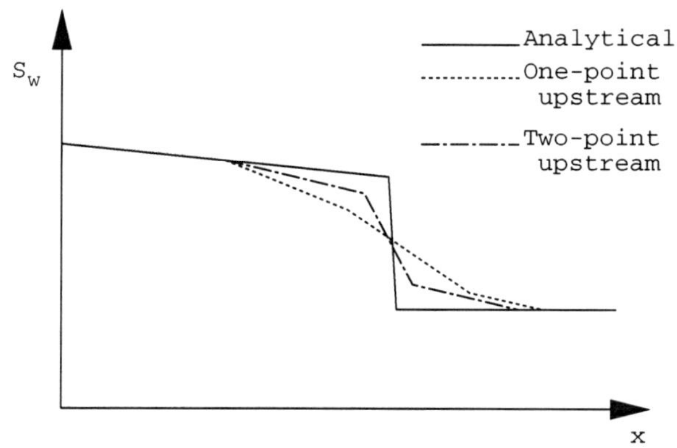

Figure 5.4. *Solutions of single-point and two-point upstream weightings.*

The single-point and two-point upstream weightings are first order and second order approximations, respectively, and both techniques lead to the same physical solution. However, the latter technique gives sharper solution fronts for the same number of gridblocks (Todd, O'Dell, and Hirasaki, 1972; cf. Fig. 5.4). If an explicit treatment of relative permeabilities is used (i.e., they are evaluated at old time levels), then the two techniques require the same computational efforts. However, if these permeabilities are evaluated implicitly, the two-point upstream weighting increases the bandwidth of the Jacobian matrix and thus the computational complexity. In general, the relative permeabilities can be approximated using either of the weightings, while the slopes of the capillary pressures in equations (5.50) and (5.51) are approximated by the single-point upstream weighting.

5.3.5 Solution Approaches in Time

Classical IMPES

Let $J = (0, T]$ ($T > 0$) be the time interval of interest, and for a positive integer N, let $0 = t^0 < t^1 < \cdots < t^N = T$ be a partition of J. For the pressure computation in the classical IMPES method, the saturation S in (5.50) is assumed to be known, and equation (5.50) is solved implicitly for p. That is, for each $n = 0, 1, \ldots,$ p^n satisfies

$$\begin{aligned}
&- T^n_{1,i+1/2,j,k}\left(p^n_{i+1,j,k} - p^n_{i,j,k}\right) + T^n_{1,i-1/2,j,k}\left(p^n_{i,j,k} - p^n_{i-1,j,k}\right)\\
&\quad - T^n_{2,i,j+1/2,k}\left(p^n_{i,j+1,k} - p^n_{i,j,k}\right) + T^n_{2,i,j-1/2,k}\left(p^n_{i,j,k} - p^n_{i,j-1,k}\right)\\
&\quad - T^n_{3,i,j,k+1/2}\left(p^n_{i,j,k+1} - p^n_{i,j,k}\right) + T^n_{3,i,j,k-1/2}\left(p^n_{i,j,k} - p^n_{i,j,k-1}\right)\\
&= -T^n_{w1,i+1/2,j,k}\left(p^n_{c,i+1,j,k} - p^n_{c,i,j,k}\right) + T^n_{w1,i-1/2,j,k}\left(p^n_{c,i,j,k} - p^n_{c,i-1,j,k}\right)\\
&\quad - T^n_{w2,i,j+1/2,k}\left(p^n_{c,i,j+1,k} - p^n_{c,i,j,k}\right) + T^n_{w2,i,j-1/2,k}\left(p^n_{c,i,j,k} - p^n_{c,i,j-1,k}\right) \qquad (5.55)\\
&\quad - T^n_{w3,i,j,k+1/2}\left(p^n_{c,i,j,k+1} - p^n_{c,i,j,k}\right) + T^n_{w3,i,j,k-1/2}\left(p^n_{c,i,j,k} - p^n_{c,i,j,k-1}\right)\\
&\quad - (T\bar{\gamma})^n_{1,i+1/2,j,k}\left(z_{i+1,j,k} - z_{i,j,k}\right) + (T\bar{\gamma})^n_{1,i-1/2,j,k}\left(z_{i,j,k} - z_{i-1,j,k}\right)
\end{aligned}$$

$$
\begin{aligned}
&- (T\bar{\gamma})^n_{2,i,j+1/2,k}\left(z_{i,j+1,k} - z_{i,j,k}\right) + (T\bar{\gamma})^n_{2,i,j-1/2,k}\left(z_{i,j,k} - z_{i,j-1,k}\right)\\
&- (T\bar{\gamma})^n_{3,i,j,k+1/2}\left(z_{i,j,k+1} - z_{i,j,k}\right) + (T\bar{\gamma})^n_{3,i,j,k-1/2}\left(z_{i,j,k} - z_{i,j,k-1}\right)\\
&+ \tilde{Q}^n_{i,j,k},
\end{aligned}
$$

where S^n is given and

$$
T_{1,i-1/2,j,k} = \left(\frac{A_1\lambda k_{11}}{h_1}\right)_{i-1/2,j,k}, \qquad (T\bar{\gamma})_{1,i-1/2,j,k} = \left(\frac{A_1 k_{11}\bar{\gamma}}{h_1}\right)_{i-1/2,j,k},
$$

and analogous notation holds for other quantities. The total source/sink term $\tilde{q}$ is evaluated as follows:

$$
\tilde{q}^n_{i,j,k} = \sum_{\nu,m}\left(\tilde{q}^{(\nu)}_{w,m} + \tilde{q}^{(\nu)}_{o,m}\right)\left(p^n_{i,j,k}, S^n_{i,j,k}\right)\delta\left(\mathbf{x}_{i,j,k} - \mathbf{x}^{(\nu)}_m\right).
$$

In IMPES, (5.51) is explicitly solved for S; i.e., for each $n = 0, 1, 2, \ldots,$ S^{n+1} satisfies

$$
\begin{aligned}
&\left(V\phi\frac{S^{n+1} - S^n}{\Delta t}\right)_{i,j,k}\\
&= T^n_{w1,i+1/2,j,k}\left(p^n_{w,i+1,j,k} - p^n_{w,i,j,k}\right) - T^n_{w1,i-1/2,j,k}\left(p^n_{w,i,j,k} - p^n_{w,i-1,j,k}\right)\\
&+ T^n_{w2,i,j+1/2,k}\left(p^n_{w,i,j+1,k} - p^n_{w,i,j,k}\right) - T^n_{w2,i,j-1/2,k}\left(p^n_{w,i,j,k} - p^n_{w,i,j-1,k}\right)\\
&+ T^n_{w3,i,j,k+1/2}\left(p^n_{w,i,j,k+1} - p^n_{w,i,j,k}\right) - T^n_{w3,i,j,k-1/2}\left(p^n_{w,i,j,k} - p^n_{w,i,j,k-1}\right)\\
&- (T_w\gamma_w)^n_{1,i+1/2,j,k}\left(z_{i+1,j,k} - z_{i,j,k}\right) + (T_w\gamma_w)^n_{1,i-1/2,j,k}\left(z_{i,j,k} - z_{i-1,j,k}\right)\\
&- (T_w\gamma_w)^n_{2,i,j+1/2,k}\left(z_{i,j+1,k} - z_{i,j,k}\right) + (T_w\gamma_w)^n_{2,i,j-1/2,k}\left(z_{i,j,k} - z_{i,j-1,k}\right)\\
&- (T_w\gamma_w)^n_{3,i,j,k+1/2}\left(z_{i,j,k+1} - z_{i,j,k}\right) + (T_w\gamma_w)^n_{3,i,j,k-1/2}\left(z_{i,j,k} - z_{i,j,k-1}\right)\\
&+ \tilde{Q}^n_{w,i,j,k},
\end{aligned}
\tag{5.56}
$$

where $p^n_w = p^n - p_c(S^n)$ and

$$
\tilde{q}^n_{w,i,j,k} = \sum_{\nu,m}\tilde{q}^{(\nu)}_{w,m}\left(p^n_{i,j,k}, S^n_{i,j,k}\right)\delta\left(\mathbf{x}_{i,j,k} - \mathbf{x}^{(\nu)}_m\right).
$$

The IMPES method goes as follows: After startup, for $n = 0, 1, \ldots,$ we use (5.56) and S^n to evaluate p^n; next, we utilize S^n, p^n, and (5.56) to compute S^{n+1}. As noted, the time step Δt must be sufficiently small for this method to be stable.

To control the variation of saturation, we need to find a suitable time step Δt before we solve equation (5.56) for S^{n+1} for each $n = 0, 1, \ldots.$ The control strategy is defined as follows: We calculate the maximum value of $\partial S^{n+1}/\partial t$ at all computational nodes, denoted by $(\partial S^{n+1}/\partial t)_{max}$, which is, by (5.56),

$$
\left(\frac{\partial S^{n+1}}{\partial t}\right)_{max} = \max_{i,j,k}\left(\frac{G(p^n, S^n)}{V\phi}\right)_{i,j,k},
$$

where $G(p, s)$ represents the right-hand side of equation (5.56) and the maximum is taken over all gridblocks. Then we apply the following formula to find Δt:

$$
\Delta t = DS_{max}/\left(\frac{\partial S^{n+1}}{\partial t}\right)_{max}, \tag{5.57}
$$

where DS_{max} is the maximum variation of the saturation to be allowed (set by the user). Now, we use this time step in (5.56) to obtain S^{n+1}. This approach guarantees that the saturation variation does not exceed DS_{max}. Note that DS_{max} can depend on the time level n.

For incompressible flow and constant viscosities, the seven-point stencil system (5.56) is a linear system and can be solved using an iterative algorithm (Chen, Huan, and Ma, 2006). The saturation system (5.56) does not require any iterative algorithm.

Improved IMPES

Most of the computational time in the classical IMPES method is spent on the implicit calculation of pressure (Chen, Huan, and Ma, 2006). It follows from the mechanics of fluid flow in porous media that pressure changes less rapidly in time than saturation. Furthermore, the constraint on time steps is primarily used in the explicit calculation of saturation. For all these reasons, it is appropriate to take a much larger time step for the pressure than for the saturation.

Again, for a positive integer N, let $0 = t^0 < t^1 < \cdots < t^N = T$ be a partition of J into subintervals $J^n = (t^{n-1}, t^n]$, with length $\Delta t_p^n = t^n - t^{n-1}$. This time partition is used for pressure. For saturation, each subinterval J^n is divided into subsubintervals $J^{n,m} = (t^{n-1,m-1}, t^{n-1,m}]$:

$$t^{n-1,m} = t^{n-1} + m\Delta t_p^n/M^n, \quad m = 1, 2, \ldots, M^n.$$

The length of $J^{n,m}$ is denoted by $\Delta t_S^{n,m} = t^{n-1,m} - t^{n-1,m-1}$, $m = 1, \ldots, M^n, n = 0, 1, \ldots$. The number of steps, M^n, can depend on n. Below we simply write $t^{n-1,0} = t^{n-1}$ and set $v^{n,m} = v(\cdot, t^{n,m})$.

In the improved IMPES method, the computation of pressure is the same as in system (5.56): For each $n = 0, 1, \ldots,$ p^n satisfies

$$\begin{aligned}
&- T^n_{1,i+1/2,j,k}\left(p^n_{i+1,j,k} - p^n_{i,j,k}\right) + T^n_{1,i-1/2,j,k}\left(p^n_{i,j,k} - p^n_{i-1,j,k}\right) \\
&- T^n_{2,i,j+1/2,k}\left(p^n_{i,j+1,k} - p^n_{i,j,k}\right) + T^n_{2,i,j-1/2,k}\left(p^n_{i,j,k} - p^n_{i,j-1,k}\right) \\
&- T^n_{3,i,j,k+1/2}\left(p^n_{i,j,k+1} - p^n_{i,j,k}\right) + T^n_{3,i,j,k-1/2}\left(p^n_{i,j,k} - p^n_{i,j,k-1}\right) \\
= &- T^n_{w1,i+1/2,j,k}\left(p^n_{c,i+1,j,k} - p^n_{c,i,j,k}\right) + T^n_{w1,i-1/2,j,k}\left(p^n_{c,i,j,k} - p^n_{c,i-1,j,k}\right) \\
&- T^n_{w2,i,j+1/2,k}\left(p^n_{c,i,j+1,k} - p^n_{c,i,j,k}\right) + T^n_{w2,i,j-1/2,k}\left(p^n_{c,i,j,k} - p^n_{c,i,j-1,k}\right) \\
&- T^n_{w3,i,j,k+1/2}\left(p^n_{c,i,j,k+1} - p^n_{c,i,j,k}\right) + T^n_{w3,i,j,k-1/2}\left(p^n_{c,i,j,k} - p^n_{c,i,j,k-1}\right) \\
&- (T\bar{\gamma})^n_{1,i+1/2,j,k}\left(z_{i+1,j,k} - z_{i,j,k}\right) + (T\bar{\gamma})^n_{1,i-1/2,j,k}\left(z_{i,j,k} - z_{i-1,j,k}\right) \\
&- (T\bar{\gamma})^n_{2,i,j+1/2,k}\left(z_{i,j+1,k} - z_{i,j,k}\right) + (T\bar{\gamma})^n_{2,i,j-1/2,k}\left(z_{i,j,k} - z_{i,j-1,k}\right) \\
&- (T\bar{\gamma})^n_{3,i,j,k+1/2}\left(z_{i,j,k+1} - z_{i,j,k}\right) + (T\bar{\gamma})^n_{3,i,j,k-1/2}\left(z_{i,j,k} - z_{i,j,k-1}\right) \\
&+ \tilde{Q}^n_{i,j,k}.
\end{aligned} \tag{5.58}$$

The difference is in the calculation of saturation: For $m = 1, \dots, M^n$, $n = 0, 1, \dots$, find $S^{n+1,m}$ such that

$$
\begin{aligned}
&\left(V\phi \frac{S^{n+1,m} - S^{n+1,m-1}}{\Delta t_S^{n+1,m}} \right)_{i,j,k} \\
&= T_{w1,i+1/2,j,k}^{n+1,m-1} \left(p_{w,i+1,j,k}^{n+1,m-1} - p_{w,i,j,k}^{n+1,m-1} \right) - T_{w1,i-1/2,j,k}^{n+1,m-1} \left(p_{w,i,j,k}^{n+1,m-1} - p_{w,i-1,j,k}^{n+1,m-1} \right) \\
&+ T_{w2,i,j+1/2,k}^{n+1,m-1} \left(p_{w,i,j+1,k}^{n+1,m-1} - p_{w,i,j,k}^{n+1,m-1} \right) - T_{w2,i,j-1/2,k}^{n+1,m-1} \left(p_{w,i,j,k}^{n+1,m-1} - p_{w,i,j-1,k}^{n+1,m-1} \right) \\
&+ T_{w3,i,j,k+1/2}^{n+1,m-1} \left(p_{w,i,j,k+1}^{n+1,m-1} - p_{w,i,j,k}^{n+1,m-1} \right) - T_{w3,i,j,k-1/2}^{n+1,m-1} \left(p_{w,i,j,k}^{n+1,m-1} - p_{w,i,j,k-1}^{n+1,m-1} \right) \\
&- (T_w \gamma_w)_{1,i+1/2,j,k}^{n+1,m-1} \left(z_{i+1,j,k} - z_{i,j,k} \right) + (T_w \gamma_w)_{1,i-1/2,j,k}^{n+1,m-1} \left(z_{i,j,k} - z_{i-1,j,k} \right) \\
&- (T_w \gamma_w)_{2,i,j+1/2,k}^{n+1,m-1} \left(z_{i,j+1,k} - z_{i,j,k} \right) + (T_w \gamma_w)_{2,i,j-1/2,k}^{n+1,m-1} \left(z_{i,j,k} - z_{i,j-1,k} \right) \\
&- (T_w \gamma_w)_{3,i,j,k+1/2}^{n+1,m-1} \left(z_{i,j,k+1} - z_{i,j,k} \right) + (T_w \gamma_w)_{3,i,j,k-1/2}^{n+1,m-1} \left(z_{i,j,k} - z_{i,j,k-1} \right) \\
&+ \tilde{Q}_{w,i,j,k}^{n+1,m-1},
\end{aligned}
\tag{5.59}
$$

where $p_w^{n+1,m-1} = p^n - p_c(S^{n+1,m-1})$ and in all the transmissibility coefficients and the source/sink term $\tilde{Q}_w$ pressure and saturation are evaluated at time levels t^n and $t^{n+1,m-1}$, respectively.

The time step $\Delta t_S^{n+1,m}$ in (5.59) is chosen as follows: Set

$$
\left(\frac{\partial S^{n+1,m}}{\partial t} \right)_{max} = \max_{i,j,k} \left(\frac{G(p^n, S^{n+1,m-1})}{V\phi} \right)_{i,j,k}, \tag{5.60}
$$

and then calculate

$$
\Delta t_S^{n+1,m} = DS_{max} / \left(\frac{\partial S^{n+1,m}}{\partial t} \right)_{max}, \qquad m = 1, 2, \dots, M^n, \ n = 0, 1, \dots, \tag{5.61}
$$

where $G(p^n, S^{n+1,m-1})$ denotes the right-hand side of equation (5.59).

For a comparison between the classical and improved IMPES methods, the reader can refer to Chen, Huan, and Ma (2006). It was shown that the latter is much more efficient. Furthermore, the classical IMPES method has not successfully been applied to the solution of two-phase *coning problems*, but the improved IMPES method is capable of solving them.

Chapter 6

The Black Oil Model and Numerical Solution

In this chapter, we develop basic equations for simultaneous flow of three phases: water (aqueous), oil (oleic), and gas phases through a porous medium. Previously, we assumed that mass does not transfer between phases. The *black oil model* relaxes this assumption. It is now assumed that the hydrocarbon components are divided into a gas component and an oil component in a stock tank at standard pressure and temperature, and that no mass transfer occurs between the water phase and the other two phases (oil and gas). The gas component mainly consists of methane and ethane. Rock and fluid properties given in Chapter 2 are briefly reviewed, and three solution techniques (simultaneous solution, sequential, and IMPES) for the black oil model are studied.

6.1 Basic Differential Equations

6.1.1 Mass Conservation and Darcy's Law

To reduce confusion, we carefully distinguish between phases and components. We use lowercase and uppercase letter subscripts to denote the phases and components, respectively. Note that the water phase is just the water component. The subscript s indicates standard conditions. The mass conservation equations stated in (5.3) apply here. However, because of mass interchange between the oil and gas phases, mass is not conserved within each phase, but rather the total mass of each component must be conserved:

$$\frac{\partial(\phi\rho_w S_w)}{\partial t} = -\nabla \cdot (\rho_w \mathbf{u}_w) + q_W \tag{6.1}$$

for the water component,

$$\frac{\partial(\phi\rho_{Oo} S_o)}{\partial t} = -\nabla \cdot (\rho_{Oo} \mathbf{u}_o) + q_O \tag{6.2}$$

for the oil component, and

$$\frac{\partial}{\partial t}\big(\phi(\rho_{Go} S_o + \rho_g S_g)\big) = -\nabla \cdot (\rho_{Go} \mathbf{u}_o + \rho_g \mathbf{u}_g) + q_G \tag{6.3}$$

for the gas component, where ρ_{Oo} and ρ_{Go} indicate the *partial densities* of the oil and gas components in the oil phase, respectively. Equation (6.3) implies that the gas component may exist in both the oil and gas phases.

Darcy's law for each phase is written in the usual form (cf. (5.6)):

$$\mathbf{u}_\alpha = -\frac{1}{\mu_\alpha}\mathbf{k}_\alpha\left(\nabla p_\alpha - \rho_\alpha \wp \nabla z\right), \qquad \alpha = w,\ o,\ g. \tag{6.4}$$

The fact that the three phases jointly fill the void space is given by the equation

$$S_w + S_o + S_g = 1. \tag{6.5}$$

Finally, the phase pressures are related by capillary pressures

$$p_{cow} = p_o - p_w, \qquad p_{cgo} = p_g - p_o. \tag{6.6}$$

It is not necessary to define a third capillary pressure since it can be defined in terms of p_{cow} and p_{cgo}.

The alternative differential equations developed for two phases in Chapter 5 can be adapted for the three-phase black oil model in a similar fashion (Chen, 2000). That is, equations (6.1)–(6.6) can be rewritten in the three-pressure formulation, in a pressure and two-saturation formulation, or in a global pressure and two-saturation formulation. In the global formulation, the pressure equation is elliptic or parabolic depending on the effects of densities. The two-saturation equations are parabolic if the capillary pressure effects exist; otherwise, they are hyperbolic (Chen, 2000).

For the black oil model, it is often convenient to work with the conservation equations on "standard volumes," instead of the conservation equations on "mass" (6.1)–(6.3). The mass fractions of the oil and gas components in the oil phase can be determined by *gas solubility*, R_{so} (also called *dissolved gas/oil ratio*), which is the volume of gas (measured at standard conditions) dissolved at a given pressure and reservoir temperature in a unit volume of stock tank oil:

$$R_{so}(p, T) = V_{Gs}/V_{Os}. \tag{6.7}$$

Note that

$$V_{Os} = W_O/\rho_{Os}, \qquad V_{Gs} = W_G/\rho_{Gs}, \tag{6.8}$$

where W_O and W_G are the weights of the oil and gas components, respectively. Then equation (6.7) becomes

$$R_{so} = \frac{W_G \rho_{Os}}{W_O \rho_{Gs}}. \tag{6.9}$$

The *oil formation volume factor* B_o is the ratio of the volume V_o of the oil phase (measured at reservoir conditions) to the volume V_{Os} of the oil component measured at standard conditions:

$$B_o(p, T) = V_o(p, T)/V_{Os}, \tag{6.10}$$

where

$$V_o = \frac{W_O + W_G}{\rho_o}. \tag{6.11}$$

Consequently, combining (6.8), (6.10), and (6.11), we have

$$B_o = \frac{(W_O + W_G)\rho_{Os}}{W_O \rho_o}. \tag{6.12}$$

Now, using (6.9) and (6.12), the *mass fractions* of the oil and gas components in the oil phase are, respectively,

$$C_{Oo} = \frac{W_O}{W_O + W_G} = \frac{\rho_{Os}}{B_o \rho_o},$$

$$C_{Go} = \frac{W_G}{W_O + W_G} = \frac{R_{so}\rho_{Gs}}{B_o \rho_o},$$

which, together with $C_{Oo} + C_{Go} = 1$, yield

$$\rho_o = \frac{R_{so}\rho_{Gs} + \rho_{Os}}{B_o}. \tag{6.13}$$

The *gas formation volume factor* B_g is the ratio of the volume of the gas phase measured at reservoir conditions to the volume of the gas component measured at standard conditions:

$$B_g(p, T) = V_g(p, T)/V_{Gs}.$$

Let $W_g = W_G$ be the weight of free gas. Because $V_g = W_G/\rho_g$ and $V_{Gs} = W_G/\rho_{Gs}$, we see that

$$\rho_g = \frac{\rho_{Gs}}{B_g}. \tag{6.14}$$

For completeness, the *water formation volume factor*, B_w, is defined by

$$\rho_w = \frac{\rho_{Ws}}{B_w}. \tag{6.15}$$

The flow rates are defined by

$$\begin{aligned} q_W &= \frac{q_{Ws}\rho_{Ws}}{B_w}, \quad q_O = \frac{q_{Os}\rho_{Os}}{B_o}, \\ q_G &= \frac{q_{Gs}\rho_{Gs}}{B_g} + \frac{q_{Os}R_{so}\rho_{Gs}}{B_o}, \end{aligned} \tag{6.16}$$

where q_{Ws}, q_{Os}, and q_{Gs} are the rates at standard conditions. We introduce the fluid gravities

$$\gamma_\alpha = \rho_\alpha \wp, \qquad \alpha = w, o, g. \tag{6.17}$$

Moreover, we define the *transmissibility*

$$\mathbf{T}_\alpha = \frac{k_{r\alpha}}{\mu_\alpha B_\alpha}\mathbf{k}, \qquad \alpha = w, o, g. \tag{6.18}$$

Substituting (6.16)–(6.18) into (6.1)–(6.3) and dividing the resulting equations by ρ_{Ws}, ρ_{Os}, and ρ_{Gs}, respectively, we obtain

$$\begin{aligned}
&\frac{\partial}{\partial t}\left(\frac{\phi S_w}{B_w}\right) = \nabla \cdot (\mathbf{T}_w [\nabla p_w - \gamma_w \nabla z]) + \frac{q_{Ws}}{B_w},\\
&\frac{\partial}{\partial t}\left(\frac{\phi S_o}{B_o}\right) = \nabla \cdot (\mathbf{T}_o [\nabla p_o - \gamma_o \nabla z]) + \frac{q_{Os}}{B_o},\\
&\frac{\partial}{\partial t}\left[\phi\left(\frac{S_g}{B_g} + \frac{R_{so} S_o}{B_o}\right)\right]\\
&\quad = \nabla \cdot \left(\mathbf{T}_g \left[\nabla p_g - \gamma_g \nabla z\right] + R_{so}\mathbf{T}_o [\nabla p_o - \gamma_o \nabla z]\right) + \frac{q_{Gs}}{B_g} + \frac{q_{Os} R_{so}}{B_o}.
\end{aligned} \tag{6.19}$$

These are the conservation equations on "standard volumes."

The volumetric flow rates at wells (at standard conditions) are (cf. Chapter 4)

$$\begin{aligned}
q_{Ws} &= \sum_{\nu=1}^{N_w}\sum_{m=1}^{M_{w\nu}} WI_m^{(\nu)} \frac{k_{rw}}{\mu_w}\left[p_{bh}^{(\nu)} - p_w - \gamma_w (z_{bh}^{(\nu)} - z)\right] \delta(\mathbf{x} - \mathbf{x}_m^{(\nu)}),\\
q_{Os} &= \sum_{\nu=1}^{N_w}\sum_{m=1}^{M_{w\nu}} WI_m^{(\nu)} \frac{k_{ro}}{\mu_o}\left[p_{bh}^{(\nu)} - p_o - \gamma_o (z_{bh}^{(\nu)} - z)\right] \delta(\mathbf{x} - \mathbf{x}_m^{(\nu)}),\\
q_{Gs} &= \sum_{\nu=1}^{N_w}\sum_{m=1}^{M_{w\nu}} WI_m^{(\nu)} \frac{k_{rg}}{\mu_g}\left[p_{bh}^{(\nu)} - p_g - \gamma_g (z_{bh}^{(\nu)} - z)\right] \delta(\mathbf{x} - \mathbf{x}_m^{(\nu)}),
\end{aligned} \tag{6.20}$$

where the well index is

$$WI_m^{(\nu)} = \left.\frac{2\pi \bar{k} \Delta h}{\ln(r_e/r_w)}\right|_m^{(\nu)},$$

$\delta(\mathbf{x})$ is the Dirac delta function, N_w is the total number of wells, $M_{w\nu}$ is the total number of perforated zones of the νth well, $\Delta h_m^{(\nu)}$ and $\mathbf{x}_m^{(\nu)}$ are the segment length and central location of the mth perforated zone of the νth well, the quantity $\bar{k}$ is an average of $\mathbf{k}$ at the wells (cf. Section 4.3.2), $r_w^{(\nu)}$ denotes the wellbore radius of the νth well, $r_{e,m}^{(\nu)}$ is the drainage radius of the νth well at the gridblock in which $\mathbf{x}_m^{(\nu)}$ is located, and $p_{bh}^{(\nu)}$ is the bottom hole pressure of the νth well at the well datum $z_{bh}^{(\nu)}$.

Typical expressions of p_{cow}, p_{cgo}, and $k_{r\alpha}$ as functions of S_w and S_g were introduced in Chapter 2. Equations (6.5), (6.6), and (6.19) provide six equations for the six unknowns p_α and S_α, $\alpha = w, o, g$. If the bottom hole pressure $p_{bh}^{(\nu)}$ is not given, the source/sink term defining this pressure introduces one more unknown (i.e., $p_{bh}^{(\nu)}$). With appropriate boundary and initial conditions, this is a closed differential system for these unknowns.

6.1.2 Rock/Fluid Properties

The rock-fluid interaction properties were considered in Chapter 2 for three-phase flow; for completeness, we state them briefly. The oil pressure is one of the primary variables to be used:

$$p = p_o. \tag{6.21}$$

The capillary pressures p_{cw} and p_{cg} are assumed to be functions of the saturations only (Leverett and Lewis, 1941):

$$p_{cow} = p_{cow}(S_w), \quad p_{cgo} = p_{cgo}(S_g). \tag{6.22}$$

The relative permeabilities for water and gas are assumed to be of the form

$$\begin{aligned} k_{rw} &= k_{rw}(S_w), & k_{row} &= k_{row}(S_w), \\ k_{rg} &= k_{rg}(S_g), & k_{rog} &= k_{rog}(S_g). \end{aligned} \tag{6.23}$$

As an example, Stone's model II for the oil relative permeability is used (cf. Section 2.5.2),

$$\begin{aligned} &k_{ro}(S_w, S_g) \\ &= k_{rc}\left\{\left[\frac{k_{row}(S_w)}{k_{rc}} + k_{rw}(S_w)\right]\left[\frac{k_{rog}(S_g)}{k_{rc}} + k_{rg}(S_g)\right] - k_{rw}(S_w) - k_{rg}(S_g)\right\}, \end{aligned} \tag{6.24}$$

where $k_{rc} = k_{row}(S_{wc})$ and S_{wc} is the critical saturation (cf. Chapter 2). Finally, the porosity ϕ is assumed to have the form

$$\phi = \phi^o\left(1 + c_R(p - p^o)\right), \tag{6.25}$$

where ϕ^o is the porosity at a reference pressure p^o and c_R is the rock compressibility.

6.1.3 Fluid Properties

The fluid properties were also stated in Chapter 2; we briefly review the definitions of densities and viscosities. The water density ρ_{Ws} at standard conditions is determined using water salinities (Chen, Huan, and Ma, 2006), while the water phase density ρ_w is determined by

$$\rho_w = \frac{\rho_{Ws}}{B_{wi}}\left(1 + c_w(p - p^0)\right), \tag{6.26}$$

where B_{wi} is the water formation volume factor at the initial formation pressure p^0, and c_w is the water compressibility. The water viscosity μ_w is taken to be constant.

The black oil model involves three phases and three components: water, oil, and gas. The relationship between the phases and components is that the water component is all the water phase with density ρ_w, the oil component exists solely in the oil phase with density ρ_{Oo}, and the gas component is divided into two parts: one part in the gas phase that is called *free gas* with density ρ_g, and the other part in the oil phase that is termed the *solution gas* with density ρ_{Go}. Thus the oil phase density ρ_o is given by

$$\rho_o = \rho_{Oo} + \rho_{Go}. \tag{6.27}$$

The oil component density ρ_{Oo} is evaluated from

$$\rho_{Oo} = \frac{\rho_{Os}}{B_o}, \tag{6.28}$$

where the oil formation volume factor B_o is

$$B_o = B_{ob}(p_b)\big(1 - c_o(p - p_b)\big), \tag{6.29}$$

with B_{ob} being the formation volume factor at the bubble point pressure p_b and c_o the oil compressibility. The solution gas density ρ_{Go} is computed by

$$\rho_{Go} = \frac{R_{so}\rho_{Gs}}{B_o}. \tag{6.30}$$

The free gas density ρ_g is defined by

$$\rho_g = \frac{\rho_{Gs}}{B_g}, \tag{6.31}$$

where

$$\rho_{Gs} = Y_G \rho_{air}, \quad B_g = \frac{ZT}{p}\frac{p_s}{T_s}, \tag{6.32}$$

with Y_G being the raw gas density (which is unity for air), ρ_{air} the air density, Z the gas deviation factor, T the temperature, and p_s and T_s the formation pressure and temperature at standard conditions.

The oil viscosity μ_o is given by

$$\mu_o = \mu_{ob}(p_b)\big(1 + c_\mu(p - p_b)\big), \tag{6.33}$$

where μ_{ob} is the oil viscosity at p_b and c_μ is the *oil viscosity compressibility*. The gas viscosity μ_g is a function of p:

$$\mu_g = \mu_g(p). \tag{6.34}$$

For more information on the fluid properties for the black oil model, the reader may refer to Chen, Huan, and Ma (2006).

6.1.4 Phase States

In the secondary recovery of oil, if the reservoir pressure is above the bubble point pressure of the oil phase, the flow is two-phase; if the pressure drops below the bubble point pressure, then the flow is of black oil type. Because of the frequent changes in injection and production in a reservoir, the bubble point pressure varies. If all three phases coexist, the reservoir is referred to as being in the *saturated state*. When all gas dissolves into the oil phase, there is no gas phase present (no free gas); i.e., $S_g = 0$. In this case, the reservoir is said to be in the *undersaturated state*. The critical pressure at which the saturated state becomes the undersaturated state, or vice versa, is the *bubble point pressure*. In the saturated state, $S_g \neq 0$ and $p_b = p$; the densities and viscosities depend only on pressure p:

$$\begin{aligned} &\rho_{Oo}(p) = \frac{\rho_{Os}}{B_{ob}(p)}, \quad \rho_{Go}(p) = \frac{R_{so}(p)\rho_{Gs}}{B_{ob}(p)}, \quad \rho_g(p) = \frac{\rho_{Gs}}{B_g(p)}, \\ &\mu_o = \mu_o(p), \qquad \mu_g = \mu_g(p). \end{aligned} \tag{6.35}$$

In the undersaturated state, $S_g = 0$ and $p_b < p$. The densities and viscosity in the oil phase depend on both p and p_b:

$$\begin{aligned}
\rho_{Oo}(p, p_b) &= \frac{\rho_{Os}}{B_{ob}(p_b)}\big(1 + c_o(p - p_b)\big), \\
\rho_{Go}(p, p_b) &= \frac{R_{so}(p_b)\rho_{Gs}}{B_{ob}(p_b)}\big(1 + c_o(p - p_b)\big), \quad \rho_g(p) = \frac{\rho_{Gs}}{B_g(p)}, \\
\mu_o(p, p_b) &= \mu_{ob}(p_b)\big(1 + c_\mu(p - p_b)\big), \quad \mu_g = \mu_g(p).
\end{aligned} \tag{6.36}$$

For numerical solutions of the black oil model, the choice of the primary unknowns depends on the states. In the saturated state, $p = p_o$, S_w, and S_o are the primary unknowns; in the undersaturated state, $p = p_o$, p_b, and S_w are the primary unknowns. Consequently, the initial conditions are either

$$p(\mathbf{x}, 0) = p^0(\mathbf{x}), \quad S_w(\mathbf{x}, 0) = S_w^0(\mathbf{x}), \quad S_o(\mathbf{x}, 0) = S_o^0(\mathbf{x}), \quad \mathbf{x} \in \Omega, \tag{6.37}$$

or

$$p(\mathbf{x}, 0) = p^0(\mathbf{x}), \quad S_w(\mathbf{x}, 0) = S_w^0(\mathbf{x}), \quad p_b(\mathbf{x}, 0) = p_b^0(\mathbf{x}), \quad \mathbf{x} \in \Omega, \tag{6.38}$$

depending on the initial state of a reservoir. More details on the specification of initial data will be given below.

6.2 Numerical Solution of the Black Oil Model

We consider the case where the permeability tensor $\mathbf{k} = \text{diag}(k_{11}, k_{22}, k_{33})$ is diagonal. Consequently, the transmissibility tensor $\mathbf{T}_\alpha$ is also diagonal, and we write $\mathbf{T}_\alpha = \text{diag}(T_{\alpha 1}, T_{\alpha 2}, T_{\alpha 3})$, $\alpha = w, o, g$. Therefore, in terms of each coordinate direction, equations (6.19) are

$$\begin{aligned}
\frac{\partial}{\partial t}\left(\frac{\phi S_w}{B_w}\right) &= \frac{\partial}{\partial x_1}\left(T_{w1}\left(\frac{\partial p_w}{\partial x_1} - \gamma_w \frac{\partial z}{\partial x_1}\right)\right) + \frac{\partial}{\partial x_2}\left(T_{w2}\left(\frac{\partial p_w}{\partial x_2} - \gamma_w \frac{\partial z}{\partial x_2}\right)\right) \\
&\quad + \frac{\partial}{\partial x_3}\left(T_{w3}\left(\frac{\partial p_w}{\partial x_3} - \gamma_w \frac{\partial z}{\partial x_3}\right)\right) + \tilde{q}_{Ws},
\end{aligned} \tag{6.39}$$

$$\begin{aligned}
\frac{\partial}{\partial t}\left(\frac{\phi S_o}{B_o}\right) &= \frac{\partial}{\partial x_1}\left(T_{o1}\left(\frac{\partial p_o}{\partial x_1} - \gamma_o \frac{\partial z}{\partial x_1}\right)\right) + \frac{\partial}{\partial x_2}\left(T_{o2}\left(\frac{\partial p_o}{\partial x_2} - \gamma_o \frac{\partial z}{\partial x_2}\right)\right) \\
&\quad + \frac{\partial}{\partial x_3}\left(T_{o3}\left(\frac{\partial p_o}{\partial x_3} - \gamma_o \frac{\partial z}{\partial x_3}\right)\right) + \tilde{q}_{Os},
\end{aligned} \tag{6.40}$$

and

$$\begin{aligned}
\frac{\partial}{\partial t}\left[\phi\left(\frac{S_g}{B_g} + \frac{R_{so} S_o}{B_o}\right)\right] &= \frac{\partial}{\partial x_1}\left(T_{g1}\left(\frac{\partial p_g}{\partial x_1} - \gamma_g \frac{\partial z}{\partial x_1}\right)\right) \\
+ \frac{\partial}{\partial x_2}\left(T_{g2}\left(\frac{\partial p_g}{\partial x_2} - \gamma_g \frac{\partial z}{\partial x_2}\right)\right) &+ \frac{\partial}{\partial x_3}\left(T_{g3}\left(\frac{\partial p_g}{\partial x_3} - \gamma_g \frac{\partial z}{\partial x_3}\right)\right)
\end{aligned}$$

$$+\frac{\partial}{\partial x_1}\left(R_{so}T_{o1}\left(\frac{\partial p_o}{\partial x_1}-\gamma_o\frac{\partial z}{\partial x_1}\right)\right)+\frac{\partial}{\partial x_2}\left(R_{so}T_{o2}\left(\frac{\partial p_o}{\partial x_2}-\gamma_o\frac{\partial z}{\partial x_2}\right)\right) \tag{6.41}$$
$$+\frac{\partial}{\partial x_3}\left(R_{so}T_{o3}\left(\frac{\partial p_o}{\partial x_3}-\gamma_o\frac{\partial z}{\partial x_3}\right)\right)+\tilde{q}_{Gs},$$

where

$$\tilde{q}_{Ws}=\frac{q_{Ws}}{B_w},\quad \tilde{q}_{Os}=\frac{q_{Os}}{B_o},\quad \tilde{q}_{Gs}=\frac{q_{Gs}}{B_g}+\frac{q_{Os}R_{so}}{B_o}.$$

Numerical solutions of equations (6.39)–(6.41) will be obtained in this section.

6.2.1 Treatment of Initial Conditions

Initial conditions for the black oil model involve the specification of phase pressures and/or saturations for each grid block in the simulation model at the beginning of each simulation. Differences in phase gravities and capillary pressures cause fluids to segregate until the reservoir system reaches gravity/capillary equilibrium. There exist up to five different fluid zones vertically from the top of the reservoir to its bottom: gas cap, gas/oil transition, oil, oil/water transition, and water zone. Different initial data can be specified in each of these five zones.

In general, the specification of initial data depends on the gravity/capillary equilibrium and the nature of the fluids that occupy the different zones. For a continuous phase, the initial pressure is directly calculated from a hydrostatic relation, while for a discontinuous phase, the initial pressure is determined from the capillary pressure function evaluated at the endpoint saturation. The initial saturation of a continuous phase is calculated from either the capillary pressure function (6.6) or the saturation relation (6.5), and the initial saturation of a discontinuous phase is given at the endpoint saturation.

Gas cap zone

Initially, in the gas cap zone, only the gas phase is continuous. Thus, the vertical distribution of the gas pressure can be computed from the *hydrostatic relation*

$$\frac{dp_g}{dz}=\gamma_g. \tag{6.42}$$

In addition,

$$S_w=S_{iw},\quad S_o=0, \tag{6.43}$$

where S_{iw} is the irreducible water saturation. From these known variables, other variables can be deduced:

$$S_g=1-S_w-S_o,\quad p_o=p_g-p_{cgo}(S_{g,max}),\quad p_w=p_o-p_{cow}(S_{iw}),$$

where $S_{g,max}$ is the maximum gas saturation in the original gas cap.

Gas/oil transition zone

In the gas/oil transition zone, both the gas and oil phases are continuous, so the vertical pressure distribution of these phases can be directly obtained from the *hydrostatic relations*

$$\frac{dp_g}{dz} = \gamma_g, \quad \frac{dp_o}{dz} = \gamma_o. \tag{6.44}$$

An additional condition is

$$S_w = S_{iw}. \tag{6.45}$$

From these conditions, we see that

$$S_g = p_{cgo}^{-1}(p_g - p_o), \quad p_w = p_o - p_{cow}(S_{iw}), \quad S_o = 1 - S_g - S_w, \tag{6.46}$$

where we assume that p_{cgo} has an inverse p_{cgo}^{-1}.

Oil zone

The oil phase is the only continuous phase in the oil zone:

$$\frac{dp_o}{dz} = \gamma_o. \tag{6.47}$$

Additionally, we have

$$S_w = S_{iw}, \quad S_g = 0. \tag{6.48}$$

It follows from equations (6.47) and (6.48) that

$$p_g = p_o + p_{cgo}(0), \quad p_w = p_o - p_{cow}(S_{iw}), \quad S_o = 1 - S_g - S_w. \tag{6.49}$$

Oil/water zone

Both the oil and water phases are continuous in the oil/water zone:

$$\frac{dp_o}{dz} = \gamma_o, \quad \frac{dp_w}{dz} = \gamma_w. \tag{6.50}$$

In addition,

$$S_g = 0. \tag{6.51}$$

From these two conditions, we see that

$$S_w = p_{cow}^{-1}(p_o - p_w), \quad p_g = p_o + p_{cgo}(0), \quad S_o = 1 - S_g - S_w, \tag{6.52}$$

where the capillary pressure p_{cow} is assumed to be invertible.

Water zone

Finally, in the water zone, only the water phase is continuous:

$$\frac{dp_w}{dz} = \gamma_w. \tag{6.53}$$

In addition,

$$S_g = S_o = 0. \tag{6.54}$$

From these initial data, we have

$$p_o = p_w + p_{cow}(S_{wmax}), \quad p_g = p_o + p_{cgo}(0), \quad S_w = 1, \tag{6.55}$$

where S_{wmax} is the maximum water saturation in the original water zone.

In reservoir simulation, the depths of the water/oil contact and the oil/gas contact are given. Then the initial pressure and saturation at all gridblocks can be uniquely determined if a reference pressure (e.g., datum pressure) and a reference depth (e.g., datum depth) are given (cf. Section 3.4.1). For an undersaturated reservoir, the reference depth and pressure are arbitrary and can be specified in any of the five fluid zones. For a saturated reservoir, the reference depth must be the depth of the oil/gas contact, and the reference pressure must be the initial bubble point (saturation) pressure.

When the hydrostatic conditions are used to obtain the initial pressure, the simulation model will initialize to equilibrium if the depths in the initialization part are the same as those in the reservoir layers. However, if a simulation model is not in an initial hydrostatic equilibrium, an initialization algorithm should be performed in several time steps (without source/sink terms) to allow the model to reach the equilibrium state.

If capillary pressures (p_{cow} and/or p_{cgo}) are ignored, the initial phase saturations (often the endpoint saturations) must be imposed, but the pressures can be obtained from the reference pressure. In this situation, no transition zone is generally assumed to exist in the reservoir.

6.2.2 Simultaneous Solution Techniques

The most natural solution technique for the black oil system is to solve the three equations simultaneously, which suggests the *simultaneous solution* (SS) technique. This technique was initially introduced by Douglas, Peaceman, and Rachford (1959) and is still widely used in black oil reservoir simulation.

Saturated state

To abuse the transmissibility notation, the numerical transmissibilities at the gridblock boundaries

$$\frac{A_m T_{\alpha m}}{h_m}, \quad m = 1, 2, 3, \ \alpha = w, o, g,$$

are still indicated by $T_{\alpha m}$, where A_m is the cross-sectional area normal to the x_m-direction.

In the saturated state, the primary unknowns are ($p = p_o, S_w, S_o$), and all other variables are obtained from them. Particularly, $S_g = 1 - S_w - S_o$, $p_w = p - p_{cow}$, and $p_g = p + p_{cgo}$. In terms of these primary unknowns, the finite difference counterparts of equations (6.39)–(6.41) are

$$
\begin{aligned}
&\frac{1}{\Delta t}\left(V\left[\left(\frac{\phi S_w}{B_w}\right)^{n+1}-\left(\frac{\phi S_w}{B_w}\right)^{n}\right]\right)_{i,j,k}\\
&= T^{n+1}_{w1,i+1/2,j,k}\left(p^{n+1}_{w,i+1,j,k}-p^{n+1}_{w,i,j,k}\right)-T^{n+1}_{w1,i-1/2,j,k}\left(p^{n+1}_{w,i,j,k}-p^{n+1}_{w,i-1,j,k}\right)\\
&\quad+T^{n+1}_{w2,i,j+1/2,k}\left(p^{n+1}_{w,i,j+1,k}-p^{n+1}_{w,i,j,k}\right)-T^{n+1}_{w2,i,j-1/2,k}\left(p^{n+1}_{w,i,j,k}-p^{n+1}_{w,i,j-1,k}\right)\\
&\quad+T^{n+1}_{w3,i,j,k+1/2}\left(p^{n+1}_{w,i,j,k+1}-p^{n+1}_{w,i,j,k}\right)-T^{n+1}_{w3,i,j,k-1/2}\left(p^{n+1}_{w,i,j,k}-p^{n+1}_{w,i,j,k-1}\right)\\
&\quad-(T_w\gamma_w)^{n+1}_{1,i+1/2,j,k}\left(z_{i+1,j,k}-z_{i,j,k}\right)+(T_w\gamma_w)^{n+1}_{1,i-1/2,j,k}\left(z_{i,j,k}-z_{i-1,j,k}\right)\\
&\quad-(T_w\gamma_w)^{n+1}_{2,i,j+1/2,k}\left(z_{i,j+1,k}-z_{i,j,k}\right)+(T_w\gamma_w)^{n+1}_{2,i,j-1/2,k}\left(z_{i,j,k}-z_{i,j-1,k}\right)\\
&\quad-(T_w\gamma_w)^{n+1}_{3,i,j,k+1/2}\left(z_{i,j,k+1}-z_{i,j,k}\right)+(T_w\gamma_w)^{n+1}_{3,i,j,k-1/2}\left(z_{i,j,k}-z_{i,j,k-1}\right)\\
&\quad+\tilde{Q}^{n+1}_{Ws,i,j,k},
\end{aligned}
\tag{6.56}
$$

$$
\begin{aligned}
&\frac{1}{\Delta t}\left(V\left[\left(\frac{\phi S_o}{B_o}\right)^{n+1}-\left(\frac{\phi S_o}{B_o}\right)^{n}\right]\right)_{i,j,k}\\
&= T^{n+1}_{o1,i+1/2,j,k}\left(p^{n+1}_{o,i+1,j,k}-p^{n+1}_{o,i,j,k}\right)-T^{n+1}_{o1,i-1/2,j,k}\left(p^{n+1}_{o,i,j,k}-p^{n+1}_{o,i-1,j,k}\right)\\
&\quad+T^{n+1}_{o2,i,j+1/2,k}\left(p^{n+1}_{o,i,j+1,k}-p^{n+1}_{o,i,j,k}\right)-T^{n+1}_{o2,i,j-1/2,k}\left(p^{n+1}_{o,i,j,k}-p^{n+1}_{o,i,j-1,k}\right)\\
&\quad+T^{n+1}_{o3,i,j,k+1/2}\left(p^{n+1}_{o,i,j,k+1}-p^{n+1}_{o,i,j,k}\right)-T^{n+1}_{o3,i,j,k-1/2}\left(p^{n+1}_{o,i,j,k}-p^{n+1}_{o,i,j,k-1}\right)\\
&\quad-(T_o\gamma_o)^{n+1}_{1,i+1/2,j,k}\left(z_{i+1,j,k}-z_{i,j,k}\right)+(T_o\gamma_o)^{n+1}_{1,i-1/2,j,k}\left(z_{i,j,k}-z_{i-1,j,k}\right)\\
&\quad-(T_o\gamma_o)^{n+1}_{2,i,j+1/2,k}\left(z_{i,j+1,k}-z_{i,j,k}\right)+(T_o\gamma_o)^{n+1}_{2,i,j-1/2,k}\left(z_{i,j,k}-z_{i,j-1,k}\right)\\
&\quad-(T_o\gamma_o)^{n+1}_{3,i,j,k+1/2}\left(z_{i,j,k+1}-z_{i,j,k}\right)+(T_o\gamma_o)^{n+1}_{3,i,j,k-1/2}\left(z_{i,j,k}-z_{i,j,k-1}\right)\\
&\quad+\tilde{Q}^{n+1}_{Os,i,j,k},
\end{aligned}
\tag{6.57}
$$

and

$$
\begin{aligned}
&\frac{1}{\Delta t}\left(V\left[\left(\phi\left(\frac{S_g}{B_g}+\frac{R_{so}S_o}{B_o}\right)\right)^{n+1}-\left(\phi\left(\frac{S_g}{B_g}+\frac{R_{so}S_o}{B_o}\right)\right)^{n}\right]\right)_{i,j,k}\\
&= T^{n+1}_{g1,i+1/2,j,k}\left(p^{n+1}_{g,i+1,j,k}-p^{n+1}_{g,i,j,k}\right)-T^{n+1}_{g1,i-1/2,j,k}\left(p^{n+1}_{g,i,j,k}-p^{n+1}_{g,i-1,j,k}\right)\\
&\quad+T^{n+1}_{g2,i,j+1/2,k}\left(p^{n+1}_{g,i,j+1,k}-p^{n+1}_{g,i,j,k}\right)-T^{n+1}_{g2,i,j-1/2,k}\left(p^{n+1}_{g,i,j,k}-p^{n+1}_{g,i,j-1,k}\right)\\
&\quad+T^{n+1}_{g3,i,j,k+1/2}\left(p^{n+1}_{g,i,j,k+1}-p^{n+1}_{g,i,j,k}\right)-T^{n+1}_{g3,i,j,k-1/2}\left(p^{n+1}_{g,i,j,k}-p^{n+1}_{g,i,j,k-1}\right)\\
&\quad-(T_g\gamma_g)^{n+1}_{1,i+1/2,j,k}\left(z_{i+1,j,k}-z_{i,j,k}\right)+(T_g\gamma_g)^{n+1}_{1,i-1/2,j,k}\left(z_{i,j,k}-z_{i-1,j,k}\right)\\
&\quad-(T_g\gamma_g)^{n+1}_{2,i,j+1/2,k}\left(z_{i,j+1,k}-z_{i,j,k}\right)+(T_g\gamma_g)^{n+1}_{2,i,j-1/2,k}\left(z_{i,j,k}-z_{i,j-1,k}\right)\\
&\quad-(T_g\gamma_g)^{n+1}_{3,i,j,k+1/2}\left(z_{i,j,k+1}-z_{i,j,k}\right)+(T_g\gamma_g)^{n+1}_{3,i,j,k-1/2}\left(z_{i,j,k}-z_{i,j,k-1}\right)\\
&\quad+(R_{so}T_o)^{n+1}_{1,i+1/2,j,k}\left(p^{n+1}_{o,i+1,j,k}-p^{n+1}_{o,i,j,k}\right)-(R_{so}T_o)^{n+1}_{1,i-1/2,j,k}\left(p^{n+1}_{o,i,j,k}-p^{n+1}_{o,i-1,j,k}\right)
\end{aligned}
$$

$$
\begin{aligned}
&+ (R_{so}T_o)^{n+1}_{2,i,j+1/2,k}\left(p^{n+1}_{o,i,j+1,k} - p^{n+1}_{o,i,j,k}\right) - (R_{so}T_o)^{n+1}_{2,i,j-1/2,k}\left(p^{n+1}_{o,i,j,k} - p^{n+1}_{o,i,j-1,k}\right)\\
&+ (R_{so}T_o)^{n+1}_{3,i,j,k+1/2}\left(p^{n+1}_{o,i,j,k+1} - p^{n+1}_{o,i,j,k}\right) - (R_{so}T_o)^{n+1}_{3,i,j,k-1/2}\left(p^{n+1}_{o,i,j,k} - p^{n+1}_{o,i,j,k-1}\right)\\
&- (R_{so}T_o\gamma_o)^{n+1}_{1,i+1/2,j,k}\left(z_{i+1,j,k} - z_{i,j,k}\right) + (R_{so}T_o\gamma_o)^{n+1}_{1,i-1/2,j,k}\left(z_{i,j,k} - z_{i-1,j,k}\right)\\
&- (R_{so}T_o\gamma_o)^{n+1}_{2,i,j+1/2,k}\left(z_{i,j+1,k} - z_{i,j,k}\right) + (R_{so}T_o\gamma_o)^{n+1}_{2,i,j-1/2,k}\left(z_{i,j,k} - z_{i,j-1,k}\right)\\
&- (R_{so}T_o\gamma_o)^{n+1}_{3,i,j,k+1/2}\left(z_{i,j,k+1} - z_{i,j,k}\right) + (R_{so}T_o\gamma_o)^{n+1}_{3,i,j,k-1/2}\left(z_{i,j,k} - z_{i,j,k-1}\right)\\
&+ \tilde{Q}^{n+1}_{Gs,i,j,k},
\end{aligned}
\tag{6.58}
$$

where $\tilde{Q}_{\alpha s,i,j,k} = (V\tilde{q}_{\alpha s})_{i,j,k}$, $\alpha = W, O, G$.

The transmissibility terms in equations (6.56)–(6.58) can be treated as in the two-phase flow equations. That is, the rock property (i.e., absolute permeability), fluid properties (i.e., viscosities and formation volume factors), and rock/fluid properties (i.e., relative permeabilities and capillary pressures) at internal boundaries of gridblocks should be evaluated using the harmonic averaging, (weighted) arithmetic averaging, and upstream weighting techniques (cf. Section 5.3.4), respectively.

Equations (6.56)–(6.58) are nonlinear in terms of the primary unknowns (p^{n+1}, S_w^{n+1}, S_o^{n+1}) and can be linearized via the Newton–Raphson iteration introduced in Chapter 3. At each gridblock, there are three unknowns $(p^{n+1}, S_w^{n+1}, S_o^{n+1})_{i,j,k}$. In each Newton–Raphson iteration step, they are written as

$$
\begin{aligned}
p^{n+1,l+1} &= p^{n+1,l} + \delta p^{n+1,l+1}, \qquad S_w^{n+1,l+1} = S_w^{n+1,l} + \delta S_w^{n+1,l+1},\\
S_o^{n+1,l+1} &= S_o^{n+1,l} + \delta S_o^{n+1,l+1}.
\end{aligned}
$$

Below, the superscript $n+1$ will be dropped.

The residuals of equations (6.56)–(6.58) at the Newton–Raphson iteration level l are

$$
\begin{aligned}
R^l_{w,i,j,k} = {}& \frac{1}{\Delta t}\left(V\left[\left(\frac{\phi S_w}{B_w}\right)^l - \left(\frac{\phi S_w}{B_w}\right)^n\right]\right)_{i,j,k}\\
&- T^l_{w1,i+1/2,j,k}\left(p^l_{w,i+1,j,k} - p^l_{w,i,j,k}\right) + T^l_{w1,i-1/2,j,k}\left(p^l_{w,i,j,k} - p^l_{w,i-1,j,k}\right)\\
&- T^l_{w2,i,j+1/2,k}\left(p^l_{w,i,j+1,k} - p^l_{w,i,j,k}\right) + T^l_{w2,i,j-1/2,k}\left(p^l_{w,i,j,k} - p^l_{w,i,j-1,k}\right)\\
&- T^l_{w3,i,j,k+1/2}\left(p^l_{w,i,j,k+1} - p^l_{w,i,j,k}\right) + T^l_{w3,i,j,k-1/2}\left(p^l_{w,i,j,k} - p^l_{w,i,j,k-1}\right)\\
&+ (T_w\gamma_w)^l_{1,i+1/2,j,k}\left(z_{i+1,j,k} - z_{i,j,k}\right) - (T_w\gamma_w)^l_{1,i-1/2,j,k}\left(z_{i,j,k} - z_{i-1,j,k}\right)\\
&+ (T_w\gamma_w)^l_{2,i,j+1/2,k}\left(z_{i,j+1,k} - z_{i,j,k}\right) - (T_w\gamma_w)^l_{2,i,j-1/2,k}\left(z_{i,j,k} - z_{i,j-1,k}\right)\\
&+ (T_w\gamma_w)^l_{3,i,j,k+1/2}\left(z_{i,j,k+1} - z_{i,j,k}\right) - (T_w\gamma_w)^l_{3,i,j,k-1/2}\left(z_{i,j,k} - z_{i,j,k-1}\right)\\
&- \tilde{Q}^l_{Ws,i,j,k},
\end{aligned}
\tag{6.59}
$$

$$
\begin{aligned}
R^l_{o,i,j,k} = {}& \frac{1}{\Delta t}\left(V\left[\left(\frac{\phi S_o}{B_o}\right)^l - \left(\frac{\phi S_o}{B_o}\right)^n\right]\right)_{i,j,k}\\
&- T^l_{o1,i+1/2,j,k}\left(p^l_{o,i+1,j,k} - p^l_{o,i,j,k}\right) + T^l_{o1,i-1/2,j,k}\left(p^l_{o,i,j,k} - p^l_{o,i-1,j,k}\right)\\
&- T^l_{o2,i,j+1/2,k}\left(p^l_{o,i,j+1,k} - p^l_{o,i,j,k}\right) + T^l_{o2,i,j-1/2,k}\left(p^l_{o,i,j,k} - p^l_{o,i,j-1,k}\right)
\end{aligned}
$$

$$
\begin{aligned}
&- T^l_{o3,i,j,k+1/2}\left(p^l_{o,i,j,k+1} - p^l_{o,i,j,k}\right) + T^l_{o3,i,j,k-1/2}\left(p^l_{o,i,j,k} - p^l_{o,i,j,k-1}\right)\\
&+ (T_o\gamma_o)^l_{1,i+1/2,j,k}\left(z_{i+1,j,k} - z_{i,j,k}\right) - (T_o\gamma_o)^l_{1,i-1/2,j,k}\left(z_{i,j,k} - z_{i-1,j,k}\right)\\
&+ (T_o\gamma_o)^l_{2,i,j+1/2,k}\left(z_{i,j+1,k} - z_{i,j,k}\right) - (T_o\gamma_o)^l_{2,i,j-1/2,k}\left(z_{i,j,k} - z_{i,j-1,k}\right)\\
&+ (T_o\gamma_o)^l_{3,i,j,k+1/2}\left(z_{i,j,k+1} - z_{i,j,k}\right) - (T_o\gamma_o)^l_{3,i,j,k-1/2}\left(z_{i,j,k} - z_{i,j,k-1}\right)\\
&- \tilde{Q}^l_{Os,i,j,k},
\end{aligned}
\tag{6.60}
$$

and

$$
\begin{aligned}
R^l_{g,i,j,k} = &\frac{1}{\Delta t}\left(V\left[\left(\phi\left(\frac{S_g}{B_g} + \frac{R_{so}S_o}{B_o}\right)\right)^l - \left(\phi\left(\frac{S_g}{B_g} + \frac{R_{so}S_o}{B_o}\right)\right)^n\right]\right)_{i,j,k}\\
&- T^l_{g1,i+1/2,j,k}\left(p^l_{g,i+1,j,k} - p^l_{g,i,j,k}\right) + T^l_{g1,i-1/2,j,k}\left(p^l_{g,i,j,k} - p^l_{g,i-1,j,k}\right)\\
&- T^l_{g2,i,j+1/2,k}\left(p^l_{g,i,j+1,k} - p^l_{g,i,j,k}\right) + T^l_{g2,i,j-1/2,k}\left(p^l_{g,i,j,k} - p^l_{g,i,j-1,k}\right)\\
&- T^l_{g3,i,j,k+1/2}\left(p^l_{g,i,j,k+1} - p^l_{g,i,j,k}\right) + T^l_{g3,i,j,k-1/2}\left(p^l_{g,i,j,k} - p^l_{g,i,j,k-1}\right)\\
&+ (T_g\gamma_g)^l_{1,i+1/2,j,k}\left(z_{i+1,j,k} - z_{i,j,k}\right) - (T_g\gamma_g)^l_{1,i-1/2,j,k}\left(z_{i,j,k} - z_{i-1,j,k}\right)\\
&+ (T_g\gamma_g)^l_{2,i,j+1/2,k}\left(z_{i,j+1,k} - z_{i,j,k}\right) - (T_g\gamma_g)^l_{2,i,j-1/2,k}\left(z_{i,j,k} - z_{i,j-1,k}\right)\\
&+ (T_g\gamma_g)^l_{3,i,j,k+1/2}\left(z_{i,j,k+1} - z_{i,j,k}\right) - (T_g\gamma_g)^l_{3,i,j,k-1/2}\left(z_{i,j,k} - z_{i,j,k-1}\right)\\
&- (R_{so}T_o)^l_{1,i+1/2,j,k}\left(p^l_{o,i+1,j,k} - p^l_{o,i,j,k}\right) + (R_{so}T_o)^l_{1,i-1/2,j,k}\left(p^l_{o,i,j,k} - p^l_{o,i-1,j,k}\right)\\
&- (R_{so}T_o)^l_{2,i,j+1/2,k}\left(p^l_{o,i,j+1,k} - p^l_{o,i,j,k}\right) + (R_{so}T_o)^l_{2,i,j-1/2,k}\left(p^l_{o,i,j,k} - p^l_{o,i,j-1,k}\right)\\
&- (R_{so}T_o)^l_{3,i,j,k+1/2}\left(p^l_{o,i,j,k+1} - p^l_{o,i,j,k}\right) + (R_{so}T_o)^l_{3,i,j,k-1/2}\left(p^l_{o,i,j,k} - p^l_{o,i,j,k-1}\right)\\
&+ (R_{so}T_o\gamma_o)^l_{1,i+1/2,j,k}\left(z_{i+1,j,k} - z_{i,j,k}\right) - (R_{so}T_o\gamma_o)^l_{1,i-1/2,j,k}\left(z_{i,j,k} - z_{i-1,j,k}\right)\\
&+ (R_{so}T_o\gamma_o)^l_{2,i,j+1/2,k}\left(z_{i,j+1,k} - z_{i,j,k}\right) - (R_{so}T_o\gamma_o)^l_{2,i,j-1/2,k}\left(z_{i,j,k} - z_{i,j-1,k}\right)\\
&+ (R_{so}T_o\gamma_o)^l_{3,i,j,k+1/2}\left(z_{i,j,k+1} - z_{i,j,k}\right) - (R_{so}T_o\gamma_o)^l_{3,i,j,k-1/2}\left(z_{i,j,k} - z_{i,j,k-1}\right)\\
&- \tilde{Q}^l_{Gs,i,j,k}.
\end{aligned}
\tag{6.61}
$$

Now, we define the unknown and residual vectors

$$
\mathbf{y} = (p, S_w, S_o)^T, \quad \mathbf{R}^l_{i,j,k} = \left(R^l_{w,i,j,k}, R^l_{o,i,j,k}, R^l_{g,i,j,k}\right)^T,
$$

where the superscript T indicates the transpose. Then application of the Newton–Raphson iteration to equations (6.56)–(6.58) yields a linear system of equations in terms of $\delta\mathbf{y}^{l+1}$:

$$
\begin{aligned}
&\frac{\partial \mathbf{R}^l_{i,j,k}}{\partial \mathbf{y}_{i,j,k-1}}\delta\mathbf{y}^{l+1}_{i,j,k-1} + \frac{\partial \mathbf{R}^l_{i,j,k}}{\partial \mathbf{y}_{i,j-1,k}}\delta\mathbf{y}^{l+1}_{i,j-1,k} + \frac{\partial \mathbf{R}^l_{i,j,k}}{\partial \mathbf{y}_{i-1,j,k}}\delta\mathbf{y}^{l+1}_{i-1,j,k}\\
&\quad + \frac{\partial \mathbf{R}^l_{i,j,k}}{\partial \mathbf{y}_{i,j,k}}\delta\mathbf{y}^{l+1}_{i,j,k} + \frac{\partial \mathbf{R}^l_{i,j,k}}{\partial \mathbf{y}_{i+1,j,k}}\delta\mathbf{y}^{l+1}_{i+1,j,k} + \frac{\partial \mathbf{R}^l_{i,j,k}}{\partial \mathbf{y}_{i,j+1,k}}\delta\mathbf{y}^{l+1}_{i,j+1,k}\\
&\quad + \frac{\partial \mathbf{R}^l_{i,j,k}}{\partial \mathbf{y}_{i,j,k+1}}\delta\mathbf{y}^{l+1}_{i,j,k+1} = -\mathbf{R}^l_{i,j,k},
\end{aligned}
\tag{6.62}
$$

which is the *block seven-point stencil* in the increment $\delta \mathbf{y}^{l+1}$, and

$$\frac{\partial \mathbf{R}}{\partial \mathbf{y}} = \begin{pmatrix} \frac{\partial R_w}{\partial p} & \frac{\partial R_w}{\partial S_o} & \frac{\partial R_w}{\partial S_w} \\ \frac{\partial R_o}{\partial p} & \frac{\partial R_o}{\partial S_o} & \frac{\partial R_o}{\partial S_w} \\ \frac{\partial R_g}{\partial p} & \frac{\partial R_g}{\partial S_o} & \frac{\partial R_g}{\partial S_w} \end{pmatrix}.$$

The elements of the coefficient matrix in system (6.62) are submatrices, and thus this matrix is of a block form (cf. Fig. 6.1). After the increment $\delta \mathbf{y}^{l+1}$ is obtained, the solution is updated until a convergence criterion is achieved:

$$\mathbf{y}^{l+1} = \mathbf{y}^l + \delta \mathbf{y}^{l+1}.$$

The source/sink terms are determined by

$$\begin{aligned} q_{Ws}^{l+1} &= \sum_{\nu=1}^{N_w} \sum_{m=1}^{M_{w\nu}} WI_m^{(\nu)} \frac{k_{rw}^{l+1}}{\mu_w} \left[\left(p_{bh}^{(\nu)}\right)^{l+1} - p^{l+1} - p_{cw}^{l+1} - \rho_w^{l+1} \wp (z_{bh}^{(\nu)} - z) \right] \delta(\mathbf{x} - \mathbf{x}_m^{(\nu)}), \\ q_{Os}^{l+1} &= \sum_{\nu=1}^{N_w} \sum_{m=1}^{M_{w\nu}} WI_m^{(\nu)} \frac{k_{ro}^{l+1}}{\mu_o^{l+1}} \left[\left(p_{bh}^{(\nu)}\right)^{l+1} - p^{l+1} - \rho_o^{l+1} \wp (z_{bh}^{(\nu)} - z) \right] \delta(\mathbf{x} - \mathbf{x}_m^{(\nu)}), \\ q_{Gs}^{l+1} &= \sum_{\nu=1}^{N_w} \sum_{m=1}^{M_{w\nu}} WI_m^{(\nu)} \frac{k_{rg}^{l+1}}{\mu_g^{l+1}} \left[\left(p_{bh}^{(\nu)}\right)^{l+1} - p^{l+1} - p_{cg}^{l+1} - \rho_g^{l+1} \wp (z_{bh}^{(\nu)} - z) \right] \delta(\mathbf{x} - \mathbf{x}_m^{(\nu)}). \end{aligned} \tag{6.63}$$

When the bottom hole pressure p_{bh} is given, the equations in (6.63) can be substituted into the residuals of equations (6.56)–(6.58) to solve for the primary unknowns ($\delta p, \delta S_w, \delta S_o$). On the other hand, if the rates are given, system (6.63) is coupled to the flow equations (6.56)–(6.58) for the unknowns ($\delta p, \delta S_w, \delta S_o, \delta p_{bh}$) (cf. Section 6.2.6).

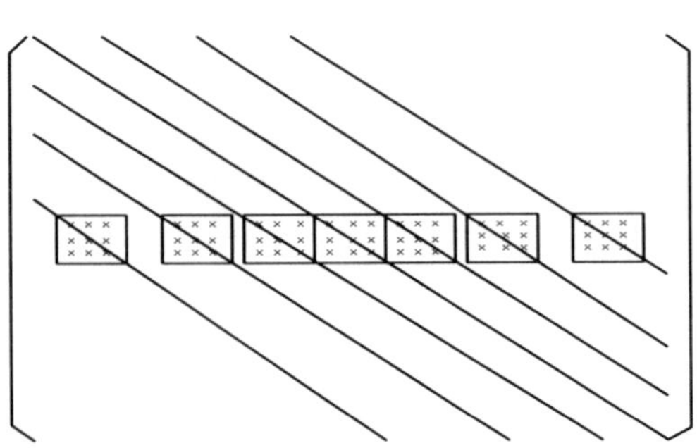

Figure 6.1. *Block matrix.*

Undersaturated state

In the undersaturated state, the primary unknowns are $(p = p_o, S_w, p_b)$. In the present case, $S_g = 0$ and $p_w = p - p_{cow}$. Analogous equations can be obtained as in the saturated state:

$$
\begin{aligned}
&\frac{1}{\Delta t}\left(V\left[\left(\frac{\phi S_w}{B_w}\right)^{n+1} - \left(\frac{\phi S_w}{B_w}\right)^{n}\right]\right)_{i,j,k} \\
&= T^{n+1}_{w1,i+1/2,j,k}\left(p^{n+1}_{w,i+1,j,k} - p^{n+1}_{w,i,j,k}\right) - T^{n+1}_{w1,i-1/2,j,k}\left(p^{n+1}_{w,i,j,k} - p^{n+1}_{w,i-1,j,k}\right) \\
&\quad + T^{n+1}_{w2,i,j+1/2,k}\left(p^{n+1}_{w,i,j+1,k} - p^{n+1}_{w,i,j,k}\right) - T^{n+1}_{w2,i,j-1/2,k}\left(p^{n+1}_{w,i,j,k} - p^{n+1}_{w,i,j-1,k}\right) \\
&\quad + T^{n+1}_{w3,i,j,k+1/2}\left(p^{n+1}_{w,i,j,k+1} - p^{n+1}_{w,i,j,k}\right) - T^{n+1}_{w3,i,j,k-1/2}\left(p^{n+1}_{w,i,j,k} - p^{n+1}_{w,i,j,k-1}\right) \\
&\quad - (T_w\gamma_w)^{n+1}_{1,i+1/2,j,k}\left(z_{i+1,j,k} - z_{i,j,k}\right) + (T_w\gamma_w)^{n+1}_{1,i-1/2,j,k}\left(z_{i,j,k} - z_{i-1,j,k}\right) \\
&\quad - (T_w\gamma_w)^{n+1}_{2,i,j+1/2,k}\left(z_{i,j+1,k} - z_{i,j,k}\right) + (T_w\gamma_w)^{n+1}_{2,i,j-1/2,k}\left(z_{i,j,k} - z_{i,j-1,k}\right) \\
&\quad - (T_w\gamma_w)^{n+1}_{3,i,j,k+1/2}\left(z_{i,j,k+1} - z_{i,j,k}\right) + (T_w\gamma_w)^{n+1}_{3,i,j,k-1/2}\left(z_{i,j,k} - z_{i,j,k-1}\right) \\
&\quad + \tilde{Q}^{n+1}_{Ws,i,j,k},
\end{aligned}
\tag{6.64}
$$

$$
\begin{aligned}
&\frac{1}{\Delta t}\left(V\left[\left(\frac{\phi S_o}{B_o}\right)^{n+1} - \left(\frac{\phi S_o}{B_o}\right)^{n}\right]\right)_{i,j,k} \\
&= T^{n+1}_{o1,i+1/2,j,k}\left(p^{n+1}_{o,i+1,j,k} - p^{n+1}_{o,i,j,k}\right) - T^{n+1}_{o1,i-1/2,j,k}\left(p^{n+1}_{o,i,j,k} - p^{n+1}_{o,i-1,j,k}\right) \\
&\quad + T^{n+1}_{o2,i,j+1/2,k}\left(p^{n+1}_{o,i,j+1,k} - p^{n+1}_{o,i,j,k}\right) - T^{n+1}_{o2,i,j-1/2,k}\left(p^{n+1}_{o,i,j,k} - p^{n+1}_{o,i,j-1,k}\right) \\
&\quad + T^{n+1}_{o3,i,j,k+1/2}\left(p^{n+1}_{o,i,j,k+1} - p^{n+1}_{o,i,j,k}\right) - T^{n+1}_{o3,i,j,k-1/2}\left(p^{n+1}_{o,i,j,k} - p^{n+1}_{o,i,j,k-1}\right) \\
&\quad - (T_o\gamma_o)^{n+1}_{1,i+1/2,j,k}\left(z_{i+1,j,k} - z_{i,j,k}\right) + (T_o\gamma_o)^{n+1}_{1,i-1/2,j,k}\left(z_{i,j,k} - z_{i-1,j,k}\right) \\
&\quad - (T_o\gamma_o)^{n+1}_{2,i,j+1/2,k}\left(z_{i,j+1,k} - z_{i,j,k}\right) + (T_o\gamma_o)^{n+1}_{2,i,j-1/2,k}\left(z_{i,j,k} - z_{i,j-1,k}\right) \\
&\quad - (T_o\gamma_o)^{n+1}_{3,i,j,k+1/2}\left(z_{i,j,k+1} - z_{i,j,k}\right) + (T_o\gamma_o)^{n+1}_{3,i,j,k-1/2}\left(z_{i,j,k} - z_{i,j,k-1}\right) \\
&\quad + \tilde{Q}^{n+1}_{Os,i,j,k},
\end{aligned}
\tag{6.65}
$$

and

$$
\begin{aligned}
&\frac{1}{\Delta t}\left(V\left[\left(\frac{\phi R_{so} S_o}{B_o}\right)^{n+1} - \left(\frac{\phi R_{so} S_o}{B_o}\right)^{n}\right]\right)_{i,j,k} \\
&= (R_{so}T_o)^{n+1}_{1,i+1/2,j,k}\left(p^{n+1}_{o,i+1,j,k} - p^{n+1}_{o,i,j,k}\right) - (R_{so}T_o)^{n+1}_{1,i-1/2,j,k}\left(p^{n+1}_{o,i,j,k} - p^{n+1}_{o,i-1,j,k}\right) \\
&\quad + (R_{so}T_o)^{n+1}_{2,i,j+1/2,k}\left(p^{n+1}_{o,i,j+1,k} - p^{n+1}_{o,i,j,k}\right) - (R_{so}T_o)^{n+1}_{2,i,j-1/2,k}\left(p^{n+1}_{o,i,j,k} - p^{n+1}_{o,i,j-1,k}\right) \\
&\quad + (R_{so}T_o)^{n+1}_{3,i,j,k+1/2}\left(p^{n+1}_{o,i,j,k+1} - p^{n+1}_{o,i,j,k}\right) - (R_{so}T_o)^{n+1}_{3,i,j,k-1/2}\left(p^{n+1}_{o,i,j,k} - p^{n+1}_{o,i,j,k-1}\right)
\end{aligned}
$$

$$
\begin{aligned}
&- (R_{so}T_o\gamma_o)^{n+1}_{1,i+1/2,j,k}\left(z_{i+1,j,k} - z_{i,j,k}\right) + (R_{so}T_o\gamma_o)^{n+1}_{1,i-1/2,j,k}\left(z_{i,j,k} - z_{i-1,j,k}\right) \quad (6.66)\\
&- (R_{so}T_o\gamma_o)^{n+1}_{2,i,j+1/2,k}\left(z_{i,j+1,k} - z_{i,j,k}\right) + (R_{so}T_o\gamma_o)^{n+1}_{2,i,j-1/2,k}\left(z_{i,j,k} - z_{i,j-1,k}\right)\\
&- (R_{so}T_o\gamma_o)^{n+1}_{3,i,j,k+1/2}\left(z_{i,j,k+1} - z_{i,j,k}\right) + (R_{so}T_o\gamma_o)^{n+1}_{3,i,j,k-1/2}\left(z_{i,j,k} - z_{i,j,k-1}\right)\\
&+ \left(\frac{Vq_{Os}R_{so}}{B_o}\right)^{n+1}_{i,j,k}.
\end{aligned}
$$

Again, (6.64)–(6.66) are nonlinear in terms of the primary unknowns $(p^{n+1}, S_w^{n+1}, p_b^{n+1})$ and can be linearized via the Newton–Raphson iteration:

$$
\begin{aligned}
p^{n+1,l+1} &= p^{n+1,l} + \delta p^{n+1,l+1}, \qquad S_w^{n+1,l+1} = S_w^{n+1,l} + \delta S_w^{n+1,l+1},\\
p_b^{n+1,l+1} &= p_b^{n+1,l} + \delta p_b^{n+1,l+1}.
\end{aligned}
$$

The resulting linear system can be expressed in the increments $\delta p^{n+1,l+1}$, $\delta S_w^{n+1,l+1}$, and $\delta p_b^{n+1,l+1}$ as in the saturated case; we omit the details.

Termination of the Newton–Raphson iteration

To terminate a Newton–Raphson iteration, some important factors should be considered. First, the iteration number should be smaller than a given maximum number. Second, the iteration values of the unknowns and the right-hand vectors of the linear equation systems to be solved are used as part of the termination condition. The absolute iteration values of the increments of pressure, water saturation, oil saturation (respectively, bubble point pressure), and the bottom hole pressure of wells must be less than their respective allowable maximum limits. Third, from our simulation experience the ratio of the infinite norm of the right-hand-side vector of a linear system of equations to the maximum absolute value of the sum of the oil and gas component flow rates of perforated zones of wells must be less than a certain given limit. Mass balance errors are not used as part of the termination condition of the Newton–Raphson iteration, but are monitored during a simulation. *Material balance* means that the cumulative component mass production equals the initial component mass in place minus the current component mass in place.

Treatment of bubble point problems

It is very important to deal properly with the bubble point problem to control convergence of a Newton–Raphson iteration. The state of a reservoir can change from saturated to undersaturated, or vice versa. Determining a proper state during the state transition is the *bubble point problem*. If the bubble point problem can be promptly recognized and reasonable unknowns can be selected for different states of a reservoir, convergence of the Newton–Raphson iteration can be better monitored and sped up.

To handle the bubble point problem properly, we must figure out the *trigger* that causes the transition of states of a reservoir using the state machine (Booch, Rumbaugh, and Jacobson, 1998) shown in Fig. 6.2. A location in the reservoir can stay in either the saturated state or the undersaturated state. Furthermore, from the lth iteration to the $(l+1)$th iteration in a Newton–Raphson iteration at the $(n+1)$th time step, the location can stay

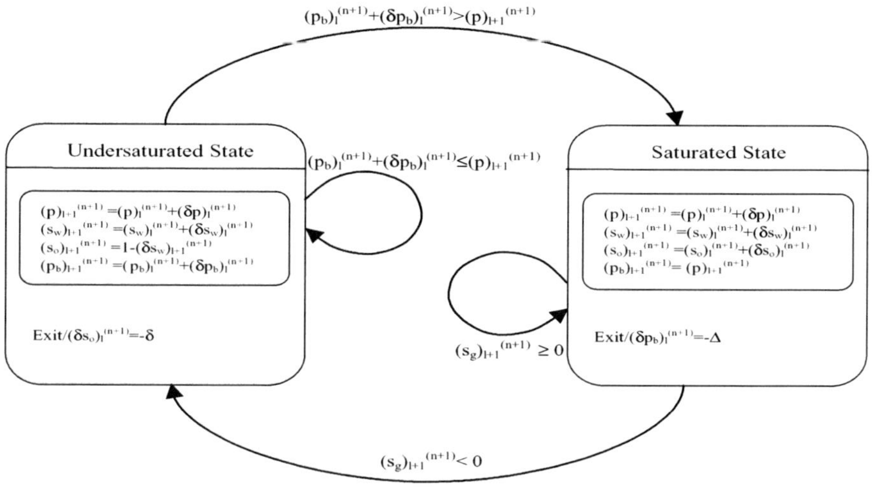

Figure 6.2. *A state machine.*

in the same state or transfer to another state. The constraint conditions and triggers are different in different states. In the undersaturated state, the constraint conditions are

$$\begin{aligned} S_w^{n+1,l} + S_o^{n+1,l} &= 1, \\ p^{n+1,l} &> p_b^{n+1,l}. \end{aligned} \tag{6.67}$$

On the other hand, in the saturated state, the constraint conditions are

$$\begin{aligned} S_w^{n+1,l} + S_o^{n+1,l} + S_g^{n+1,l} &= 1, \\ p^{n+1,l} &= p_b^{n+1,l}. \end{aligned} \tag{6.68}$$

The trigger that causes the transition from the undersaturated state to the saturated state is

$$p_b^{n+1,l} + \delta p_b > p^{n+1,l+1}, \tag{6.69}$$

and the trigger that causes the transition from the saturated state to the undersaturated state is

$$S_g^{n+1,l+1} < 0. \tag{6.70}$$

To deal with the bubble point problem properly, we must check the triggers to determine whether a location in a reservoir stays in the old state or transfers to a new state. Then we let the unknowns satisfy the constraint conditions of the corresponding state. When the reservoir pressure at a location in a reservoir drops below the bubble point pressure, then $(p_b)^{n+1,l} + \delta p_b > p^{n+1,l+1}$, the dissolved gas comes out from the oil phase, and the oil saturation decreases. It triggers the state to transfer from the undersaturated state to the saturated state at this location. In order to enter the new state, δS_o is set with a small negative value so that the gas saturation is greater than zero and the dissolved gas is released. When the reservoir at this location is in the saturated state, the unknowns corresponding to the grid point of the location are updated to satisfy the constraint conditions (6.68). Similarly, if the reservoir pressure at a location increases to the point that all the gas dissolves into the oil phase, then the state changes from the saturated state to the undersaturated state at this location and $S_g^{n+1,l+1} < 0$, which triggers the state to transfer from the saturated state to the undersaturated state. In order to guarantee that the oil phase pressure will be greater than the bubble point pressure in the new state, δp_b is set with a small negative value. After the reservoir at the location enters this new state, the unknowns are updated to meet the constraint conditions (6.67) in the undersaturated state.

6.2.3 Sequential Solution Techniques

The *sequential solution technique* (MacDonald and Coats, 1970) is similar to the SS technique discussed in the previous subsection. The difference is that the three equations in the black oil system are now solved separately and *sequentially*.

Saturated state

In the sequential technique, all the finite difference equations are the same as in (6.56)–(6.58). However, in the Newton–Raphson iteration, all the saturation functions k_{rw}, k_{ro}, k_{rg}, p_{cow}, and p_{cgo} use the previous iteration values of saturations; i.e., they are explicitly handled in this iteration:

$$
\begin{aligned}
&p_w^{(l+1)*} = p^{l+1} - p_{cow}(S_w^l), \quad p_g^{(l+1)*} = p^{l+1} + p_{cgo}(S_g^l),\\
&\mathbf{T}_\alpha^{(l+1)*} = \frac{k_{r\alpha}^l}{\mu_\alpha^{l+1} B_\alpha^{l+1}}\mathbf{k}, \qquad \alpha = w, o, g.
\end{aligned}
\tag{6.71}
$$

The residuals of equations (6.56)–(6.58) at the Newton–Raphson iteration level l are now

$$
\begin{aligned}
R_{w,i,j,k}^{l*} = {} & \frac{1}{\Delta t}\left(V\left[\left(\frac{\phi S_w}{B_w}\right)^l - \left(\frac{\phi S_w}{B_w}\right)^n\right]\right)_{i,j,k}\\
& - T_{w1,i+1/2,j,k}^{l*}\left(p_{w,i+1,j,k}^{l*} - p_{w,i,j,k}^{l*}\right) + T_{w1,i-1/2,j,k}^{l*}\left(p_{w,i,j,k}^{l*} - p_{w,i-1,j,k}^{l*}\right)\\
& - T_{w2,i,j+1/2,k}^{l*}\left(p_{w,i,j+1,k}^{l*} - p_{w,i,j,k}^{l*}\right) + T_{w2,i,j-1/2,k}^{l*}\left(p_{w,i,j,k}^{l*} - p_{w,i,j-1,k}^{l*}\right)\\
& - T_{w3,i,j,k+1/2}^{l*}\left(p_{w,i,j,k+1}^{l*} - p_{w,i,j,k}^{l*}\right) + T_{w3,i,j,k-1/2}^{l*}\left(p_{w,i,j,k}^{l*} - p_{w,i,j,k-1}^{l*}\right)\\
& + (T_w\gamma_w)_{1,i+1/2,j,k}^{l*}\left(z_{i+1,j,k} - z_{i,j,k}\right) - (T_w\gamma_w)_{1,i-1/2,j,k}^{l*}\left(z_{i,j,k} - z_{i-1,j,k}\right)\\
& + (T_w\gamma_w)_{2,i,j+1/2,k}^{l*}\left(z_{i,j+1,k} - z_{i,j,k}\right) - (T_w\gamma_w)_{2,i,j-1/2,k}^{l*}\left(z_{i,j,k} - z_{i,j-1,k}\right)\\
& + (T_w\gamma_w)_{3,i,j,k+1/2}^{l*}\left(z_{i,j,k+1} - z_{i,j,k}\right) - (T_w\gamma_w)_{3,i,j,k-1/2}^{l*}\left(z_{i,j,k} - z_{i,j,k-1}\right)\\
& - \tilde{Q}_{Ws,i,j,k}^{l*},
\end{aligned}
\tag{6.72}
$$

and $R^{l*}_{o,i,j,k}$ and $R^{l*}_{g,i,j,k}$ can be similarly defined. By the definition of the transmissibility coefficients $\mathbf{T}^{l*}_{\alpha}$, the Newton–Raphson iteration for $R^{l*}_{w,i,j,k}$ has the following form:

$$\begin{aligned} &\frac{\partial R^{l*}_{w,i,j,k}}{\partial p_{i,j,k-1}}\delta p^{l+1}_{i,j,k-1} + \frac{\partial R^{l*}_{w,i,j,k}}{\partial p_{i,j-1,k}}\delta p^{l+1}_{i,j-1,k} + \frac{\partial R^{l*}_{w,i,j,k}}{\partial p_{i-1,j,k}}\delta p^{l+1}_{i-1,j,k} \\ &\quad + \frac{\partial R^{l*}_{w,i,j,k}}{\partial p_{i,j,k}}\delta p^{l+1}_{i,j,k} + \frac{\partial R^{l*}_{w,i,j,k}}{\partial p_{i+1,j,k}}\delta p^{l+1}_{i+1,j,k} + \frac{\partial R^{l*}_{w,i,j,k}}{\partial p_{i,j+1,k}}\delta p^{l+1}_{i,j+1,k} \\ &\quad + \frac{\partial R^{l*}_{w,i,j,k}}{\partial p_{i,j,k+1}}\delta p^{l+1}_{i,j,k+1} + \frac{\partial R^{l*}_{w,i,j,k}}{\partial S_{w,i,j,k}}\delta S^{l+1}_{w,i,j,k} = -R^{l*}_{w,i,j,k}. \end{aligned} \tag{6.73}$$

For the oil and gas components, analogous equations hold:

$$\begin{aligned} &\frac{\partial R^{l*}_{o,i,j,k}}{\partial p_{i,j,k-1}}\delta p^{l+1}_{i,j,k-1} + \frac{\partial R^{l*}_{o,i,j,k}}{\partial p_{i,j-1,k}}\delta p^{l+1}_{i,j-1,k} + \frac{\partial R^{l*}_{o,i,j,k}}{\partial p_{i-1,j,k}}\delta p^{l+1}_{i-1,j,k} \\ &\quad + \frac{\partial R^{l*}_{o,i,j,k}}{\partial p_{i,j,k}}\delta p^{l+1}_{i,j,k} + \frac{\partial R^{l*}_{o,i,j,k}}{\partial p_{i+1,j,k}}\delta p^{l+1}_{i+1,j,k} + \frac{\partial R^{l*}_{o,i,j,k}}{\partial p_{i,j+1,k}}\delta p^{l+1}_{i,j+1,k} \\ &\quad + \frac{\partial R^{l*}_{w,i,j,k}}{\partial p_{i,j,k+1}}\delta p^{l+1}_{i,j,k+1} + \frac{\partial R^{l*}_{o,i,j,k}}{\partial S_{o,i,j,k}}\delta S^{l+1}_{o,i,j,k} = -R^{l*}_{o,i,j,k} \end{aligned} \tag{6.74}$$

and

$$\begin{aligned} &\frac{\partial R^{l*}_{g,i,j,k}}{\partial p_{i,j,k-1}}\delta p^{l+1}_{i,j,k-1} + \frac{\partial R^{l*}_{g,i,j,k}}{\partial p_{i,j-1,k}}a\delta p^{l+1}_{i,j-1,k} + \frac{\partial R^{l*}_{g,i,j,k}}{\partial p_{i-1,j,k}}\delta p^{l+1}_{i-1,j,k} \\ &\quad + \frac{\partial R^{l*}_{g,i,j,k}}{\partial p_{i,j,k}}\delta p^{l+1}_{i,j,k} + \frac{\partial R^{l*}_{g,i,j,k}}{\partial p_{i+1,j,k}}\delta p^{l+1}_{i+1,j,k} + \frac{\partial R^{l*}_{g,i,j,k}}{\partial p_{i,j+1,k}}\delta p^{l+1}_{i,j+1,k} \\ &\quad + \frac{\partial R^{l*}_{g,i,j,k}}{\partial p_{i,j,k+1}}\delta p^{l+1}_{i,j,k+1} + \frac{\partial R^{l*}_{g,i,j,k}}{\partial S_{w,i,j,k}}\delta S^{l+1}_{w,i,j,k} + \frac{\partial R^{l*}_{g,i,j,k}}{\partial S_{o,i,j,k}}\delta S^{l+1}_{o,i,j,k} = -R^{l*}_{g,i,j,k}. \end{aligned} \tag{6.75}$$

We express these three equations, respectively, as

$$\begin{aligned} &\frac{\partial R^{l*}_{w,i,j,k}}{\partial S_{w,i,j,k}}\delta S^{l+1}_{w,i,j,k} = F_{w,i,j,k}(\delta p^{l+1}), \\ &\frac{\partial R^{l*}_{o,i,j,k}}{\partial S_{o,i,j,k}}\delta S^{l+1}_{o,i,j,k} = F_{o,i,j,k}(\delta p^{l+1}), \\ &\frac{\partial R^{l*}_{g,i,j,k}}{\partial S_{w,i,j,k}}\delta S^{l+1}_{w,i,j,k} + \frac{\partial R^{l*}_{g,i,j,k}}{\partial S_{o,i,j,k}}\delta S^{l+1}_{o,i,j,k} = F_{g,i,j,k}(\delta p^{l+1}), \end{aligned} \tag{6.76}$$

where F_w, F_o, and F_g denote all the terms involving δp^{l+1} and the right-hand terms in their respective equations (6.73)–(6.75). System (6.76) can be further reduced to

$$\begin{aligned} &\frac{\partial R^{l*}_{g,i,j,k}}{\partial S_{w,i,j,k}}\frac{F_{w,i,j,k}(\delta p^{l+1})}{\partial R^{l*}_{w,i,j,k}/\partial S_{w,i,j,k}} \\ &\quad + \frac{\partial R^{l*}_{g,i,j,k}}{\partial S_{o,i,j,k}}\frac{F_{o,i,j,k}(\delta p^{l+1})}{\partial R^{l*}_{o,i,j,k}/\partial S_{o,i,j,k}} = -F_{g,i,j,k}(\delta p^{l+1}). \end{aligned} \tag{6.77}$$

This equation is used to solve for δp^{l+1}. Once δp^{l+1} is calculated, we can obtain δS_w^{l+1} and δS_o^{l+1} from the first two equations of (6.76).

Undersaturated state

In the undersaturated state, the residual $R^{l*}_{g,i,j,k}$ becomes

$$\begin{aligned} R^{l*}_{g,i,j,k} = &\frac{1}{\Delta t}\left(V\left[\left(\frac{\phi R_{so}S_o}{B_o}\right)^l - \left(\frac{\phi R_{so}S_o}{B_o}\right)^n\right]\right)_{i,j,k} \\ &- (R_{so}T_o)^{l*}_{1,i+1/2,j,k}\left(p^{l*}_{o,i+1,j,k} - p^{l*}_{o,i,j,k}\right) + (R_{so}T_o)^{l*}_{1,i-1/2,j,k}\left(p^{l*}_{o,i,j,k} - p^{l*}_{o,i-1,j,k}\right) \\ &- (R_{so}T_o)^{l*}_{2,i,j+1/2,k}\left(p^{l*}_{o,i,j+1,k} - p^{l*}_{o,i,j,k}\right) + (R_{so}T_o)^{l*}_{2,i,j-1/2,k}\left(p^{l*}_{o,i,j,k} - p^{l*}_{o,i,j-1,k}\right) \\ &- (R_{so}T_o)^{l*}_{3,i,j,k+1/2}\left(p^{l*}_{o,i,j,k+1} - p^{l*}_{o,i,j,k}\right) + (R_{so}T_o)^{l*}_{3,i,j,k-1/2}\left(p^{l*}_{o,i,j,k} - p^{l*}_{o,i,j,k-1}\right) \\ &+ (R_{so}T_o\gamma_o)^{l*}_{1,i+1/2,j,k}\left(z_{i+1,j,k} - z_{i,j,k}\right) - (R_{so}T_o\gamma_o)^{l*}_{1,i-1/2,j,k}\left(z_{i,j,k} - z_{i-1,j,k}\right) \\ &+ (R_{so}T_o\gamma_o)^{l*}_{2,i,j+1/2,k}\left(z_{i,j+1,k} - z_{i,j,k}\right) - (R_{so}T_o\gamma_o)^{l*}_{2,i,j-1/2,k}\left(z_{i,j,k} - z_{i,j-1,k}\right) \\ &+ (R_{so}T_o\gamma_o)^{l*}_{3,i,j,k+1/2}\left(z_{i,j,k+1} - z_{i,j,k}\right) - (R_{so}T_o\gamma_o)^{l*}_{3,i,j,k-1/2}\left(z_{i,j,k} - z_{i,j,k-1}\right) \\ &- \left(\frac{Vq_{Os}R_{so}}{B_o}\right)^{l*}_{i,j,k}. \end{aligned} \tag{6.78}$$

Because the nonlinearity in the undersaturated state is weaker than that in the saturated state, the transmissibilities can be treated explicitly:

$$\mathbf{T}_\alpha^{(l+1)*} = \frac{k^l_{r\alpha}}{\mu^l_\alpha B^l_\alpha}\mathbf{k}, \qquad \alpha = w, o, g. \tag{6.79}$$

With this treatment, the residual equations for the water, oil, and gas components are

$$\begin{aligned} &\frac{\partial R^{l*}_{w,i,j,k}}{\partial p_{i,j,k-1}}\delta p^{l+1}_{i,j,k-1} + \frac{\partial R^{l*}_{w,i,j,k}}{\partial p_{i,j-1,k}}\delta p^{l+1}_{i,j-1,k} + \frac{\partial R^{l*}_{iw,,j,k}}{\partial p_{i-1,j,k}}\delta p^{l+1}_{i-1,j,k} \\ &+\frac{\partial R^{l*}_{w,i,j,k}}{\partial p_{i,j,k}}\delta p^{l+1}_{i,j,k} + \frac{\partial R^{l*}_{w,i,j,k}}{\partial p_{i+1,j,k}}\delta p^{l+1}_{i+1,j,k} + \frac{\partial R^{l*}_{w,i,j,k}}{\partial p_{i,j+1,k}}\delta p^{l+1}_{i,j+1,k} \\ &+\frac{\partial R^{l*}_{w,i,j,k}}{\partial p_{i,j,k+1}}\delta p^{l+1}_{i,j,k+1} + \frac{\partial R^{l*}_{w,i,j,k}}{\partial S_{w,i,j,k}}\delta S^{l+1}_{w,i,j,k} = -R^{l*}_{w,i,j,k}, \end{aligned} \tag{6.80}$$

$$\begin{aligned} &\frac{\partial R^{l*}_{o,i,j,k}}{\partial p_{i,j,k-1}}\delta p^{l+1}_{i,j,k-1} + \frac{\partial R^{l*}_{o,i,j,k}}{\partial p_{i,j-1,k}}\delta p^{l+1}_{i,j-1,k} + \frac{\partial R^{l*}_{o,i,j,k}}{\partial p_{i-1,j,k}}\delta p^{l+1}_{i-1,j,k} \\ &+\frac{\partial R^{l*}_{o,i,j,k}}{\partial p_{i,j,k}}\delta p^{l+1}_{i,j,k} + \frac{\partial R^{l*}_{o,i,j,k}}{\partial p_{i+1,j,k}}\delta p^{l+1}_{i+1,j,k} + \frac{\partial R^{l*}_{o,i,j,k}}{\partial p_{i,j+1,k}}\delta p^{l+1}_{i,j+1,k} \\ &+\frac{\partial R^{l*}_{w,i,j,k}}{\partial p_{i,j,k+1}}\delta p^{l+1}_{i,j,k+1} + \frac{\partial R^{l*}_{o,i,j,k}}{\partial S_{w,i,j,k}}\delta S^{l+1}_{w,i,j,k} + \frac{\partial R^{l*}_{o,i,j,k}}{\partial (p_b)_{i,j,k}}\delta (p_b)^{l+1}_{i,j,k} = -R^{l*}_{o,i,j,k}, \end{aligned} \tag{6.81}$$

and

$$\begin{aligned}
&\frac{\partial R^{l*}_{g,i,j,k}}{\partial p_{i,j,k-1}}\delta p^{l+1}_{i,j,k-1} + \frac{\partial R^{l*}_{g,i,j,k}}{\partial p_{i,j-1,k}}\delta p^{l+1}_{i,j-1,k} + \frac{\partial R^{l*}_{g,i,j,k}}{\partial p_{i-1,j,k}}\delta p^{l+1}_{i-1,j,k} \\
&\quad + \frac{\partial R^{l*}_{g,i,j,k}}{\partial p_{i,j,k}}\delta p^{l+1}_{i,j,k} + \frac{\partial R^{l*}_{g,i,j,k}}{\partial p_{i+1,j,k}}\delta p^{l+1}_{i+1,j,k} + \frac{\partial R^{l*}_{g,i,j,k}}{\partial p_{i,j+1,k}}\delta p^{l+1}_{i,j+1,k} \\
&\quad + \frac{\partial R^{l*}_{g,i,j,k}}{\partial p_{i,j,k+1}}\delta p^{l+1}_{i,j,k+1} + \frac{\partial R^{l*}_{g,i,j,k}}{\partial S_{w,i,j,k}}\delta S^{l+1}_{w,i,j,k} + \frac{\partial R^{l*}_{g,i,j,k}}{\partial (p_b)_{i,j,k}}\delta (p_b)^{l+1}_{i,j,k} = -R^{l*}_{g,i,j,k}.
\end{aligned} \tag{6.82}$$

Equations (6.80)–(6.82) are rewritten, respectively, as

$$\begin{aligned}
&\frac{\partial R^{l*}_{w,i,j,k}}{\partial S_{w,i,j,k}}\delta S^{l+1}_{w,i,j,k} = F_{w,i,j,k}(\delta p^{l+1}), \\
&\frac{\partial R^{l*}_{o,i,j,k}}{\partial S_{w,i,j,k}}\delta S^{l+1}_{w,i,j,k} + \frac{\partial R^{l*}_{o,i,j,k}}{\partial (p_b)_{i,j,k}}\delta (p_b)^{l+1}_{i,j,k} = F_{o,i,j,k}(\delta p^{l+1}), \\
&\frac{\partial R^{l*}_{g,i,j,k}}{\partial S_{w,i,j,k}}\delta S^{l+1}_{w,i,j,k} + \frac{\partial R^{l*}_{g,i,j,k}}{\partial (p_b)_{i,j,k}}\delta (p_b)^{l+1}_{i,j,k} = F_{g,i,j,k}(\delta p^{l+1}),
\end{aligned} \tag{6.83}$$

where F_w, F_o, and F_g indicate all the terms involving δp^{l+1} and the right-hand term in their respective equations (6.80)–(6.82). In this system, we first eliminate δS_w and δp_b to obtain a reduced system only for δp, which is solved. Then, after δp is obtained, any two equations in system (6.83) can be used to find δS_w and δp_b.

In summary, the sequential technique has the following features:

- The difference between the SS and sequential techniques is that the three differential equations are solved simultaneously in the former, while these equations are solved sequentially at each grid node in the latter.
- All the saturation functions k_{rw}, k_{ro}, k_{rg}, p_{cw}, and p_{cg} use the previous Newton–Raphson iteration values of saturations in the sequential technique.

Selection of time steps

The bubble point problem in the sequential technique can be treated in the same way as in Section 6.2.2 for the SS technique. Compared with the SS technique, the implicitness of the sequential technique is lower. Selecting reasonable time steps is key to controlling convergence of a Newton–Raphson iteration and speeding up a simulation procedure. If the time steps are too small, too much computational time will be consumed; if they are too large, a Newton–Raphson iteration may diverge.

To select suitable time steps, from our experimental experience we have adopted the following empirical rules:

- With a given maximum time step Δt_{max} and a minimum allowable time step size Δt_{min}, the time step Δt should satisfy $\Delta t_{min} < \Delta t \leq \Delta t_{max}$.

- In the saturated state, Δt is bounded by

$$\Delta t \leq \Delta t^n \min\left\{R_{\Delta t}, \frac{(dp)_{max}}{(\delta p)^n_{max}}, \frac{(dS_w)_{max}}{(\delta S_w)^n_{max}}, \frac{(dS_o)_{max}}{(\delta S_o)^n_{max}}\right\}, \tag{6.84}$$

where Δt^n is the previous time step size; $R_{\Delta t}$ is a maximum allowable ratio of $\Delta t/\Delta t^n$ ($R_{\Delta t} > 1$ and generally is between 2 and 3); $(dp)_{max}$, $(dS_w)_{max}$, and $(dS_o)_{max}$ are the allowable maximum values of the pressure, water saturation, and oil saturation increments, respectively; and $(\delta p)^n_{max}$, $(\delta S_w)^n_{max}$, $(\delta S_o)^n_{max}$ are the maximum values of these increments at the nth time step. In the undersaturated state, (6.84) becomes

$$\Delta t \leq \Delta t^n \min\left\{R_{\Delta t}, \frac{(dp)_{max}}{(\delta p)^n_{max}}, \frac{(dS_w)_{max}}{(\delta S_w)^n_{max}}, \frac{(dp_b)_{max}}{(\delta p_b)^n_{max}}\right\}, \tag{6.85}$$

where $(dp_b)_{max}$ is the allowable maximum value of the bubble point pressure increment.

- For a given time period, Δt should guarantee that the simulation time reaches the period time.

With these rules, a time step Δt can be automatically selected. Its choice must also take into account the convergence of a Newton–Raphson iteration. If the number of iterations is larger than a given maximum number when Δt is selected according to these rules, then the selected time step may be too large and must be reduced. First, we reduce Δt by $\Delta t/R_{\Delta t}$ because of the occurrence of $R_{\Delta t}$ in (6.84) and (6.85). Then the oil phase and bubble point pressures and water and oil saturations at the nth time step are taken as the first iteration values of the Newton–Raphson iteration at the $(n+1)$th time step.

6.2.4 Iterative IMPES Solution Techniques

The IMPES method was discussed in the preceding chapter for two-phase flow and is a very useful technique for flow of this type. Particularly, the improved IMPES method introduced in Section 5.3.5 is very powerful for solving two-phase flow. We now discuss the IMPES method for the solution of the black oil model. When IMPES is used within a Newton–Raphson iteration, it is called *iterative IMPES*. In iterative IMPES, only the pressure equation is computed implicitly, and the other two (saturation and bubble point pressure) equations are evaluated explicitly.

In iterative IMPES, all the saturation functions k_{rw}, k_{ro}, k_{rg}, p_{cw}, and p_{cg} are evaluated at the saturation values of the previous time step in a Newton–Raphson iteration, and the fluid formation volume factors and viscosities in the transmissibilities, phase potentials, and well terms are computed using the previous Newton–Raphson iteration values. Thus we define

$$\begin{aligned} p_w^{(l+1)\star} &= p^{l+1} - p_{cow}(S_w^n), \quad p_g^{(l+1)\star} = p^{l+1} + p_{cgo}(S_g^n), \\ \mathbf{T}_\alpha^{(l+1)\star} &= \frac{k^n_{r\alpha}}{\mu^l_\alpha B^l_\alpha}\mathbf{k}, \qquad \alpha = w, o, g. \end{aligned} \tag{6.86}$$

Saturated state

Equations (6.56)–(6.58) remain the same. Their residuals at the Newton–Raphson iteration level l in the iterative IMPES technique are

$$\begin{aligned}
R^{l\star}_{w,i,j,k} = & \frac{1}{\Delta t}\left(V\left[\left(\frac{\phi S_w}{B_w}\right)^l - \left(\frac{\phi S_w}{B_w}\right)^n\right]\right)_{i,j,k} \\
& - T^{l\star}_{w1,i+1/2,j,k}\left(p^{l\star}_{w,i+1,j,k} - p^{l\star}_{w,i,j,k}\right) + T^{l\star}_{w1,i-1/2,j,k}\left(p^{l\star}_{w,i,j,k} - p^{l\star}_{w,i-1,j,k}\right) \\
& - T^{l\star}_{w2,i,j+1/2,k}\left(p^{l\star}_{w,i,j+1,k} - p^{l\star}_{w,i,j,k}\right) + T^{l\star}_{w2,i,j-1/2,k}\left(p^{l\star}_{w,i,j,k} - p^{l\star}_{w,i,j-1,k}\right) \\
& - T^{l\star}_{w3,i,j,k+1/2}\left(p^{l\star}_{w,i,j,k+1} - p^{l\star}_{w,i,j,k}\right) + T^{l\star}_{w3,i,j,k-1/2}\left(p^{l\star}_{w,i,j,k} - p^{l\star}_{w,i,j,k-1}\right) \\
& + (T_w\gamma_w)^{l\star}_{1,i+1/2,j,k}\left(z_{i+1,j,k} - z_{i,j,k}\right) - (T_w\gamma_w)^{l\star}_{1,i-1/2,j,k}\left(z_{i,j,k} - z_{i-1,j,k}\right) \\
& + (T_w\gamma_w)^{l\star}_{2,i,j+1/2,k}\left(z_{i,j+1,k} - z_{i,j,k}\right) - (T_w\gamma_w)^{l\star}_{2,i,j-1/2,k}\left(z_{i,j,k} - z_{i,j-1,k}\right) \\
& + (T_w\gamma_w)^{l\star}_{3,i,j,k+1/2}\left(z_{i,j,k+1} - z_{i,j,k}\right) - (T_w\gamma_w)^{l\star}_{3,i,j,k-1/2}\left(z_{i,j,k} - z_{i,j,k-1}\right) \\
& - \tilde{Q}^{l\star}_{Ws,i,j,k},
\end{aligned} \tag{6.87}$$

and $R^{l\star}_{o,i,j,k}$ and $R^{l\star}_{g,i,j,k}$ can be similarly defined. It follows from the definition of the transmissibility coefficients $\mathbf{T}^{l\star}_\alpha$ that the Newton–Raphson iteration for $R^{l\star}_{w,i,j,k}$ gives

$$\begin{aligned}
& \frac{\partial R^{l\star}_{w,i,j,k}}{\partial p_{i,j,k-1}}\delta p^{l+1}_{i,j,k-1} + \frac{\partial R^{l\star}_{w,i,j,k}}{\partial p_{i,j-1,k}}\delta p^{l+1}_{i,j-1,k} + \frac{\partial R^{l\star}_{iw,,j,k}}{\partial p_{i-1,j,k}}\delta p^{l+1}_{i-1,j,k} \\
& \quad + \frac{\partial R^{l\star}_{w,i,j,k}}{\partial p_{i,j,k}}\delta p^{l+1}_{i,j,k} + \frac{\partial R^{l\star}_{w,i,j,k}}{\partial p_{i+1,j,k}}\delta p^{l+1}_{i+1,j,k} + \frac{\partial R^{l\star}_{w,i,j,k}}{\partial p_{i,j+1,k}}\delta p^{l+1}_{i,j+1,k} \\
& \quad + \frac{\partial R^{l\star}_{w,i,j,k}}{\partial p_{i,j,k+1}}\delta p^{l+1}_{i,j,k+1} + \frac{\partial R^{l\star}_{w,i,j,k}}{\partial S_{w,i,j,k}}\delta S^{l+1}_{w,i,j,k} = -R^{l\star}_{w,i,j,k}.
\end{aligned} \tag{6.88}$$

For the oil and gas components, analogous equations hold:

$$\begin{aligned}
& \frac{\partial R^{l\star}_{o,i,j,k}}{\partial p_{i,j,k-1}}\delta p^{l+1}_{i,j,k-1} + \frac{\partial R^{l\star}_{o,i,j,k}}{\partial p_{i,j-1,k}}\delta p^{l+1}_{i,j-1,k} + \frac{\partial R^{l\star}_{o,i,j,k}}{\partial p_{i-1,j,k}}\delta p^{l+1}_{i-1,j,k} \\
& \quad + \frac{\partial R^{l\star}_{o,i,j,k}}{\partial p_{i,j,k}}\delta p^{l+1}_{i,j,k} + \frac{\partial R^{l\star}_{o,i,j,k}}{\partial p_{i+1,j,k}}\delta p^{l+1}_{i+1,j,k} + \frac{\partial R^{l\star}_{o,i,j,k}}{\partial p_{i,j+1,k}}\delta p^{l+1}_{i,j+1,k} \\
& \quad + \frac{\partial R^{l\star}_{w,i,j,k}}{\partial p_{i,j,k+1}}\delta p^{l+1}_{i,j,k+1} + \frac{\partial R^{l\star}_{o,i,j,k}}{\partial S_{o,i,j,k}}\delta S^{l+1}_{o,i,j,k} = -R^{l\star}_{o,i,j,k}
\end{aligned} \tag{6.89}$$

and

$$\begin{aligned}
& \frac{\partial R^{l\star}_{g,i,j,k}}{\partial p_{i,j,k-1}}\delta p^{l+1}_{i,j,k-1} + \frac{\partial R^{l\star}_{g,i,j,k}}{\partial p_{i,j-1,k}}\delta p^{l+1}_{i,j-1,k} + \frac{\partial R^{l*}_{g,i,j,k}}{\partial p_{i-1,j,k}}\delta p^{l+1}_{i-1,j,k} \\
& \quad + \frac{\partial R^{l\star}_{g,i,j,k}}{\partial p_{i,j,k}}\delta p^{l+1}_{i,j,k} + \frac{\partial R^{l\star}_{g,i,j,k}}{\partial p_{i+1,j,k}}\delta p^{l+1}_{i+1,j,k} + \frac{\partial R^{l\star}_{g,i,j,k}}{\partial p_{i,j+1,k}}\delta p^{l+1}_{i,j+1,k} \\
& \quad + \frac{\partial R^{l\star}_{g,i,j,k}}{\partial p_{i,j,k+1}}\delta p^{l+1}_{i,j,k+1} + \frac{\partial R^{l\star}_{g,i,j,k}}{\partial S_{w,i,j,k}}\delta S^{l+1}_{w,i,j,k} + \frac{\partial R^{l\star}_{g,i,j,k}}{\partial S_{o,i,j,k}}\delta S^{l+1}_{o,i,j,k} = -R^{l\star}_{g,i,j,k}.
\end{aligned} \tag{6.90}$$

Equations (6.88)–(6.90) can be solved, in a decoupled manner, as in the sequential solution technique.

Undersaturated state

In the undersaturated state, the residual $R^{l\star}_{g,i,j,k}$ becomes

$$
\begin{aligned}
R^{l\star}_{g,i,j,k} = {} & \frac{1}{\Delta t}\left(V\left[\left(\frac{\phi R_{so}S_o}{B_o}\right)^l - \left(\frac{\phi R_{so}S_o}{B_o}\right)^n\right]\right)_{i,j,k} \\
& - (R_{so}T_o)^{l\star}_{1,i+1/2,j,k}\left(p^{l\star}_{o,i+1,j,k} - p^{l\star}_{o,i,j,k}\right) + (R_{so}T_o)^{l\star}_{1,i-1/2,j,k}\left(p^{l\star}_{o,i,j,k} - p^{l\star}_{o,i-1,j,k}\right) \\
& - (R_{so}T_o)^{l\star}_{2,i,j+1/2,k}\left(p^{l\star}_{o,i,j+1,k} - p^{l\star}_{o,i,j,k}\right) + (R_{so}T_o)^{l\star}_{2,i,j-1/2,k}\left(p^{l\star}_{o,i,j,k} - p^{l\star}_{o,i,j-1,k}\right) \\
& - (R_{so}T_o)^{l\star}_{3,i,j,k+1/2}\left(p^{l\star}_{o,i,j,k+1} - p^{l\star}_{o,i,j,k}\right) + (R_{so}T_o)^{l\star}_{3,i,j,k-1/2}\left(p^{l\star}_{o,i,j,k} - p^{l\star}_{o,i,j,k-1}\right) \\
& + (R_{so}T_o\gamma_o)^{l\star}_{1,i+1/2,j,k}\left(z_{i+1,j,k} - z_{i,j,k}\right) - (R_{so}T_o\gamma_o)^{l\star}_{1,i-1/2,j,k}\left(z_{i,j,k} - z_{i-1,j,k}\right) \\
& + (R_{so}T_o\gamma_o)^{l\star}_{2,i,j+1/2,k}\left(z_{i,j+1,k} - z_{i,j,k}\right) - (R_{so}T_o\gamma_o)^{l\star}_{2,i,j-1/2,k}\left(z_{i,j,k} - z_{i,j-1,k}\right) \\
& + (R_{so}T_o\gamma_o)^{l\star}_{3,i,j,k+1/2}\left(z_{i,j,k+1} - z_{i,j,k}\right) - (R_{so}T_o\gamma_o)^{l\star}_{3,i,j,k-1/2}\left(z_{i,j,k} - z_{i,j,k-1}\right) \\
& - \left(\frac{Vq_{Os}R_{so}}{B_o}\right)^{l\star}_{i,j,k}.
\end{aligned}
\tag{6.91}
$$

Now, the residual equations for the water, oil, and gas components are

$$
\begin{aligned}
& \frac{\partial R^{l\star}_{w,i,j,k}}{\partial p_{i,j,k-1}}\delta p^{l+1}_{i,j,k-1} + \frac{\partial R^{l\star}_{w,i,j,k}}{\partial p_{i,j-1,k}}\delta p^{l+1}_{i,j-1,k} + \frac{\partial R^{l\star}_{iw,,j,k}}{\partial p_{i-1,j,k}}\delta p^{l+1}_{i-1,j,k} \\
& \quad + \frac{\partial R^{l\star}_{w,i,j,k}}{\partial p_{i,j,k}}\delta p^{l+1}_{i,j,k} + \frac{\partial R^{l\star}_{w,i,j,k}}{\partial p_{i+1,j,k}}\delta p^{l+1}_{i+1,j,k} + \frac{\partial R^{l\star}_{w,i,j,k}}{\partial p_{i,j+1,k}}\delta p^{l+1}_{i,j+1,k} \\
& \quad + \frac{\partial R^{l\star}_{w,i,j,k}}{\partial p_{i,j,k+1}}\delta p^{l+1}_{i,j,k+1} + \frac{\partial R^{l\star}_{w,i,j,k}}{\partial S_{w,i,j,k}}\delta S^{l+1}_{w,i,j,k} = -R^{l\star}_{w,i,j,k},
\end{aligned}
\tag{6.92}
$$

$$
\begin{aligned}
& \frac{\partial R^{l\star}_{o,i,j,k}}{\partial p_{i,j,k-1}}\delta p^{l+1}_{i,j,k-1} + \frac{\partial R^{l\star}_{o,i,j,k}}{\partial p_{i,j-1,k}}\delta p^{l+1}_{i,j-1,k} + \frac{\partial R^{l\star}_{o,i,j,k}}{\partial p_{i-1,j,k}}\delta p^{l+1}_{i-1,j,k} \\
& \quad + \frac{\partial R^{l\star}_{o,i,j,k}}{\partial p_{i,j,k}}\delta p^{l+1}_{i,j,k} + \frac{\partial R^{l\star}_{o,i,j,k}}{\partial p_{i+1,j,k}}\delta p^{l+1}_{i+1,j,k} + \frac{\partial R^{l\star}_{o,i,j,k}}{\partial p_{i,j+1,k}}\delta p^{l+1}_{i,j+1,k} \\
& \quad + \frac{\partial R^{l\star}_{o,i,j,k}}{\partial p_{i,j,k+1}}\delta p^{l+1}_{i,j,k+1} + \frac{\partial R^{l\star}_{o,i,j,k}}{\partial S_{w,i,j,k}}\delta S^{l+1}_{w,i,j,k} + \frac{\partial R^{l\star}_{o,i,j,k}}{\partial (p_b)_{i,j,k}}\delta (p_b)^{l+1}_{i,j,k} = -R^{l\star}_{o,i,j,k},
\end{aligned}
\tag{6.93}
$$

and

$$
\begin{aligned}
& \frac{\partial R^{l\star}_{g,i,j,k}}{\partial p_{i,j,k-1}}\delta p^{l+1}_{i,j,k-1} + \frac{\partial R^{l\star}_{g,i,j,k}}{\partial p_{i,j-1,k}}\delta p^{l+1}_{i,j-1,k} + \frac{\partial R^{l*}_{g,i,j,k}}{\partial p_{i-1,j,k}}\delta p^{l+1}_{i-1,j,k} \\
& \quad + \frac{\partial R^{l\star}_{g,i,j,k}}{\partial p_{i,j,k}}\delta p^{l+1}_{i,j,k} + \frac{\partial R^{l\star}_{g,i,j,k}}{\partial p_{i+1,j,k}}\delta p^{l+1}_{i+1,j,k} + \frac{\partial R^{l\star}_{g,i,j,k}}{\partial p_{i,j+1,k}}\delta p^{l+1}_{i,j+1,k} \\
& \quad + \frac{\partial R^{l\star}_{g,i,j,k}}{\partial p_{i,j,k+1}}\delta p^{l+1}_{i,j,k+1} + \frac{\partial R^{l\star}_{g,i,j,k}}{\partial S_{w,i,j,k}}\delta S^{l+1}_{w,i,j,k} + \frac{\partial R^{l\star}_{g,i,j,k}}{\partial (p_b)_{i,j,k}}\delta (p_b)^{l+1}_{i,j,k} = -R^{l\star}_{g,i,j,k}.
\end{aligned}
\tag{6.94}
$$

Equations (6.92)–(6.94) can also be solved, in a decoupled fashion, as in the sequential solution technique.

In summary, the iterative IMPES method has the following features:

- The difference between iterative IMPES and classical IMPES is that the iterative technique is used within each Newton–Raphson iteration loop, while the classical one is exploited before a Newton–Raphson iteration.
- All the saturation functions k_{rw}, k_{ro}, k_{rg}, p_{cw}, and p_{cg} use the previous time step values of saturations in a Newton–Raphson iteration.
- The fluid formation volume factors and viscosities in the transmissibilities, phase potentials, and well terms are computed using the previous Newton–Raphson iteration values.
- The pressure unknown is obtained implicitly, and the other two unknowns are obtained explicitly.

As in the sequential technique, the saturation functions k_{rw}, k_{ro}, k_{rg}, p_{cw}, and p_{cg} may use the previous Newton–Raphson iteration values of saturations, instead of the previous time step values of saturation. The bubble point problem in iterative IMPES can be treated in the same manner as in the SS technique, and the time steps can be controlled in a similar way as in the sequential technique. The improved IMPES method developed in the preceding chapter for two-phase flow can be extended to iterative IMPES for the black oil model. In particular, the time steps can be different for pressure than for saturations.

Numerical comparisons between the SS, sequential, and iterative IMPES solution techniques for solving the black oil differential equations were performed by Chen, Huan, and Ma (2006). Field-scale simulation models of oil reservoirs were used to test these solution schemes for both the saturated and undersaturated states of these reservoirs. From the comparisons, the following observations were obtained:

- The iterative IMPES technique is not a suitable choice for the three-phase black oil simulation.
- The SS technique is the most stable and robust, but it requires the highest memory and computational cost.
- The sequential technique is convergent and stable for an undersaturated reservoir, and it can significantly reduce memory and computational cost compared with the SS technique. For a saturated reservoir the accuracy of the sequential scheme depends on whether free gas is injected. For no gas injection, this scheme is convergent and accurate and can reduce computational cost. However, for gas injection, the pressures and gas/oil ratios obtained from this technique differ from those from the SS technique, even though it seems convergent.

6.2.5 Adaptive Implicit Techniques

An *adaptive implicit technique* was introduced in reservoir simulation by Thomas and Thurnau (1983). The principal idea of this technique is to seek an efficient middle ground between the IMPES (or sequential) and SS techniques. That is, at a given time step, the expensive SS technique is confined to those gridblocks that require it, while on the remaining gridblocks the IMPES technique is implemented. In this technique, pressure is computed implicitly everywhere in a porous medium (as in the IMPES, sequential, and SS techniques), but the computation of saturation is implicit in selected gridblocks and explicit elsewhere. This division into implicit and explicit gridblocks may be different from one time step to the next. The principal issue in implementation of this technique is a *switching criterion* that determines whether the saturation equation should be considered implicit or explicit.

In the original work (Thomas and Thurnau, 1983), the switching criterion is based on solution variable changes. When a change at an IMPES gridblock exceeds a specified threshold value, the gridblock switches to the SS treatment. This criterion has the drawback that although instability leads to large solution changes, small changes do not guarantee stability. This drawback has led to the development of other criteria, such as those based on eigenvalues (Fung, Collins, and Nghiem, 1989) and hyperbolic equation stability analysis (i.e., the well-known CFL stability analysis; cf. Section 3.3.7).

Research on parallel computation in reservoir simulation was extensively carried out in the late 1980s, particularly due to the introduction of shared and distributed memory computers. For example, Scott, Wainwright, and Raghavan (1987) presented a multiple instruction multiple data (MIMD) approach to reservoir simulation, and Chien et al. (1987) described parallel processing on distributed memory machines. Several methods are available in the literature for parallelization of reservoir codes. Most of them are based on message passing techniques such as PVM (parallel virtual machine) and MPI (message passing interface) and *domain decomposition* methods. In most parallel approaches, a reservoir is split into a number of subdomains, and a processor is assigned to each subdomain problem (Killough and Wheeler, 1987); the *Schur complement* method can be used to solve interface problems (Smith, Bjorstad, and Gropp, 1996). Parallel computing for reservoir simulation was discussed in detail by Ma and Chen (2004) and Chen, Huan, and Ma (2006).

Parallel algorithms have been used in the SS (Mayer, 1989), IMPES (Rutledge et al., 1991), and adaptive implicit (Verdière et al., 1999) solution techniques for various multiphase flows. That is, in each of these solution techniques, both the pressure and saturation equations are solved in a parallel fashion. Benchmark computations have indicated that linear (or nearly linear) speedup in CPU time can be obtained with an increasing number of processors. The parallel idea can also be used as a solution technique for multiphase flow. In the IMPES, sequential, and SS techniques, the pressure and saturation equations are solved either separately or simultaneously on the same processor. However, these two equations can be solved in parallel; i.e., their solution can be assigned to different processors at the same time point. This idea seems very useful for multicomponent, multiphase flow where the equations for different components (or phases) can be assigned to different processors. This research direction has yet to be investigated.

6.2.6 Well Coupling

Various *well constraints* must be taken into account for the black oil model. Two kinds of well constraints are used for an injection well: Either the well bottom hole pressure p_{bh} is given or a phase injection rate is fixed. In the former case,

$$p_{bh}^{(\nu)} = P_{bh}^{(\nu)}, \tag{6.95}$$

where ν is the number of the well with this kind of well control and $P_{bh}^{(\nu)}$ is the given bottom hole pressure at this well. In this case,

$$\delta p_{bh}^{(\nu)} = 0. \tag{6.96}$$

In the latter case, it follows from (6.20) that the injection rate controls for water and gas injection wells are, respectively,

$$Q_{Ws}^{(\nu)} = \sum_{m=1}^{M_{w\nu}} WI_m^{(\nu)} \frac{k_{rwmax}}{\mu_w} \left[p_{bh}^{(\nu)} - p_w - \gamma_w (z_{bh}^{(\nu)} - z) \right] \delta(\mathbf{x} - \mathbf{x}_m^{(\nu)}) \tag{6.97}$$

and

$$Q_{Gs}^{(\nu)} = \sum_{m=1}^{M_{w\nu}} WI_m^{(\nu)} \frac{k_{rgmax}}{\mu_g} \left[p_{bh}^{(\nu)} - p_g - \gamma_g (z_{bh}^{(\nu)} - z) \right] \delta(\mathbf{x} - \mathbf{x}_m^{(\nu)}), \tag{6.98}$$

where $Q_{Ws}^{(\nu)}$ and $Q_{Gs}^{(\nu)}$ are the given water and gas injection rates, respectively, at the νth well and $k_{r\alpha max}$ is the maximum relative permeability of the α phase, $\alpha = w, g$. The well control equations (6.97) and (6.98) are coupled to the flow equations, and the coupled system can be linearized by the Newton–Raphson iteration method as in the SS technique.

For a production well, there are three kinds of well constraints: a constant bottom hole pressure, a constant total liquid production rate, and a constant total flow rate. The constant bottom hole pressure constraint has the form (6.95), and thus equation (6.96) holds. The constant total liquid production rate control takes the form

$$\begin{aligned} Q_{Ls}^{(\nu)} = & \sum_{m=1}^{M_{w\nu}} WI_m^{(\nu)} \frac{k_{rw}}{\mu_w} \left[p_{bh}^{(\nu)} - p_w - \gamma_w (z_{bh}^{(\nu)} - z) \right] \delta(\mathbf{x} - \mathbf{x}_m^{(\nu)}) \\ & + \sum_{m=1}^{M_{w\nu}} WI_m^{(\nu)} \frac{k_{ro}}{\mu_o} \left[p_{bh}^{(\nu)} - p_o - \gamma_o \wp (z_{bh}^{(\nu)} - z) \right] \delta(\mathbf{x} - \mathbf{x}_m^{(\nu)}), \end{aligned} \tag{6.99}$$

where $Q_{Ls}^{(\nu)}$ is the given total liquid production rate at the νth well. The *water cut*, defined as the ratio of water production to the sum of water and oil production, at a perforated zone of a well with this kind of well constraint must be less than a certain limit; over this limit, that perforated zone must be shut down. The constant total flow rate control can be defined similarly; in this case, gas production is added, and the gas/oil ratio at a perforated zone of a well must be less than a certain limit (over this limit, that perforated zone needs to be shut down). These well controls are also coupled to the flow equations.

Chapter 7

Transport of Multicomponents in a Fluid and Numerical Solution

Before we describe a full compositional model for multicomponent, multiphase flow in the next chapter, we consider the differential equations used to model the transport of multicomponents in a fluid phase in a porous medium. Perhaps the most important unconstrained aspect of petroleum systems analysis concerns the charging and placement of petroleum to a structure or prospect. Petroleum charge rates, leakage, and spill control petroleum residence time in a reservoir, which is fundamental for prediction of biodegradation rates, seal integrity (failure), and oil quality, all of which are affected by fluid mixing processes. While forward models can estimate plausible charge rates based on thermal histories, there are no field data proxies for the charge rates that are necessary to constrain migration and charge models. To solve this problem, we can couple high resolution, full physics reservoir simulation protocols to full 4D basin models such that gradients in petroleum compositions from models and from chemical analysis can be used to constrain charges rates. This new generation of hybrid reservoir simulator/basin models necessitates rapid high-resolution fluid mixing solvers and multicomponent fluids.

This chapter deals with the development of compositional fluid mixing simulators that enable the forward simulation of detailed reservoir charging and fluid property evolution, coupling the effects of advection, diffusion, and gravity segregation to predict the development of compositional gradients in petroleum columns that can be used to constrain reservoir charging and alteration processes. The traditional simulator for solving the isothermal gravity/chemical equilibrium problem is deduced as a special example of the simulators presented here. We do not consider the effects of chemical reactions between the components in the fluid phase, radioactive decay, biodegradation, and growth due to bacterial activities that cause the quantity of this component to increase or decrease.

Most of the present formulations for the mass conservation of species do not correctly represent changes in the time scale of diffusion as a function of porosity in a system of porous media. In sediments, or in any porous system, the presence of solid particles causes the diffusion paths of species to deviate from straight lines. To represent the role of porosity on diffusion, the diffusion coefficient must be scaled with tortuosity. In this chapter we also review available formulations for the scaled diffusion coefficient with tortuosity.

7.1 Basic Differential Equations

In this and the next two chapters, instead of using mass concentrations, we use *mole fractions* in the description of mass conservation, because the phase equilibrium relations are usually defined in terms of mole fractions. We consider a gas or liquid mixture that consists of N_c chemical species. Let n_m be the number of moles for the mth species in the mixture, and then the total moles of the mixture are

$$n = \sum_{m=1}^{N_c} n_m. \tag{7.1}$$

The mole fraction of each component is defined by

$$x_m = \frac{n_m}{n}, \qquad m = 1, 2, \ldots, N_c,$$

so

$$\sum_{m=1}^{N_c} x_m = 1. \tag{7.2}$$

Let ξ_m be the *molar density* of component m in the mixture, with physical dimensions of moles per pore volume:

$$\xi_m = \frac{n_m}{v_m}, \qquad m = 1, 2, \ldots, N_c,$$

where v_m is the *molar volume* of component m. The molar density of the mixture is

$$\xi = \sum_{m=1}^{N_c} \xi_m.$$

Conservation of mass of each component in the fluid mixture is

$$\frac{\partial(\phi x_m \xi)}{\partial t} = -\nabla \cdot (x_m \xi \mathbf{u} + \mathbf{J}_m) + q_m, \qquad m = 1, 2, \ldots, N_c, \tag{7.3}$$

where ϕ is the porosity, $\mathbf{u}$ is the fluid velocity, and q_m and $\mathbf{J}_m$ are the source/sink term and the diffusive mass flux of the mth component, respectively. Darcy's law for the fluid is

$$\mathbf{u} = -\frac{1}{\mu}\mathbf{k}\left(\nabla p - \rho \wp \nabla z\right), \tag{7.4}$$

where $\mathbf{k}$ is the permeability tensor, μ, p, and ρ are the fluid viscosity, pressure, and mass density, respectively, $\wp$ is the gravitational constant, and z is the depth. The fluid mass density is related to its molar density by

$$\rho = \xi W \equiv \xi \sum_{m=1}^{N_c} x_m W_m, \tag{7.5}$$

where W is the total molecular weight and W_m is the molecular weight of the mth component.

The primary variables in (7.3) and (7.4) are the mole fractions $\mathbf{x} = (x_1, x_2, \ldots, x_{N_c})$ and pressure p. The fluid viscosity has the following dependence:

$$\mu = \mu(p, T, x_1, x_2, \ldots, x_{N_c}),$$

which can be obtained from the correlation of Lohrenz, Bray, and Clark (1964), for example, with the input data: the critical pressure, critical temperature, critical volume, and molecular weight of each component (cf. Section 7.2). Here T denotes temperature. The molar density

$$\xi = \xi(p, T, x_1, x_2, \ldots, x_{N_c})$$

can be calculated using the equations of state, such as Peng and Robinson (PR, 1976) and Redlich, Kwong, and Soave (RKS, 1972) (cf. Section 7.3). The method of volume translation is widely used for correcting volumetric deficiencies of the original PR and RKS equations (Peneloux, Rauzy, and Freze, 1982). The least well understood term in equation (7.3) is that involving the diffusive flux $\mathbf{J}$, which will be discussed in detail in Section 7.4. Finally, the source/sink term q_m can be defined as (cf. Chapter 4)

$$q_m = \sum_{\nu=1}^{N_w}\sum_{l=1}^{M_{w\nu}} \left.\frac{2\pi x_m \xi \bar{k} \Delta h}{\mu \ln(r_e/r_w)}\right|_l^{(\nu)} \left(p_{bh}^{(\nu)} - p - \rho\wp(z_{bh}^{(\nu)} - z)\right)\delta\left(\mathbf{x} - \mathbf{x}_l^{(\nu)}\right), \qquad (7.6)$$

where $\delta(\mathbf{x})$ is the Dirac delta function, N_w is the total number of wells, $M_{w\nu}$ is the total number of perforated zones of the νth well, $\Delta h_l^{(\nu)}$ and $\mathbf{x}_l^{(\nu)}$ are the segment length and central location of the lth perforated zone of the νth well, $p_{bh}^{(\nu)}$ is the *bottom hole pressure* at the datum level depth $z_{bh}^{(\nu)}$, $r_{l,e}^{(\nu)}$ is the equivalent well radius, $r_w^{(\nu)}$ is the radius of the νth well, and $\bar{k}$ is some average of $\mathbf{k}$ at the wells.

7.2 Computation of Fluid Viscosity

Depending on the characteristics of the reservoir fluid (light, medium, or heavy), there exist many different analytical formulas for the computation of the fluid viscosity. In this section we briefly review a correlation formula due to Lohrenz, Bray, and Clark (1964), which is widely used in commercial reservoir simulators for light and medium oils.

For each component, or pseudocomponent, the following input data are required:

- critical pressure, p_{cm};
- critical temperature, T_{cm};
- critical volume, V_{cm};
- molecular weight, W_m, $\qquad m = 1, 2, \ldots, N_c$.

They are available for each component from laboratory experiments. The reduced temperature for component m is

$$T_{rm} = \frac{T}{T_{cm}}, \qquad m = 1, 2, \ldots, N_c.$$

Now, the steps to calculate μ are as follows:

- Computation of the component viscosity at low pressure, μ_m^* (cp):

$$\mu_m^* = \begin{cases} 34(10^{-5})T_{rm}^{0.94}/\eta_m & \text{if } T_{rm} < 1.5, \\ 17.78(10^{-5})(4.58T_{rm} - 1.67)^{5/8}/\eta_m & \text{if } T_{rm} \geq 1.5, \end{cases}$$

where the units are K for temperature and atm for pressure, and

$$\eta_m = \frac{T_{cm}^{1/6}}{W_m^{1/2} p_{cm}^{2/3}}.$$

- Computation of the fluid viscosity at low pressure, μ^* (cp):

$$\mu^* = \frac{\sum_{m=1}^{N_c} x_m \mu_m^* W_m^{1/2}}{\sum_{m=1}^{N_c} x_m W_m^{1/2}}.$$

- Computation of the reduced density:

$$\xi_r = \frac{\xi}{\xi_c},$$

where

$$\xi_c = \frac{1}{\sum_{m=1}^{N_c} x_m V_{cm}}.$$

- Computation of the fluid viscosity μ (cp):

$$\left[(\mu - \mu^*)\eta + 10^{-4}\right]^{1/4} = 0.1023 + 0.023364\xi_r + 0.058533\xi_r^2 - 0.40758\xi_r^3 + 0.0093324\xi_r^4,$$

where

$$\eta = \frac{\left(\sum_{m=1}^{N_c} x_m T_{cm}\right)^{1/6}}{\left(\sum_{m=1}^{N_c} x_m W_m\right)^{1/2} \left(\sum_{m=1}^{N_c} x_m p_{cm}\right)^{2/3}}.$$

7.3 Equations of State

In recent years, the *equations of state* (EOS) have been widely employed in reservoir simulation because they produce consistent compositions, densities, and molar volumes. There are thousands of EOS. The simplest EOS is that of van der Waals (Reid, Prausnitz, and Sherwood, 1977). Other EOS include PR, Redlich–Kwong (RK), and RKS. Here we describe only the PR EOS; others will be given in the next chapter. The method of volume translation is widely used for correcting volumetric deficiencies of the original EOS (Peneloux et al., 1982).

The mixing principle for the PR EOS is

$$a = \sum_{m=1}^{N_c} \sum_{\nu=1}^{N_c} x_m x_\nu (1 - k_{m\nu}) \sqrt{a_m a_\nu}, \quad b = \sum_{m=1}^{N_c} x_m b_m,$$

where $k_{m\nu}$ is a binary interaction parameter between components m and ν, and a_m and b_m are empirical factors for the pure component m. The interaction parameters account for molecular interactions between two unlike molecules. By definition, $k_{m\nu}$ is zero when m and ν represent the same component, small when m and ν represent components that do not differ much (e.g., when components m and ν are both alkanes), and large when m and ν represent components that are substantially different. Ideally, $k_{m\nu}$ depends on pressure and temperature and only on the identities of components m and ν (Zudkevitch and Joffe, 1970; Whitson, 1982).

The factors a_m and b_m can be computed from

$$a_m = \Omega_{ma} \alpha_m \frac{R^2 T_{cm}^2}{p_{cm}}, \quad b_m = \Omega_{mb} \frac{R T_{cm}}{p_{cm}},$$

where R is the universal gas constant, the EOS parameters Ω_{ma} and Ω_{mb} are given by

$$\begin{aligned} \Omega_{ma} &= 0.45724, \quad \Omega_{mb} = 0.077796, \\ \alpha_m &= \left(1 + \lambda_m \left[1 - \sqrt{T/T_{cm}}\right]\right)^2, \\ \lambda_m &= 0.37464 + 1.5423\omega_m - 0.26992\omega_m^2, \end{aligned}$$

and ω_m is the acentric factor for component m. The acentric factors roughly express the deviation of the shape of a molecule from a sphere. Define

$$A = \frac{ap}{R^2 T^2}, \quad B = \frac{bp}{RT}, \tag{7.7}$$

where the pressure p is given by the PR two-parameter EOS

$$p = \frac{RT}{V - b} - \frac{a}{V(V + b) + b(V - b)},$$

and V is the molar volume of the fluid phase. Introduce the compressibility factor

$$Z = \frac{pV}{RT}.$$

PR's cubic equation in Z is

$$Z^3 - (1 - B)Z^2 + (A - 2B - 3B^2)Z - (AB - B^2 - B^3) = 0. \tag{7.8}$$

This equation has three roots (Chen, Huan, and Ma, 2006). When only one root is real, it is selected. In the case where there are three real roots, say,

$$Z_1 > Z_2 > Z_3,$$

we select Z_1 if the mixture is a vapor (gas) phase. If it is a liquid (oil) phase, we select Z_1 when $Z_2 \leq 0$; select Z_2 when $Z_2 > 0$ and $Z_3 \leq 0$; select Z_3 when $Z_3 > 0$. That is, we select the smallest positive root.

The fluid molar and mass densities are

$$\xi = \frac{p}{RTZ}, \quad \rho = \xi \sum_{m=1}^{N_c} x_m W_m. \tag{7.9}$$

Note that the fluid compressibility coefficient is

$$c_f = \frac{1}{\xi}\frac{\partial \xi}{\partial p}. \tag{7.10}$$

It follows from (7.9) that

$$\frac{\partial \xi}{\partial p} = \frac{1}{RTZ} - \frac{p}{RTZ^2}\frac{\partial Z}{\partial p},$$

so, by equation (7.10),

$$c_f = \frac{1}{p} - \frac{1}{Z}\frac{\partial Z}{\partial p}. \tag{7.11}$$

Thus, to find c_f, it remains to calculate $\partial Z/\partial p$. Implicit differentiation on the cubic equation (7.8) yields

$$\begin{aligned} \frac{\partial Z}{\partial p} = -\Bigg\{ & \frac{\partial B}{\partial p} Z^2 + \left(\frac{\partial A}{\partial p} - 2\left[1 + 3B\right] \frac{\partial B}{\partial p} \right) Z \\ & - \left(\frac{\partial A}{\partial p} B + \left[A - 2B - 3B^2\right] \frac{\partial B}{\partial p} \right) \Bigg\} \\ & / \left(3Z^2 - 2(1 - B)Z + (A - 2B - 3B^2)\right). \end{aligned} \tag{7.12}$$

In addition, it follows from (7.7) that

$$\frac{\partial A}{\partial p} = \frac{a}{R^2T^2}, \quad \frac{\partial B}{\partial p} = \frac{b}{RT}. \tag{7.13}$$

Consequently, equations (7.11)–(7.13) can be used to find the fluid compressibility coefficient c_f.

7.4 Diffusion, Dispersion, and Tortuosity

7.4.1 Fick's Law

Application of the classical single-phase *Fick's law* to the mass conservation equation (7.3) gives

$$\mathbf{J}_i = -\phi \xi \mathbf{D}_i \nabla x_i, \quad i = 1, 2, \ldots, N_c, \tag{7.14}$$

where $\mathbf{D}_i$ is the diffusion/dispersion tensor of component i in the fluid mixture. This tensor in three dimensions is defined by

$$\mathbf{D}_i(\mathbf{u}) = d_{im}\mathbf{I} + |\mathbf{u}| \left(d_{il}\mathbf{E}(\mathbf{u}) + d_{it}\mathbf{E}^{\perp}(\mathbf{u}) \right), \tag{7.15}$$

where d_{im} is the *molecular diffusion* coefficient, d_{il} and d_{it} are, respectively, the *longitudinal* and *transverse dispersion* coefficients of component i, $|\mathbf{u}|$ is the Euclidean norm of $\mathbf{u} = (u_1, u_2, u_3)$: $|\mathbf{u}| = \sqrt{u_1^2 + u_2^2 + u_3^2}$, $\mathbf{E}(\mathbf{u})$ is the orthogonal projection along the velocity,

$$\mathbf{E}(\mathbf{u}) = \frac{1}{|\mathbf{u}|^2} \begin{pmatrix} u_1^2 & u_1 u_2 & u_1 u_3 \\ u_2 u_1 & u_2^2 & u_2 u_3 \\ u_3 u_1 & u_3 u_2 & u_3^2 \end{pmatrix},$$

and $\mathbf{E}^{\perp}(\mathbf{u}) = \mathbf{I} - \mathbf{E}(\mathbf{u})$, with $\mathbf{I}$ the identity matrix. In some cases, the tensor dispersion is more significant than the molecular diffusion; also, d_{il} is usually considerably larger than d_{it}.

We consider a scenario where a species only diffuses within a constant density fluid in a homogeneous porous medium without a source or sink. In this case, it follows from (7.3) without dispersion that the conservation equation becomes

$$\frac{\partial x}{\partial t} = d_m \Delta x. \tag{7.16}$$

As anticipated by the evaluation of the governing equation, the solution of (7.16) does not depend on porosity since diffusion time and length scales are not functions of the porosity. This implies that the concentrations will be identical when the user stipulates 0% or 100% porosity, an incorrect result. To represent the role of porosity on ordinary molecular diffusion, the diffusion coefficient must be scaled with tortuosity.

7.4.2 Impact of Tortuosity on Diffusion

In sediments, or in any porous system, the presence of solid particles causes the diffusion paths of species to deviate from straight lines (cf. Fig. 7.1). Consequently, the diffusion coefficients of species must be corrected for the tortuosity. Both theory and dimensional reasoning (Petersen, 1958; van Brakel and Heertjes, 1974; Ullman and Aller, 1982) suggest that the diffusion coefficient be scaled with *tortuosity* τ as follows:

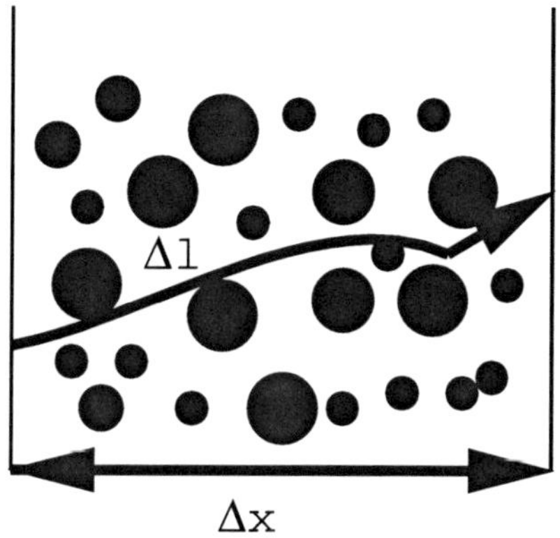

Figure 7.1. *Convolute diffusion path in a sediment.*

$$d'_m = \frac{d_m}{\tau^2}, \tag{7.17}$$

where d_m is now the diffusion coefficient of the species in question in the fluid (e.g., water) without the presence of the sediment matrix. Relations other than (7.17) will be discussed at the end of this section.

Theoretical relations

The general requirements for the tortuosity τ are as follows (Boudreau, 1996): First,

$$\tau^2 \geq 1, \tag{7.18}$$

which simply means that the actual (average) path traversed by the species while diffusing in the interstitial fluid in a porous medium is longer than in the absence of the solid. Second,

$$\lim_{\phi \to 1} \tau = 1; \tag{7.19}$$

that is, there is no hindrance to diffusion in the absence of any (impermeable) solid. Finally, $\tau \to \infty$ if and only if $\phi \to 0$; i.e., the tortuosity is finite for all nonzero values of porosity, so hindrance is complete only if the pore-space disappears. This property skips the important topic of connectivity at low porosity.

Physically, tortuosity is defined as the ratio of the actual distance Δl traveled by the species per unit length Δx of the medium (cf. Fig. 7.1):

$$\tau = \frac{\Delta l}{\Delta x}. \tag{7.20}$$

It is sometimes defined by the square of this ratio (Dullien, 1979), which, together with (7.20), will lead to an equivalent diffusion coefficient (see the discussion at the end of this section). Unlike ϕ, the parameter τ cannot be measured directly. The literature contains both theoretical and empirical relations for evaluating τ. Theoretical relations are based on a certain model of the structure of a porous medium. They have the advantage that they generally do not contain any adjustable parameter, but these relations are highly idealized. The simplest of such models is a collection of randomly oriented capillaries cutting through a solid body. Different treatments of the capillary model yield (Petersen, 1958)

$$\tau = \sqrt{2}, \tag{7.21}$$

or (Bhatia, 1985; Dykhuizen and Casey, 1989)

$$\tau = \sqrt{3}. \tag{7.22}$$

Experimental values of τ can be either smaller or larger than these numerical values (cf. Figs. 7.2–7.4). In fact, its measured values correlate with ϕ (Archie, 1942; van Brakel and Heertjes, 1974), which suggests that theoretical models for τ depend on ϕ.

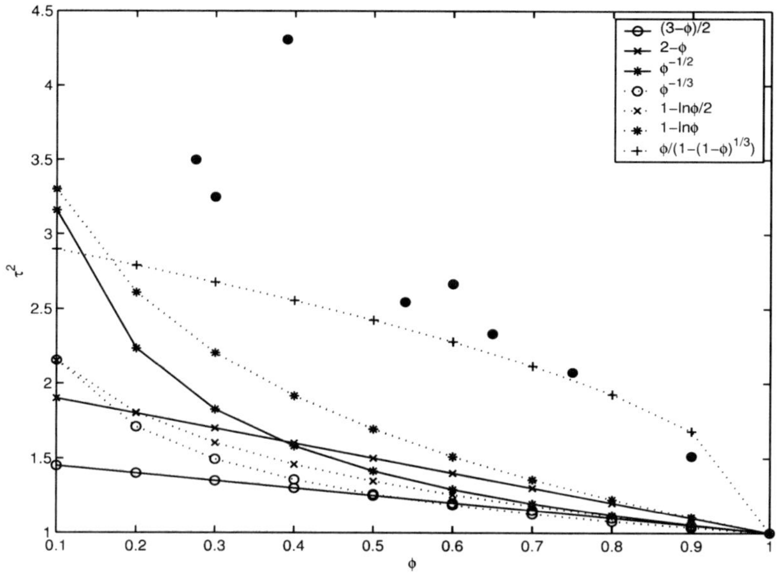

Figure 7.2. *First eight theoretical relations against measured data.*

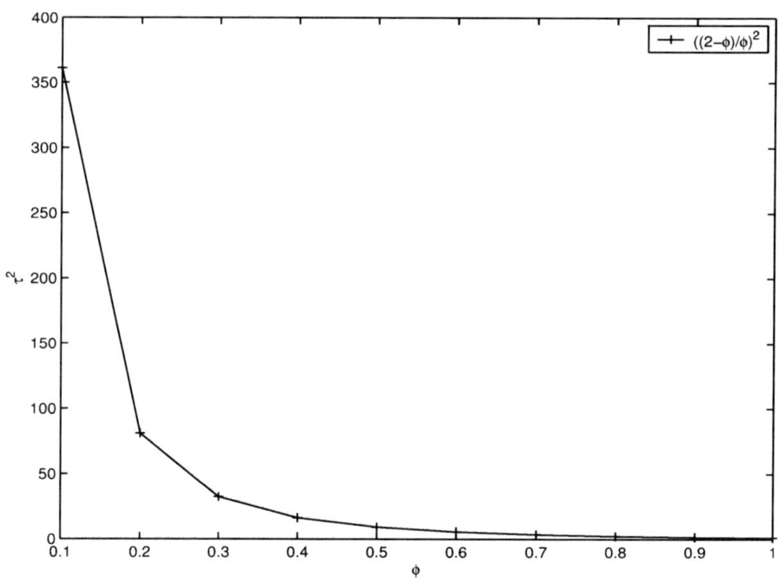

Figure 7.3. *Last theoretical tortuosity-porosity relation in Table* 7.1.

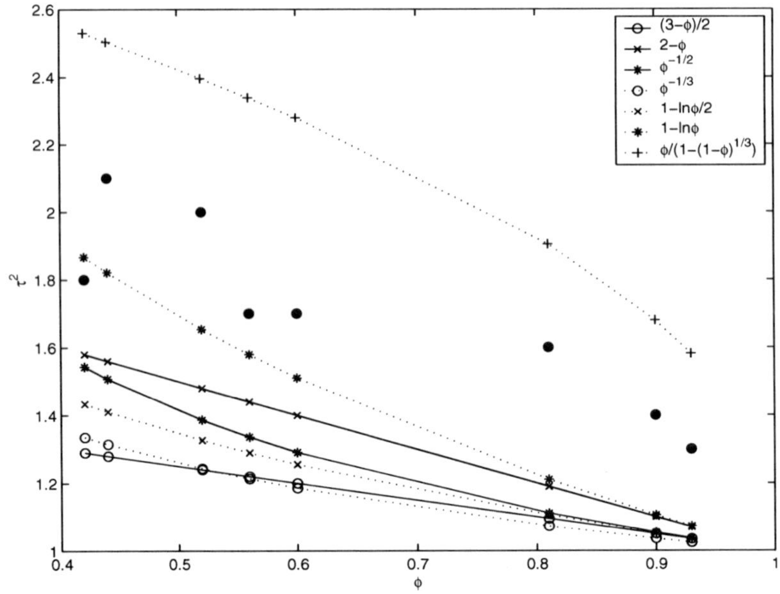

Figure 7.4. *First eight theoretical relations against different measured data.*

Table 7.1 is a collection of theoretical tortuosity-porosity relations that satisfy the above three requirements, with remarks on the physical system on which each relation is based. The important point here is whether these relations are useful in calibrating observed sedimentary tortuosity data. The data have been measured from various marine and lacustrine sediments (Manheim and Waterman, 1974; Goldhaber et al., 1977; Li and Gregory, 1977; Jorgensen, 1978; Krom and Berner, 1980; Ullman and Aller, 1982; Archer, Emerson, and Reimers, 1989; Sweerts et al., 1991). Since the data given by Iversen and Jorgensen (1993) and Sweerts et al. (1991) are well represented for the fine-grained marine and lacustrine sediments, respectively, as examples we plot the theoretical relations given in Table 7.1 against these two sets of data.

In Iversen and Jorgensen (1993) the tracer diffusion coefficients of sulfate and methane were determined in seawater, with sediments collected along the Danish coast. At water depth varying from 50 cm to 200 m, the sediment composition varied from sandy sediments to fine-grained silt and clay, with the majority of the samples from silt-clay types. A set of cores was collected from the bottom of clay beach dunes. This station cannot be classified as marine sediments, but the dune is raised sea-bottom, and the clay was in contact with seawater and closely resembled consolidated marine. The tracer diffusion measurement method by Iversen and Jorgensen (1993) was based on the instantaneous source technique by Duursma and Bosch (1970); that is, the solution to the 1D counterpart of equation (7.16) with an instantaneous source was used to calculate the tracer diffusion coefficients.

Figs. 7.2 and 7.3 plot the theoretical relations in Table 7.1 against the data from Iversen and Jorgensen (1993) (because of a great scale difference, the first eight relations are plotted in Fig. 7.2, while the last one in Table 7.1 is plotted in Fig. 7.3). These two plots indicate

Table 7.1. *Theoretical relations, their physical systems, and references.*

Relations	Remarks	References
$\tau^2 = (3 - \phi)/2$	Ordered packings	Maxwell, 1881; Akanni et al., 1987
$\tau^2 = (3 - \phi)/2$	Random homogeneous isotropic sphere packings	Neale and Nader, 1973
$\tau^2 = 2 - \phi$	A hyperbola of revolution	Rayleigh, 1892; Petersen, 1958
$\tau^2 = \phi^{-1/2}$	Not for monosized spheres	Bruggemann, 1935
$\tau^2 = \phi^{-1/3}$	Partly saturated homogeneous isotropic monodisperse sphere packings	Millington, 1959; van Brakel and Heertjes, 1974
$\tau^2 = 1 - \ln \phi/2$	Overlapping spheres	Weissberg, 1963; Ho and Strieder, 1981
$\tau^2 = 1 - \ln \phi$	Random arrays of freely overlapping cylinders	Tomadakis and Sotirchos, 1983
$\tau^2 = \phi/\left(1 - (1 - \phi)^{1/3}\right)$	Heterogeneous catalyst	Beekman, 1990
$\tau^2 = ((2 - \phi)/\phi)^2$	Cation-exchange resin membrane	Mackie and Meares, 1955

that none of the theoretical relations in Table 7.1 adequately match the measured data from Iversen and Jorgensen (1993).

We now plot the theoretical relations in Table 7.1 against the data for lacustrine sediments taken from Sweerts et al. (1991) (cf. Fig. 7.4). In Sweerts et al. (1991) the whole-sediment molecular diffusion coefficients for tritiated water in pore waters of various lakes were determined experimentally by adding ^3H_2O to the overlying water of asphyxiated (without bioirrigation) and unasphyxiated cores and measuring the resulting pore-water profiles after a period of time. The measurement method used a constant-source computer model that was based on the error function technique (Duursma and Hoede, 1967).

All the plots in Figs. 7.2–7.4 show that none of the theoretical relations in Table 7.1 accurately match the measured data. The problem is that these relations are based on mathematically idealized geometries (e.g., spherical or rectangular packings of some type, as shown in Table 7.1) of real sedimentary fabrics.

Empirical relations

As shown above, the theoretical relations do not match the data that describe sediments of very different types. Empirical relations that contain adjustable parameters are analogous to the theoretical ones but better describe the observed data.

Experimentally, the tortuosity of a sediment can be obtained by measuring the porosity ϕ and the formation resistivity factor F. These three variables are related by

$$\tau^2 = (F\phi)^n, \qquad (7.23)$$

where n is an adjustable empirical parameter. The formation factor F is determined by measuring the resistivity of a porous medium, relative to that of the free solution.

Microelectrode technology currently leads to fine-scale profiles of this factor in surface sediments (van Cappellen and Gaillard, 1996). However, this information is hardly available and must rely on an empirical relation that relates F to the sediment porosity ϕ. Following Faris et al. (1954) and Nelson and Simmons (1995), this relation is usually described by Archie's law (Archie, 1942):

$$F = \frac{A}{\phi^m}, \tag{7.24}$$

where A and m are empirical parameters. Substituting (7.23) into (7.24) gives the relation of tortuosity in terms of porosity,

$$\tau^2 = \left(A\phi^{1-m}\right)^n, \tag{7.25}$$

with three adjustable parameters A, m, and n. These parameters are lithology-dependent. Relation (7.25) was used for sands (Lerman, 1979) and muds (Ullman and Aller, 1982) with $A = n = 1$ (also see Boudreau, 1996, and van Cappellen and Gaillard, 1996), for the same choice of these parameters).

There are other types of empirical relations such as the linear function (Low, 1981; Iversen and Jorgensen, 1993)

$$\tau^2 = \phi + B(1 - \phi), \tag{7.26}$$

and the logarithmic function (Weissberg, 1963; Boudreau, 1996)

$$\tau^2 = 1 - C \ln \phi, \tag{7.27}$$

where B and C are adjustable parameters. These three types of empirical relations are summarized in Table 7.2.

Note that (7.26) with $B = 0$ becomes (7.25) with $A = n = 1$ and $m = 0$. Also, with $C = 2$ (Boudreau, 1996), relation (7.27) becomes

$$\tau^2 = 1 - 2 \ln \phi = 1 + \ln\left(\frac{1}{\phi^2}\right),$$

which can be expanded in the series of the logarithm about the value $\phi^{-2} = 1$ (i.e., $\phi = 1$):

$$\tau^2 = 1 + \left(\frac{1}{\phi^2} - 1\right) - \frac{1}{2}\left(\frac{1}{\phi^2} - 1\right)^2 + \frac{1}{3}\left(\frac{1}{\phi^2} - 1\right)^3 - \cdots.$$

Table 7.2. *Empirical relations, their physical systems, and references.*

Relations	Parameters	Remarks	References
$\tau^2 = \left(A\phi^{1-m}\right)^n$	A, m, and n	Sands, muds	Lerman, 1979; Ullman and Aller, 1982
$\tau^2 = \phi + B(1 - \phi)$	B	Soils, catalysts	Low, 1981; Iversen and Jorgensen, 1993
$\tau^2 = 1 - C \ln \phi$	C	Fine-grained unlithified sediments	Weissberg, 1963; Boudreau, 1996

For values of ϕ near 1, only the second term on the right-hand side is significant, so

$$\tau^2 \approx 1 + \left(\frac{1}{\phi^2} - 1\right) = \frac{1}{\phi^2},$$

which corresponds to (7.25) with $A = n = 1$ and $m = 3$. This case was investigated by Ullman and Aller (1982) where a ϕ^{-3}-dependence for the formation factor F was necessary to explain high porosity data.

Figs. 7.5 and 7.6 show the best least-squares fits of relations (7.25)–(7.27) to the same sets of data as given in Figs. 7.2 and 7.4, and Table 7.3 lists the corresponding parameter values and the statistics of these fits. The measured data from Iversen and Jorgensen (1993) were given only for $\phi \geq 0.275$. In Fig. 7.5, for completeness we plot the three relations for ϕ in the interval [0.1, 1.0]. All three relations provide statistically significant correlations to the data in excess of 50% of the variance for $\phi \geq 0.275$. The latter two relations seem to give better fits than the first one. This can be also seen from Fig. 7.6, where the measured data from Sweerts et al. (1991) were given only for $\phi \geq 0.4$.

The default values in equation (7.25) can be chosen as follows:

$$A = 1, \quad m = 2, \quad n = 1. \tag{7.28}$$

The tortuosity-porosity relations given by Low (1981), Boudreau (1996), and Boving and Grathwohl (2001), together with relation (7.25) with these default values, are plotted in Fig. 7.7. These relations seem matching well for $\phi \geq 0.4$.

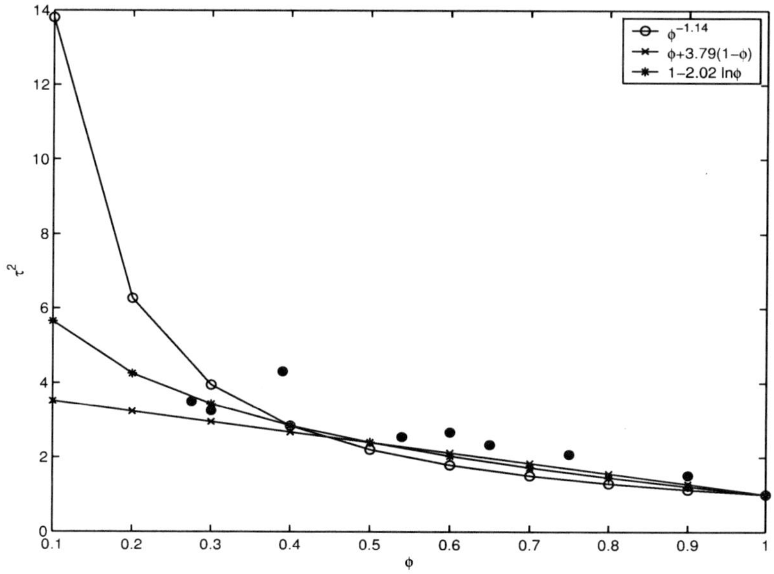

Figure 7.5. *Plot of empirical relations against measured data.*

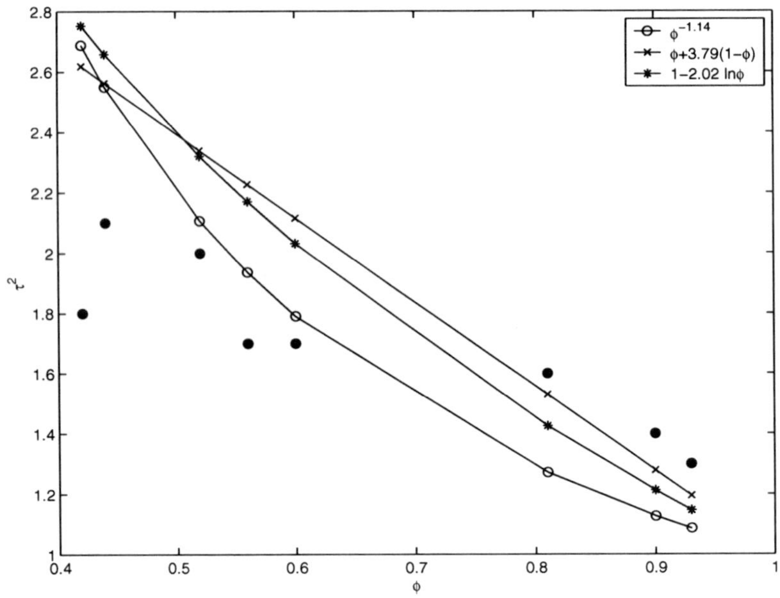

Figure 7.6. *Plot of empirical relations against different measured data.*

Table 7.3. *Best least-squares fits for empirical relations* (7.25)–(7.27).

Relations	Parameter values	r^2 (variance)
$\tau^2 = \left(A\phi^{1-m}\right)^n$	$A = n = 1$ and $m = 2.14 \pm 0.02$	0.54
$\tau^2 = \phi + B(1-\phi)$	$B = 3.79 \pm 0.10$	0.63
$\tau^2 = 1 - C \ln \phi$	$C = 2.02 \pm 0.09$	0.64

Remarks on the impact of tortuosity on diffusion

There are essentially two classes of definitions for the scaled diffusion coefficient d'_m. The first class follows equation (7.17) (Berner, 1980; Maerki et al., 2004), with the tortuosity τ given by (7.20). In McDuff and Ellis (1979), Andrews and Bennett (1981), and Boudreau (1996), the tortuosity is given by $\tau^2 = F\phi$, which corresponds to equation (7.23) with $n = 1$.

The second class defines the scaled diffusion coefficient as follows (Aris, 1975; Sherwood, Pigford, and Wilke, 1975; Dullien, 1979; Thomas, 1989; Shackelford, 1991):

$$d'_m = \frac{d_m}{\tau}, \tag{7.29}$$

with the tortuosity τ given by

$$\tau = \left(\frac{\Delta l}{\Delta x}\right)^2. \tag{7.30}$$

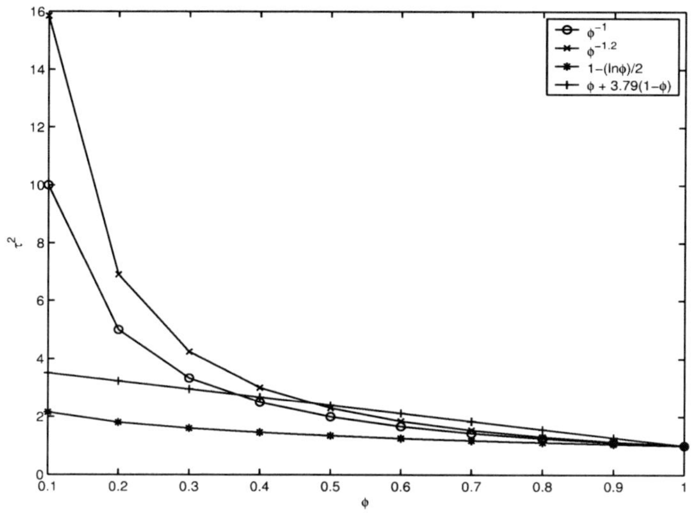

Figure 7.7. *Plot of various empirical relations and relations* (7.25)–(7.28).

We can easily see that the two definitions, (7.17) with (7.20) and (7.29) with (7.30), are equivalent.

In addition to the definitions (7.17) and (7.29), there exist other definitions for d'_m. In Popovičová and Brusseau (1997), the tortuosity τ in equation (7.29) is given by

$$\tau = \phi^{-2/3}.$$

Also, in Sherwood, Pigford, and Wilke (1975) it is defined by

$$d'_m = \frac{d_m \phi}{\tau}. \tag{7.31}$$

Finally, the scaled diffusion coefficient d'_m by Boving and Grathwohl (2001) is defined, with τ given by (7.30), as

$$d'_m = \frac{d_m \phi \delta}{\tau}, \tag{7.32}$$

where δ is the "constrictivity"—a dimensionless variable that becomes important if the size of the species becomes comparable to the size of the pore.

An question arises: Should the inverse relationship with tortuosity be also multiplied with the dispersion terms? Keeping it separate is motivated by the existing papers. Since the dispersion coefficients are also empirical, however, where the inverse relationship is placed would depend on how the dispersion coefficients are measured. Future research is needed to determine how the tortuosity affects dispersion. For a survey of the available formulations for the scaled diffusion coefficient with tortuosity, sensitivity tests with analytical solutions, and calibration of these formulations with respect to measured data, the reader should refer to Shen and Chen (2007).

7.4.3 Soret Effects and Gravity Segregation

In general, the diffusive flux $\mathbf{J}$ in equation (7.3) is much more complex than that presented in Fick's law (7.14). For example, thermal and pressure diffusions are very important for the study of compositional variations in hydrocarbon reservoirs (Firoozabadi, 1998). The ratio of molecular, thermal, and pressure diffusions to the total diffusion depends on the nonideality of a fluid mixture. An understanding of these diffusions is required to determine the distribution of various species of the fluid mixture in cavities with dimensions on the order of hundreds of meters and more.

Set

$$\mathbf{J} = \left(\mathbf{J}_1, \mathbf{J}_2, \ldots, \mathbf{J}_{N_c}\right)^T,$$

where the superscript T indicates the transpose of a vector or matrix. Note that the diffusive fluxes satisfy

$$\sum_{i=1}^{N_c} \mathbf{J}_i = \mathbf{0}. \tag{7.33}$$

The most general form (without dispersion) of each diffusive mass flux in equation (7.3) is

$$\mathbf{J}_i = -\phi\xi \left(\sum_j \mathbf{D}_{ij}^M \nabla x_j + \mathbf{D}_i^T \nabla T + \mathbf{D}_i^p \nabla p \right), \tag{7.34}$$

where $\mathbf{D}_{ij}^M$, $\mathbf{D}_i^T$, and $\mathbf{D}_i^p$ are the molecular, thermal (the *Soret effect*), and pressure (*gravity segregation*) diffusion tensors of component i, respectively. Following Ghorayeb and Firoozabadi (2000), each $\mathbf{J}_i$ is given by, $i = 1, 2, \ldots, N_c - 1$,

$$\begin{aligned}\mathbf{J}_i = &-\phi\xi a_{iN_c} D_{iN_c} \left\{ \frac{W_i x_i}{L_{ii}} \sum_{k=1}^{N_c-1} L_{ik} \sum_{j=1}^{N_c-1} \frac{W_j x_j + W_{N_c} x_{N_c} \delta_{jk}}{W_j} \sum_{l=1}^{N_c-1} \frac{\partial \ln f_j}{\partial x_l} \nabla x_l \right. \\ &\left. + \frac{W k_{Ti}}{T} \nabla T + \frac{W_i x_i}{RTL_{ii}} \sum_{k=1}^{N_c-1} L_{ik} \left(\sum_{j=1}^{N_c-1} x_j v_j + \frac{W_{N_c} x_{N_c}}{W_k} v_k - \frac{1}{\xi} \right) \nabla p \right\},\end{aligned} \tag{7.35}$$

where L_{ik} is a phenomenological coefficient, δ_{jk} denotes the Kronecker symbol, R is the gas constant, f_i and v_i are the fugacity and partial molar volume of component i, and the coefficients a_{iN_c}, D_{iN_c}, and k_{Ti} (the thermal diffusion ratio) are defined by, $i = 1, 2, \ldots, N_c - 1$,

$$a_{iN_c} = \frac{W_i W_{N_c}}{W^2}, \quad D_{iN_c} = \frac{W^2 R L_{ii}}{\xi W_i^2 W_{N_c}^2 x_i x_{N_c}},$$

$$k_{Ti} = \frac{W_i x_i W_{N_c} x_{N_c} L_i'}{WRTL_{ii}} \equiv \alpha_{Ti} x_i x_{N_c},$$

with L_i' being another phenomenological coefficient and α_{Ti} called the thermal diffusion factor of component i. For information on the phenomenological coefficients L_{ik} and L_i', the reader may refer to Onsager (1931a, 1931b). While a general formula is given for each diffusion flux $\mathbf{J}_i$, formula (7.35) is seldom used in multicomponent reservoir simulation.

Most of thermal and pressure diffusion coefficients are available only for a binary or ternary fluid mixture. The major difficulty in measuring diffusion is to determine the dependence of diffusion coefficients on compositions, temperature, and pressure. At or near critical points, these coefficients can change signs, thus changing mixing directions.

7.4.4 Isothermal Gravity/Chemical Equilibrium

Tremendous efforts have been devoted to modeling compositional variations under the force of gravity in hydrocarbon reservoirs over the past 30 years. Whitson and Belery (1994) gave a historical survey on the development of methods used to investigate gravity segregation. The formulation for computing the compositional variations under gravity for an isothermal system was first given by Gibbs (1876), using the constraint of *chemical equilibrium*,

$$d\mu_i + W_i \wp dz = 0, \qquad i = 1, 2, \ldots, N_c, \tag{7.36}$$

where μ_i is the *chemical potential* of component i. Equations (7.2) and (7.36) provide compositions $(x_1, x_2, \ldots, x_{N_c})$ and pressure p at any depth z once they are specified at a reference depth. An interesting fact is that the Gibbs equation (7.36) can be obtained using (7.35) and the condition of *mechanical equilibrium*,

$$dp = -\rho \wp dz. \tag{7.37}$$

For an isothermal system at the steady state,

$$\mathbf{J}_i = \mathbf{0}, \qquad i = 1, 2, \ldots, N_c.$$

Applying (7.35) in the z-direction, we see that

$$\begin{aligned} &\sum_{k=1}^{N_c-1} L_{ik} \sum_{j=1}^{N_c-1} \frac{W_j x_j + W_{N_c} x_{N_c} \delta_{jk}}{W_j} \sum_{l=1}^{N_c-1} \frac{\partial \ln f_j}{\partial x_l} \frac{dx_l}{dz} \\ &+ \frac{1}{RT} \sum_{k=1}^{N_c-1} L_{ik} \left(\sum_{j=1}^{N_c-1} x_j v_j + \frac{W_{N_c} x_{N_c}}{W_k} v_k - \frac{1}{\xi} \right) \frac{dp}{dz} = 0. \end{aligned} \tag{7.38}$$

This equation must hold for any combination of the compositions $(x_1, x_2, \ldots, x_{N_c})$. In particular, for a fixed $j = k$ and $x_m = 0$ for all m's except possibly for $m = j$ or N_c, equation (7.38) reduces to

$$\frac{W_j x_j + W_{N_c} x_{N_c}}{W_j} \sum_{l=1}^{N_c-1} \frac{\partial \ln f_j}{\partial x_l} \frac{dx_l}{dz} + \frac{1}{RT} \left(x_j v_j + \frac{W_{N_c} x_{N_c}}{W_j} v_j - \frac{1}{\xi} \right) \frac{dp}{dz} = 0, \tag{7.39}$$

which, together with equations (7.5) and (7.37), implies

$$RT \sum_{l=1}^{N_c-1} \frac{\partial \ln f_j}{\partial x_l} \frac{dx_l}{dz} - \wp(\rho v_j - W_j) = 0, \qquad i = 1, 2, \ldots, N_c - 1. \tag{7.40}$$

This equation is exactly the condition (7.36) by noting that

$$\frac{\partial \mu_j}{\partial x_l} = RT\frac{\partial \ln f_j}{\partial x_l}, \qquad \frac{\partial \mu_j}{\partial p} = v_j.$$

We remark that the chemical equilibrium constraint (7.36) (segregation equation) holds only for an isothermal system. For nonisothermal systems, this constraint is no longer valid because of nonzero entropy production. Furthermore, transience is not solved, so there is no driving force.

7.5 Numerical Solution

7.5.1 A Model Problem

To present the idea for numerical solution of the transport of multicomponents in a fluid phase, the diffusion fluxes are assumed to take the form (7.14) with or without a tortuosity effect. Due to the constraint (7.2), we require only $N_c - 1$ mass conservation equations for the first $N_c - 1$ mole fractions $(x_1, x_2, \ldots, x_{N_c-1})$:

$$\frac{\partial(\phi x_m \xi)}{\partial t} = \nabla \cdot \left(\frac{x_m \xi}{\mu}\mathbf{k}\,(\nabla p - \gamma \nabla z) + \phi\,\xi\,\mathbf{D}_m \nabla x_m \right) + q_m, \tag{7.41}$$

where $\gamma = \rho\wp$ (the fluid gravity) and Darcy's law (7.4) was substituted, $m = 1, 2, \ldots, N_c - 1$. Also, addition of all species equations (7.3) over m and application of constraints (7.2) and (7.33) give

$$\frac{\partial(\phi\xi)}{\partial t} = -\nabla \cdot (\xi\mathbf{u}) + q, \qquad q = \sum_{m=1}^{N_c} q_m.$$

Then, using Darcy's law (7.4) and performing time differentiation, we see that

$$c(p)\frac{\partial p}{\partial t} = \nabla \cdot \left(\frac{\xi}{\mu}\mathbf{k}\,(\nabla p - \gamma \nabla z) \right) + q, \tag{7.42}$$

where

$$c(p) = \phi\frac{\partial \xi}{\partial p} + \xi\frac{d\phi}{dp}.$$

The PR EOS can be used to evaluate ξ, ρ (γ), and $\partial\xi/\partial p$, and Lohrenz's correlation can be utilized to calculate μ. Equation (7.42) in form is exactly the same as equation (3.21) for the single-phase flow.

7.5.2 Finite Difference Equations

Equations (7.41) and (7.42), together with appropriate boundary and initial conditions, form a closed, coupled system for the primary unknowns $x_1, x_2, \ldots, x_{N_c-1}, p$. Assume that the permeability tensor $\mathbf{k}$ and the diffusion tensor $\mathbf{D}_m$ are diagonal: $\mathbf{k} = \mathrm{diag}(k_{11}, k_{22}, k_{33})$ and

$\mathbf{D}_m = \mathrm{diag}(D_{m,11}, D_{m,22}, D_{m,33})$. As in previous chapters, at time level t^n and gridblock (i, j, k) a fully implicit scheme for equations (7.41) and (7.42) is

$$
\begin{aligned}
&\left(V\frac{(\phi x_m\xi)^{n+1}-(\phi x_m\xi)^n}{\Delta t}\right)_{i,j,k}\\
&= \mathcal{D}^{n+1}_{m1,i+1/2,j,k}\left(x^{n+1}_{m,i+1,j,k}-x^{n+1}_{m,i,j,k}\right)-\mathcal{D}^{n+1}_{m1,1,i-1/2,j,k}\left(x^{n+1}_{m,i,j,k}-x^{n+1}_{m,i-1,j,k}\right)\\
&+\mathcal{D}^{n+1}_{m2,i,j+1/2,k}\left(x^{n+1}_{m,i,j+1,k}-x^{n+1}_{m,i,j,k}\right)-\mathcal{D}^{n+1}_{m2,i,j-1/2,k}\left(x^{n+1}_{m,i,j,k}-x^{n+1}_{m,i,j-1,k}\right)\\
&+\mathcal{D}^{n+1}_{m3,i,j,k+1/2}\left(x^{n+1}_{m,i,j,k+1}-x^{n+1}_{m,i,j,k}\right)-\mathcal{D}^{n+1}_{m3,i,j,k-1/2}\left(x^{n+1}_{m,i,j,k}-x^{n+1}_{m,i,j,k-1}\right)\\
&+T^{n+1}_{m1,i+1/2,j,k}\left(p^{n+1}_{i+1,j,k}-p^{n+1}_{i,j,k}\right)-T^{n+1}_{m1,i-1/2,j,k}\left(p^{n+1}_{i,j,k}-p^{n+1}_{i-1,j,k}\right)\\
&+T^{n+1}_{m2,i,j+1/2,k}\left(p^{n+1}_{i,j+1,k}-p^{n+1}_{i,j,k}\right)-T^{n+1}_{m2,i,j-1/2,k}\left(p^{n+1}_{i,j,k}-p^{n+1}_{i,j-1,k}\right)\\
&+T^{n+1}_{m3,i,j,k+1/2}\left(p^{n+1}_{i,j,k+1}-p^{n+1}_{i,j,k}\right)-T^{n+1}_{m3,i,j,k-1/2}\left(p^{n+1}_{i,j,k}-p^{n+1}_{i,j,k-1}\right)\\
&-(T_m\gamma)^{n+1}_{1,i+1/2,j,k}\left(z_{i+1,j,k}-z_{i,j,k}\right)+(T_m\gamma)^{n+1}_{1,i-1/2,j,k}\left(z_{i,j,k}-z_{i-1,j,k}\right)\\
&-(T_m\gamma)^{n+1}_{2,i,j+1/2,k}\left(z_{i,j+1,k}-z_{i,j,k}\right)+(T_m\gamma)^{n+1}_{2,i,j-1/2,k}\left(z_{i,j,k}-z_{i,j-1,k}\right)\\
&-(T_m\gamma)^{n+1}_{3,i,j,k+1/2}\left(z_{i,j,k+1}-z_{i,j,k}\right)+(T_m\gamma)^{n+1}_{3,i,j,k-1/2}\left(z_{i,j,k}-z_{i,j,k-1}\right)\\
&+Q^{n+1}_{m,i,j,k},\qquad m=1,2,\ldots,N_c-1,
\end{aligned}
\tag{7.43}
$$

and

$$
\begin{aligned}
&\left(Vc\left(p^{n+1}\right)\frac{p^{n+1}-p^n}{\Delta t}\right)_{i,j,k}\\
&= T^{n+1}_{1,i+1/2,j,k}\left(p^{n+1}_{i+1,j,k}-p^{n+1}_{i,j,k}\right)-T^{n+1}_{1,i-1/2,j,k}\left(p^{n+1}_{i,j,k}-p^{n+1}_{i-1,j,k}\right)\\
&+T^{n+1}_{2,i,j+1/2,k}\left(p^{n+1}_{i,j+1,k}-p^{n+1}_{i,j,k}\right)-T^{n+1}_{2,i,j-1/2,k}\left(p^{n+1}_{i,j,k}-p^{n+1}_{i,j-1,k}\right)\\
&+T^{n+1}_{3,i,j,k+1/2}\left(p^{n+1}_{i,j,k+1}-p^{n+1}_{i,j,k}\right)-T^{n+1}_{3,i,j,k-1/2}\left(p^{n+1}_{i,j,k}-p^{n+1}_{i,j,k-1}\right)\\
&-(T\gamma)^{n+1}_{1,i+1/2,j,k}\left(z_{i+1,j,k}-z_{i,j,k}\right)+(T\gamma)^{n+1}_{1,i-1/2,j,k}\left(z_{i,j,k}-z_{i-1,j,k}\right)\\
&-(T\gamma)^{n+1}_{2,i,j+1/2,k}\left(z_{i,j+1,k}-z_{i,j,k}\right)+(T\gamma)^{n+1}_{2,i,j-1/2,k}\left(z_{i,j,k}-z_{i,j-1,k}\right)\\
&-(T\gamma)^{n+1}_{3,i,j,k+1/2}\left(z_{i,j,k+1}-z_{i,j,k}\right)+(T\gamma)^{n+1}_{3,i,j,k-1/2}\left(z_{i,j,k}-z_{i,j,k-1}\right)\\
&+Q^{n+1}_{i,j,k},
\end{aligned}
\tag{7.44}
$$

where $Q_{m,i,j,k} = V_{i,j,k}q_{m,i,j,k}$, $Q_{i,j,k} = V_{i,j,k}q_{i,j,k}$, and the transmissibility and diffusion coefficients are

$$
\mathbf{T} = \frac{\xi}{\mu}\mathbf{k}\,\mathrm{diag}\left(\frac{A_1}{h_1}, \frac{A_2}{h_2}, \frac{A_3}{h_3}\right), \qquad \mathbf{T}_m = \frac{x_m\xi}{\mu}\mathbf{k}\,\mathrm{diag}\left(\frac{A_1}{h_1}, \frac{A_2}{h_2}, \frac{A_3}{h_3}\right),
$$

$$
\mathcal{D}_m = \phi\xi\mathbf{D}_m\,\mathrm{diag}\left(\frac{A_1}{h_1}, \frac{A_2}{h_2}, \frac{A_3}{h_3}\right), \qquad m = 1, 2, \ldots, N_c - 1,
$$

where A_i is the cross-sectional area normal to and h_i is the grid size in the x_i-direction, $i = 1, 2, 3$. Their evaluation at internal boundaries of gridblocks must be carried out properly. That is, the harmonic average, (weighted) arithmetic average, and upstream weighting technique should be used for the rock properties ($\mathbf{k}$ and ϕ), the fluid properties (ξ, μ, and γ), and the rock/fluid property ($\mathbf{D}_m$, $m = 1, 2, \ldots, N_c - 1$), respectively (cf. Sections 3.4.4 and 5.3.4).

Equations (7.43) and (7.44) form a fully implicit system for the primary unknowns $x_1^{n+1}, x_2^{n+1}, \ldots, x_{N_c-1}^{n+1}, p^{n+1}$. As for the numerical solution of the black oil model in Chapter 6, this system can be linearized via the Newton–Raphson iteration. At each iteration, there are three solution approaches available: simultaneous solution (SS), sequential, and iterative IMPES (now known as IMPEC, implicit for pressure and explicit for composition). When the number N_c of chemical species is small (e.g., two or three), the SS approach can be used. However, when N_c is large, one just cannot afford this approach, even with today's powerful computers. The other two approaches, sequential and iterative IMPEC, are more appropriate for solving this coupled system. For a slightly compressible fluid, the iterative IMPEC is a reasonable choice. In general, as shown in the previous chapter, the sequential solution approach is more stable than this approach, and it is thus used for solving the system of equations (7.43) and (7.44).

A new update using the Newton–Raphson iteration at the $(l+1)$th level for the primary unknowns is

$$\begin{aligned} p^{n+1,l+1} &= p^{n+1,l} + \delta p^{n+1,l+1}, \\ x_m^{n+1,l+1} &= x_m^{n+1,l} + \delta x_m^{n+1,l+1} \end{aligned}$$

for $m = 1, 2, \ldots, N_c - 1$. Below, the superscript $n+1$ is omitted. At each Newton–Raphson iteration step $l + 1$, the flow chart for the computation of pressure and mole fractions is as follows:

- Given p^l and x_m^l, $m = 1, 2, \ldots, N_c - 1$, apply the Lohrenz correlation to find μ^l; the PR EOS to calculate the molar density ξ^l, the mass density ρ^l, and the derivative $d\xi^l/dp$; and the rock property to evaluate ϕ^l.

- Solve for pressure p^{l+1}:

$$\begin{aligned} &\left(Vc\left(p^l\right) \frac{p^{l+1} - p^n}{\Delta t} \right)_{i,j,k} \\ &= T^l_{1,i+1/2,j,k}\left(p^{l+1}_{i+1,j,k} - p^{l+1}_{i,j,k}\right) - T^l_{1,i-1/2,j,k}\left(p^{l+1}_{i,j,k} - p^{l+1}_{i-1,j,k}\right) \\ &\quad + T^l_{2,i,j+1/2,k}\left(p^{l+1}_{i,j+1,k} - p^{l+1}_{i,j,k}\right) - T^l_{2,i,j-1/2,k}\left(p^{l+1}_{i,j,k} - p^{l+1}_{i,j-1,k}\right) \\ &\quad + T^l_{3,i,j,k+1/2}\left(p^{l}_{i,j,k+1} - p^{l+1}_{i,j,k}\right) - T^l_{3,i,j,k-1/2}\left(p^{l+1}_{i,j,k} - p^{l+1}_{i,j,k-1}\right) \\ &\quad - (T\gamma)^l_{1,i+1/2,j,k}\left(z_{i+1,j,k} - z_{i,j,k}\right) + (T\gamma)^l_{1,i-1/2,j,k}\left(z_{i,j,k} - z_{i-1,j,k}\right) \\ &\quad - (T\gamma)^l_{2,i,j+1/2,k}\left(z_{i,j+1,k} - z_{i,j,k}\right) + (T\gamma)^l_{2,i,j-1/2,k}\left(z_{i,j,k} - z_{i,j-1,k}\right) \\ &\quad - (T\gamma)^l_{3,i,j,k+1/2}\left(z_{i,j,k+1} - z_{i,j,k}\right) + (T\gamma)^l_{3,i,j,k-1/2}\left(z_{i,j,k} - z_{i,j,k-1}\right) \\ &\quad + Q^{l+1}_{i,j,k}. \end{aligned} \tag{7.45}$$

- Solve for the mole fractions x_m^{l+1}, $m = 1, 2, \ldots, N_c - 1$:

$$
\begin{aligned}
&\left(V\frac{(\phi x_m \xi)^{l+1} - (\phi x_m \xi)^n}{\Delta t}\right)_{i,j,k} \\
&= \mathcal{D}_{m1,i+1/2,j,k}^{l*}\left(x_{m,i+1,j,k}^{l+1} - x_{m,i,j,k}^{l+1}\right) - \mathcal{D}_{m1,1,i-1/2,j,k}^{l*}\left(x_{m,i,j,k}^{l+1} - x_{m,i-1,j,k}^{l+1}\right) \\
&\quad + \mathcal{D}_{m2,i,j+1/2,k}^{l*}\left(x_{m,i,j+1,k}^{l+1} - x_{m,i,j,k}^{l+1}\right) - \mathcal{D}_{m2,i,j-1/2,k}^{l*}\left(x_{m,i,j,k}^{l+1} - x_{m,i,j-1,k}^{l+1}\right) \\
&\quad + \mathcal{D}_{m3,i,j,k+1/2}^{l*}\left(x_{m,i,j,k+1}^{l+1} - x_{m,i,j,k}^{l+1}\right) - \mathcal{D}_{m3,i,j,k-1/2}^{l*}\left(x_{m,i,j,k}^{l+1} - x_{m,i,j,k-1}^{l+1}\right) \\
&\quad + T_{m1,i+1/2,j,k}^{l*}\left(p_{i+1,j,k}^{l+1} - p_{i,j,k}^{l+1}\right) - T_{m1,i-1/2,j,k}^{l*}\left(p_{i,j,k}^{l+1} - p_{i-1,j,k}^{l+1}\right) \\
&\quad + T_{m2,i,j+1/2,k}^{l*}\left(p_{i,j+1,k}^{l+1} - p_{i,j,k}^{l+1}\right) - T_{m2,i,j-1/2,k}^{l*}\left(p_{i,j,k}^{l+1} - p_{i,j-1,k}^{l+1}\right) \\
&\quad + T_{m3,i,j,k+1/2}^{l*}\left(p_{i,j,k+1}^{l+1} - p_{i,j,k}^{l+1}\right) - T_{m3,i,j,k-1/2}^{l*}\left(p_{i,j,k}^{l+1} - p_{i,j,k-1}^{l+1}\right) \\
&\quad - (T_m\gamma)_{1,i+1/2,j,k}^{l*}\left(z_{i+1,j,k} - z_{i,j,k}\right) + (T_m\gamma)_{1,i-1/2,j,k}^{l*}\left(z_{i,j,k} - z_{i-1,j,k}\right) \\
&\quad - (T_m\gamma)_{2,i,j+1/2,k}^{l*}\left(z_{i,j+1,k} - z_{i,j,k}\right) + (T_m\gamma)_{2,i,j-1/2,k}^{l*}\left(z_{i,j,k} - z_{i,j-1,k}\right) \\
&\quad - (T_m\gamma)_{3,i,j,k+1/2}^{l*}\left(z_{i,j,k+1} - z_{i,j,k}\right) + (T_m\gamma)_{3,i,j,k-1/2}^{l*}\left(z_{i,j,k} - z_{i,j,k-1}\right) \\
&\quad + Q_{m,i,j,k}^{l+1},
\end{aligned}
\tag{7.46}
$$

where $\mathcal{D}_m^{l*} = \phi^l \xi^l \mathbf{D}_m^{l+1}$ and $\mathbf{T}_m^{l*} = x_m^{l+1}\xi^l \mathbf{k}/\mu^l$.

- Iterate on l until a tolerance in error is achieved.

In the above algorithm, the pressure system (7.45) at level $l + 1$ is linear because the compressibility and transmissibility terms have been computed before its solution. Also, if the diffusion/dispersion coefficients $\mathbf{D}_m$ are independent of the mole fractions, system (7.46) is also linear in each x_m^{l+1}, $m = 1, 2, \ldots, N_c - 1$. Finally, material balance should be checked for each component over each time step. For example, for a closed reservoir (no-flow boundary), the accumulation of mass must be equal to the net mass entering and leaving the boundary:

$$
\sum_{i,j,k}\left(V\frac{(\phi x_m \xi)^{n+1} - (\phi x_m \xi)^n}{\Delta t}\right)_{i,j,k} = \sum_{i,j,k} Q_{m,i,j,k}^{n+1}, \qquad m = 1, 2, \ldots, N_c.
$$

This material balance over the entire time interval is

$$
\sum_n \sum_{i,j,k}\left(V\left((\phi x_m \xi)^{n+1} - (\phi x_m \xi)^n\right)\right)_{i,j,k} = \sum_n \sum_{i,j,k} Q_{m,i,j,k}^{n+1}\Delta t^n
$$

for $m = 1, 2, \ldots, N_c$.

7.6 Nonisothermal Flow

In the previous sections we have not touched on temperature T, which was assumed to be given in space and time. For thermal methods used in reservoir simulation, an additional energy conservation equation is required to compute the temperature. For example, a simple version of this equation is the following heat convection-conduction equation:

$$\frac{\partial}{\partial t}(\rho_b c_b T) + \nabla \cdot \left(T \rho c_p \mathbf{u}\right) = \nabla \cdot (k_T \nabla T) + q_T, \tag{7.47}$$

where ρ_b, c_b, c_p, k_T, and q_T are, respectively, the bulk density, bulk specific heat capacity, heat capacity of the fluid at constant pressure, bulk thermal conductivity, and heat source/sink term. The bulk quantities can be defined, for example, as follows:

- Bulk density: $\rho_b = \rho\phi + \rho_s(1-\phi)$, where ρ_s is the solid density.
- Bulk specific heat capacity: $c_b = c_V^{\phi} c_s^{1-\phi}$, where c_s is the rock specific heat capacity and c_V is the fluid specific heat capacity at constant volume.
- Bulk thermal conductivity: $k_T = k_f^{\phi} k_s^{1-\phi}$, where k_f and k_s are the fluid and rock thermal conductivities, respectively.

Another method to define the bulk quantity $\rho_b c_b$ is (Chen, Huan, and Ma, 2006)

$$\rho_b c_b = \phi\rho c_V + (1-\phi)\rho_s c_s.$$

For nonisothermal flow, the computation of equation (7.47) must be included in the iterative algorithm developed in the previous section. More details will be presented in the numerical solution of differential equations for nonisothermal flow in Chapter 9.

7.7 Examples

Numerical benchmark examples have been presented for all the flows under consideration by Chen, Huan, and Ma (2006) except for the single-phase, multicomponent flow. In this section we present a couple of numerical examples for this type of flow.

7.7.1 Forced Convection

This is a forced convection system where we inject a heavier component into a fluid which consists entirely of a lighter component without diffusion. The physical data consist of the following: The domain Ω is a 2D homogeneous medium with a no-flow boundary condition, the permeability is $5.6e^{-4}$ darcy, the porosity is 0.38, and the relative density and viscosity are 0.02 and 0.35, respectively. The grid-point number is of order 100 in each horizontal direction. The left and right boundaries are given with a constant velocity, and the top and bottom boundaries are of no-flow type. Three concentration contours at three different times are illustrated in Figs. 7.8–7.10. The interfacial instabilities are of the form of lobe-shaped protuberances that manifested themselves near the lower edge of the plume. As they develop spatially and temporally, they penetrate deeper and deeper into the plume, resulting in considerable modification to the overall dispersion.

Figure 7.8. *Concentration for a forced convection system at one time.*

Figure 7.9. *Concentration for a forced convection system at a second time.*

Figure 7.10. *Concentration for a forced convection system at a third time.*

7.7.2 Forced Convection Plus Dispersion

This is now a forced convection system with dispersion. The Péclet number (characteristic speed and length divided by diffusion/dispersion) equals 5,000, and the longitudinal and transverse dispersion coefficients are, respectively, $0.031e^{-2}$ and $0.00217e^{-2}$. Three concentration contours at three different times are presented in Figs. 7.11–7.13.

There are two basic processes operating to transport chemical species. Diffusion is the process by which both ionic and molecular species dissolved in a fluid phase (e.g., water) move from areas of higher concentration (i.e., chemical activity) to areas of lower concentration. Advection is the process by which the moving fluid phase carries with it dissolved species. The process of dispersion acts to dilute the species and lower its concentration of movement so that it may not move as fast as the advection rate indicates.

Figure 7.11. *Concentration for forced convection/dispersion at one time.*

Figure 7.12. *Concentration for forced convection/dispersion at a second time.*

Figure 7.13. *Concentration for forced convection/dispersion at a third time.*

If the transverse dispersion is too low, the movement remains narrow and stable. On the other hand, if the value is too high, the movement spreads rapidly and again remains stable. It is only at some intermediate values between these two extremes that instability appears. For more numerical tests, the reader may refer to Chen, Zhou, and Carruthers (2006) and Chen et al. (2009).

Chapter 8

Compositional Flow and Numerical Solution

The black oil model presented in Chapter 6 involves three phases and three components, and the transport problem studied in Chapter 7 describes the flow of multicomponents in a fluid phase. Here we consider *compositional flow* that involves multicomponents and three phases, and there is *mass transfer* between the hydrocarbon phases (i.e., the vapor and liquid phases). In a *compositional model,* a finite number of hydrocarbon components are used to represent the composition of reservoir fluids. These components associate as phases in a reservoir. We describe the model under the assumptions that the flow process is isothermal (i.e., constant temperature), the components form at most three phases (e.g., vapor, liquid, and water), and there is no mass interchange between the water phase and the hydrocarbon phases. Furthermore, the diffusion/dispersion effect is neglected. We could state a general compositional model that involves any number of phases and components, each of which may exist in any or all of these phases. While the governing differential equations for this type of model are easy to set up, they are extremely complex to solve. Therefore, we describe the compositional model that has been widely used in the petroleum industry.

8.1 Basic Differential Equations

8.1.1 Mass Conservation and Darcy's Law

Let ξ_{mo} and ξ_{mg} be the *molar densities* of component m in the liquid (e.g., oil) and vapor (e.g., gas) phases, respectively, $m = 1, 2, \ldots, N_c$, where N_c is the number of components. The *molar density* of phase α is

$$\xi_\alpha = \sum_{m=1}^{N_c} \xi_{m\alpha}, \quad \alpha = o,\ g. \tag{8.1}$$

The mole fraction of component m in phase α is

$$x_{m\alpha} = n_{m\alpha}/n_\alpha, \quad m = 1, 2, \ldots, N_c,\ \alpha = o, g, \tag{8.2}$$

where $n_{m\alpha}$ is the number of moles of component m in the α phase and n_α is the total moles of this phase (cf. Chapter 7).

Because of mass interchange between the phases, mass is not conserved within each phase; the total mass is conserved for each component:

$$\begin{aligned} &\frac{\partial(\phi\xi_w S_w)}{\partial t} + \nabla\cdot(\xi_w \mathbf{u}_w) = q_w, \\ &\frac{\partial(\phi[x_{mo}\xi_o S_o + x_{mg}\xi_g S_g])}{\partial t} + \nabla\cdot(x_{mo}\xi_o \mathbf{u}_o + x_{mg}\xi_g \mathbf{u}_g) = q_m \end{aligned} \tag{8.3}$$

for $m = 1, 2, \ldots, N_c$, where ξ_w is the molar density of water, and q_w and q_m are the molar flow rates of water and the mth component, respectively. In equation (8.3), the volumetric velocity $\mathbf{u}_\alpha$ is given by Darcy's law as in (6.4):

$$\mathbf{u}_\alpha = -\frac{k_{r\alpha}}{\mu_\alpha}\mathbf{k}(\nabla p_\alpha - \rho_\alpha \wp \nabla z), \qquad \alpha = w,\ o,\ g. \tag{8.4}$$

In addition to the differential equations (8.3) and (8.4), there are also algebraic constraints. The *mole fraction balance* implies that

$$\sum_{m=1}^{N_c} x_{mo} = 1, \quad \sum_{m=1}^{N_c} x_{mg} = 1. \tag{8.5}$$

In the transport process, the porous medium is saturated with fluids:

$$S_w + S_o + S_g = 1. \tag{8.6}$$

The phase pressures are related by capillary pressures:

$$p_{cow} = p_o - p_w, \qquad p_{cgo} = p_g - p_o. \tag{8.7}$$

These capillary pressures are assumed to be known functions of the saturations. The relative permeabilities $k_{r\alpha}$ are also assumed to be known in terms of the saturations, and the viscosities μ_α, molar densities ξ_α, and mass densities ρ_α are functions of their respective phase pressure and compositions, $\alpha = w, o, g$.

Note that there are more dependent variables than there are differential and algebraic relations combined; there are formally $2N_c + 9$ dependent variables: x_{mo}, x_{mg}, $\mathbf{u}_\alpha$, p_α, and S_α, $\alpha = w, o, g$, $m = 1, 2, \ldots, N_c$. It is then necessary to have $2N_c + 9$ independent relations to determine a solution of the system. Equations (8.3)–(8.7) provide $N_c + 9$ independent relations, differential or algebraic; the additional N_c relations are provided by the *equilibrium relations* that relate the numbers of moles.

Mass interchange between phases is characterized by the variation of mass distribution of each component in the vapor and liquid phases. As usual, these two phases are assumed to be in the *phase equilibrium state*. This is physically reasonable since the mass interchange between phases occurs much faster than the flow of porous media fluids. Consequently, the distribution of each hydrocarbon component into the two phases is subject to the condition of *stable thermodynamic equilibrium*, which is given by minimizing the *Gibbs free energy* of the compositional system (Bear, 1972; Chen, Qin, and Ewing, 2000):

$$f_{mo}(p_o, x_{1o}, x_{2o}, \ldots, x_{N_c o}) = f_{mg}(p_g, x_{1g}, x_{2g}, \ldots, x_{N_c g}), \tag{8.8}$$

where f_{mo} and f_{mg} are the fugacity functions of the mth component in the liquid and vapor phases, respectively, $m = 1, 2, \dots, N_c$.

8.1.2 Equations of State

Several mathematical techniques for handling the hydrocarbon behavior (the distribution of chemical components among phases) are available. The most common are based on (1) the *K-value approach*, (2) *equations of state* (EOS), and (3) a variety of empirical tables from experiments. In this section, we discuss the first two techniques.

Equilibrium K-values

The equilibrium *flash vaporization ratio* for each component is defined by

$$K_m = x_{mg}/x_{mo}, \qquad m = 1, 2, \dots, N_c, \tag{8.9}$$

where the quantity K_m is the *equilibrium K-value* of component m. At low pressure, these K-values are readily related to the mixture pressure and temperature (see an example in Section 9.1.4). In fact, they are easily estimated from the vapor pressure data of pure components. At high pressure, the K-values are functions of overall compositions. The introduction of the compositions into the K-value functions adds considerable complexity to the flash computation.

The Peng–Robinson EOS

While the K-value approach is easy to set up, it lacks generality and may result in inaccurate reservoir simulation. In recent years, EOS have been more widely employed because they produce more consistent compositions, densities, and molar volumes. The most famous EOS is the van der Waals EOS (Reid, Prausnitz, and Sherwood, 1977). Here we discuss three more accurate and reliable EOS: Peng–Robinson (PR), Redlich–Kwong (RK), and Redlich–Kwong–Soave (RKS).

The PR EOS was introduced for single-phase flow in Chapter 7, which is extended to multiphase flow here. The *mixing principle* for the PR EOS is

$$a_\alpha = \sum_{m=1}^{N_c} \sum_{l=1}^{N_c} x_{m\alpha} x_{l\alpha} (1 - \kappa_{ml}) \sqrt{a_m a_l},$$
$$b_\alpha = \sum_{m=1}^{N_c} x_{m\alpha} b_m, \qquad \alpha = o, g,$$

where κ_{ml} is a *binary interaction parameter* between components m and l, and a_m and b_m are empirical factors for pure component m. The factors a_m and b_m can be computed from

$$a_m = \Omega_{ma} \alpha_m \frac{R^2 T_{cm}^2}{p_{cm}}, \qquad b_m = \Omega_{mb} \frac{R\, T_{cm}}{p_{cm}},$$

where we recall that R is the *universal gas constant*, T is the temperature, T_{cm} and p_{cm} are the *critical temperature and pressure*, the EOS parameters Ω_{ma} and Ω_{mb} are given by

$$\begin{aligned}\Omega_{ma} &= 0.45724, \quad \Omega_{mb} = 0.077796,\\ \alpha_m &= \left(1 + \lambda_m\left[1 - \sqrt{T/T_{cm}}\right]\right)^2,\\ \lambda_m &= 0.37464 + 1.5423\omega_m - 0.26992\omega_m^2,\end{aligned}$$

and ω_m is the *acentric factor* of components m. The acentric factors roughly express the deviation of the shape of a molecule from a sphere (Reid, Prausnitz, and Sherwood, 1977). Define

$$A_\alpha = \frac{a_\alpha p_\alpha}{R^2T^2}, \quad B_\alpha = \frac{b_\alpha p_\alpha}{RT}, \qquad \alpha = o, g, \tag{8.10}$$

where the pressure p_α is given by the *PR two-parameter EOS* (Peng and Robinson, 1976)

$$p_\alpha = \frac{RT}{V_\alpha - b_\alpha} - \frac{a_\alpha(T)}{V_\alpha(V_\alpha + b_\alpha) + b_\alpha(V_\alpha - b_\alpha)}, \tag{8.11}$$

with V_α being the molar volume of phase α. Introduce the compressibility factor

$$Z_\alpha = \frac{p_\alpha V_\alpha}{RT}, \qquad \alpha = o, g. \tag{8.12}$$

Equation (8.11) can be expressed as a *cubic equation* in Z_α:

$$\begin{aligned}Z_\alpha^3 - (1 - B_\alpha)Z_\alpha^2 + (A_\alpha - 2B_\alpha - 3B_\alpha^2)Z_\alpha&\\ - (A_\alpha B_\alpha - B_\alpha^2 - B_\alpha^3) = 0.&\end{aligned} \tag{8.13}$$

This equation has three roots. When only one root is real, it is selected. If there are three real roots, the selection of the right one depends on the dominance of the liquid phase or the vapor phase (cf. Sections 7.3 and 8.3.4). Now, for $m = 1, 2, \ldots, N_c$ and $\alpha = o, g$, the *fugacity coefficient* of component m in a mixture can be obtained from

$$\begin{aligned}\ln \varphi_{m\alpha} = &\frac{b_m}{b_\alpha}(Z_\alpha - 1) - \ln(Z_\alpha - B_\alpha)\\ &- \frac{A_\alpha}{2\sqrt{2}B_\alpha}\left(\frac{2}{a_\alpha}\sum_{l=1}^{N_c} x_{l\alpha}(1 - \kappa_{ml})\sqrt{a_m a_l} - \frac{b_m}{b_\alpha}\right)\\ &\cdot \ln\left(\frac{Z_\alpha + (1 + \sqrt{2})B_\alpha}{Z_\alpha - (1 - \sqrt{2})B_\alpha}\right).\end{aligned} \tag{8.14}$$

The *fugacity* of component m is

$$f_{m\alpha} = p_\alpha x_{m\alpha}\varphi_{m\alpha}, \qquad m = 1, 2, \ldots, N_c, \ \alpha = o, g. \tag{8.15}$$

Finally, the distribution of each hydrocarbon component into the liquid and vapor phases is given by the thermodynamic equilibrium relation (8.8).

The Redlich–Kwong equation of state

The *RK two-parameter EOS* is given by

$$p_\alpha = \frac{RT}{V_\alpha - b_\alpha} - \frac{a_\alpha}{V_\alpha(V_\alpha + b_\alpha)}, \qquad \alpha = o, g. \tag{8.16}$$

With $Z_\alpha = p_\alpha V_\alpha/(RT)$, this equation can be written as the cubic equation

$$Z_\alpha^3 - Z_\alpha^2 + (A_\alpha - B_\alpha - B_\alpha^2)Z_\alpha - A_\alpha B_\alpha = 0, \qquad \alpha = o, g. \tag{8.17}$$

The correct choice of root can be made as in the PR two-parameter EOS. In the present case, the EOS parameters Ω_{ma}, Ω_{mb}, and α_m are

$$\Omega_{ma} = 0.42748, \qquad \Omega_{mb} = 0.08664, \qquad \alpha_m = T/T_{cm}.$$

All other quantities A_α, B_α, a_α, b_α, a_m, and b_m have the same definitions as in the PR EOS, $m = 1, 2, \ldots, N_c$, $\alpha = o, g$. The fugacity coefficient of component m in a mixture can be obtained from the equation

$$\begin{aligned} \ln \varphi_{m\alpha} &= \frac{b_m}{b_\alpha}(Z_\alpha - 1) - \ln(Z_\alpha - B_\alpha) \\ &- \frac{A_\alpha}{B_\alpha}\left(\frac{2}{a_\alpha}\sum_{l=1}^{N_c} x_{l\alpha}(1-\kappa_{ml})\sqrt{a_m a_l} - \frac{b_m}{b_\alpha}\right)\ln\left(\frac{Z_\alpha + B_\alpha}{Z_\alpha}\right). \end{aligned} \tag{8.18}$$

The Redlich–Kwong–Soave equation of state

The *Soave modification* of the RK EOS defines the EOS parameter α_m as

$$\alpha_m = \left(1 + \lambda_m\left[1 - \sqrt{T/T_{cm}}\right]\right)^2, \qquad m = 1, 2, \ldots, N_c,$$

where $\lambda_m = 0.48 + 1.574\omega_m - 0.176\omega_m^2$ and ω_m is the acentric factor for component m. The definitions of all other quantities and of the fugacity coefficients are the same as in the RK EOS. The PR EOS and RKS EOS have been extensively utilized in predicting EOR (enhanced oil recovery) phase behavior.

The method of volume translation is widely used for correcting volumetric deficiencies of the original PR and RKS (Peneloux, Rauzy, and Freze, 1982) equations. The method involves evaluating a linearly translated volume by adding a constant to the molar volume computed from the original EOS. Peneloux, Rauzy, and Freze (1982) showed that the volume shift modifies the fugacity of each component. This correction must be included in the fugacity formulas used for gradient computations.

8.2 Numerical Solution of Compositional Flow

The choice of a solution technique is crucial for a coupled system of partial differential equations. In Chapter 6, we discussed several solution techniques that are currently used in the numerical solution of the black oil model. These techniques include the iterative IMPES (or IMPEC), sequential, SS, and adaptive implicit techniques. They can be also employed for the numerical simulation of the compositional model. However, a typical compositional simulator includes about a dozen chemical components; the SS technique would be a very expensive technique for this type of flow, even with today's computing power. The iterative IMPEC and sequential techniques are widely used and are thus studied here. As an example, we develop iterative IMPEC for the compositional model in detail. An extension from this technique to the sequential technique can be carried out as in Chapter 6 for the black oil model.

8.2.1 Choice of Primary Variables

Equations (8.3)–(8.8) form a strongly coupled system of time-dependent, nonlinear differential equations and algebraic constraints. While there are $2N_c + 9$ equations for the same number of dependent variables, this system can be written in terms of $2N_c + 2$ primary variables, and other variables can be expressed as their functions. These primary variables must be carefully chosen so that the main physical properties inherent in the governing equations and constraints are preserved, the nonlinearity and coupling between the equations is weakened, and efficient numerical methods for the solution of the resulting system can be devised.

We use the *total mass variable* F of the hydrocarbon system (Nolen, 1973; Young and Stephenson, 1983),

$$F = \xi_o S_o + \xi_g S_g, \tag{8.19}$$

and the *mass fractions* of oil and gas in this system,

$$L = \frac{\xi_o S_o}{F}, \qquad V = \frac{\xi_g S_g}{F}. \tag{8.20}$$

Note that

$$L + V = 1.$$

Next, instead of exploiting the individual mole fractions, we use the *total mole fraction* of the components in the hydrocarbon system,

$$z_m = L x_{mo} + (1 - L) x_{mg}, \qquad m = 1, 2, \ldots, N_c. \tag{8.21}$$

Then we see, using (8.5), (8.19), and (8.20), that

$$\sum_{m=1}^{N_c} z_m = 1 \tag{8.22}$$

and

$$x_{mo}\xi_o S_o + x_{mg}\xi_g S_g = F z_m, \qquad m = 1, 2, \ldots, N_c. \tag{8.23}$$

Consequently, applying (8.4), the second equation in (8.3) becomes

$$\begin{aligned} &\frac{\partial(\phi F z_m)}{\partial t} - \nabla \cdot \left(\mathbf{k} \left[\frac{x_{mo}\xi_o k_{ro}}{\mu_o} (\nabla p_o - \gamma_o \nabla z) + \frac{x_{mg}\xi_g k_{rg}}{\mu_g} (\nabla p_g - \gamma_g \nabla z) \right] \right) \\ &\quad = x_{mo} q_o + x_{mg} q_g, \qquad m = 1, 2, \ldots, N_c, \end{aligned} \tag{8.24}$$

where $\gamma_\alpha = \rho_\alpha \wp$, $\alpha = o, g$. Adding the equations (8.24) over m and exploiting (8.5) and (8.22) gives

$$\frac{\partial(\phi F)}{\partial t} - \nabla \cdot \left(\mathbf{k} \left[\frac{\xi_o k_{ro}}{\mu_o} (\nabla p_o - \gamma_o \nabla z) + \frac{\xi_g k_{rg}}{\mu_g} \left(\nabla p_g - \gamma_g \nabla z \right) \right] \right) = q_o + q_g. \quad (8.25)$$

Equation (8.24) is the individual flow equation for the mth component (say, $m = 1, 2, \ldots, N_c - 1$), and equation (8.25) is the global hydrocarbon flow equation. Finally, the source/sink terms q_α can be defined as (cf. Chapter 4)

$$q_\alpha = \sum_{\nu=1}^{N_w} \sum_{l=1}^{M_{w\nu}} \frac{2\pi \bar{k} \Delta h}{\ln (r_e/r_w)} \frac{\xi_\alpha k_{r\alpha}}{\mu_\alpha} \bigg|_l^{(\nu)} \left(p_{bh}^{(\nu)} - p_\alpha - \gamma_\alpha (z_{bh}^{(\nu)} - z) \right) \delta \left(\mathbf{x} - \mathbf{x}_l^{(\nu)} \right), \quad (8.26)$$

where $\delta(\mathbf{x})$ is the Dirac delta function, N_w is the total number of wells, $M_{w\nu}$ is the total number of perforated zones of the νth well, $\Delta h_l^{(\nu)}$ and $\mathbf{x}_l^{(\nu)}$ are the segment length and central location of the lth perforated zone of the νth well, $p_{bh}^{(\nu)}$ is the *bottom hole pressure* at the datum level depth $z_{bh}^{(\nu)}$, $r_{l,e}^{(\nu)}$ is the equivalent well radius, $r_w^{(\nu)}$ is the radius of the νth well, and $\bar{k}$ is some average of $\mathbf{k}$ at the wells.

To simplify the differential equations further, we define the transmissibilities

$$\begin{aligned} \mathbf{T}_\alpha &= \frac{\xi_\alpha k_{r\alpha}}{\mu_\alpha} \mathbf{k}, & \alpha &= w, o, g, \\ \mathbf{T}_{m\alpha} &= \frac{x_{m\alpha} \xi_\alpha k_{r\alpha}}{\mu_\alpha} \mathbf{k}, & \alpha &= o, g, \ m = 1, 2, \ldots, N_c. \end{aligned} \quad (8.27)$$

We now summarize the equations needed in the sequential and iterative IMPEC solution approaches. The equilibrium relation (8.8) is recast as

$$\begin{aligned} f_{mo}(p_o, x_{1o}, x_{2o}, \ldots, x_{N_c o}) &= f_{mg}(p_o + p_{cg}, x_{1g}, x_{2g}, \ldots, x_{N_c g}), \\ & m = 1, 2, \ldots, N_c. \end{aligned} \quad (8.28)$$

Using (8.27), equation (8.24) becomes

$$\begin{aligned} \frac{\partial(\phi F z_m)}{\partial t} &= \nabla \cdot \left(\mathbf{T}_{mo} (\nabla p_o - \gamma_o \nabla z) + \mathbf{T}_{mg} \left(\nabla p_g - \gamma_g \nabla z \right) \right) \\ &\quad + x_{mo} q_o + x_{mg} q_g, \quad m = 1, 2, \ldots, N_c - 1. \end{aligned} \quad (8.29)$$

Similarly, it follows from (8.25) that

$$\frac{\partial(\phi F)}{\partial t} = \nabla \cdot \left(\mathbf{T}_o (\nabla p_o - \gamma_o \nabla z) + \mathbf{T}_g \left(\nabla p_g - \gamma_g \nabla z \right) \right) + q_o + q_g. \quad (8.30)$$

Next, applying the first equation of (8.3) and (8.27) yields

$$\frac{\partial(\phi \xi_w S_w)}{\partial t} = \nabla \cdot (\mathbf{T}_w (\nabla p_w - \gamma_w \nabla z)) + q_w. \quad (8.31)$$

Finally, using (8.19) and (8.20), the saturation state equation (8.6) becomes

$$F \left(\frac{L}{\xi_o} + \frac{1 - L}{\xi_g} \right) + S = 1. \quad (8.32)$$

The differential system consists of the $2N_c + 2$ equations (8.28)–(8.32) for the $2N_c + 2$ primary unknowns: x_{mo} (or x_{mg}), L (or V), z_m, F, $S = S_w$, and $p = p_o$, $m = 1, 2, \ldots, N_c - 1$. For rate-specified wells, equation (8.26) can be used to find p_{bh}.

8.2.2 Finite Difference Equations

Let the permeability tensor $\mathbf{k}$ be diagonal, $\mathbf{k} = \mathrm{diag}(k_{11}, k_{22}, k_{33})$, and let $n > 0$ (an integer) indicate a time step. A time approximation at the $(n+1)$th level for the system of equations (8.28)–(8.32) is

$$f_{mo}(p_o^{n+1}, x_{1o}^{n+1}, x_{2o}^{n+1}, \ldots, x_{N_c o}^{n+1}) \\ = f_{mg}(p_g^{n+1}, x_{1g}^{n+1}, x_{2g}^{n+1}, \ldots, x_{N_c g}^{n+1}), \quad m = 1, 2, \ldots, N_c, \tag{8.33}$$

$$\begin{aligned}
&\frac{1}{\Delta t}\left(V\left[(\phi F z_m)^{n+1} - (\phi F z_m)^n\right]\right)_{i,j,k} \\
&= T_{mo1,i+1/2,j,k}^{n+1}\left(p_{o,i+1,j,k}^{n+1} - p_{o,i,j,k}^{n+1}\right) - T_{mo1,i-1/2,j,k}^{n+1}\left(p_{o,i,j,k}^{n+1} - p_{o,i-1,j,k}^{n+1}\right) \\
&\quad + T_{mo2,i,j+1/2,k}^{n+1}\left(p_{o,i,j+1,k}^{n+1} - p_{o,i,j,k}^{n+1}\right) - T_{mo2,i,j-1/2,k}^{n+1}\left(p_{o,i,j,k}^{n+1} - p_{o,i,j-1,k}^{n+1}\right) \\
&\quad + T_{mo3,i,j,k+1/2}^{n+1}\left(p_{o,i,j,k+1}^{n+1} - p_{o,i,j,k}^{n+1}\right) - T_{mo3,i,j,k-1/2}^{n+1}\left(p_{o,i,j,k}^{n+1} - p_{o,i,j,k-1}^{n+1}\right) \\
&\quad - (T_{mo}\gamma_o)_{1,i+1/2,j,k}^{n+1}\left(z_{i+1,j,k} - z_{i,j,k}\right) + (T_{mo}\gamma_o)_{1,i-1/2,j,k}^{n+1}\left(z_{i,j,k} - z_{i-1,j,k}\right) \\
&\quad - (T_{mo}\gamma_o)_{2,i,j+1/2,k}^{n+1}\left(z_{i,j+1,k} - z_{i,j,k}\right) + (T_{mo}\gamma_o)_{2,i,j-1/2,k}^{n+1}\left(z_{i,j,k} - z_{i,j-1,k}\right) \\
&\quad - (T_{mo}\gamma_o)_{3,i,j,k+1/2}^{n+1}\left(z_{i,j,k+1} - z_{i,j,k}\right) + (T_{mo}\gamma_o)_{3,i,j,k-1/2}^{n+1}\left(z_{i,j,k} - z_{i,j,k-1}\right) \\
&\quad + T_{mg1,i+1/2,j,k}^{n+1}\left(p_{g,i+1,j,k}^{n+1} - p_{g,i,j,k}^{n+1}\right) - T_{mg1,i-1/2,j,k}^{n+1}\left(p_{g,i,j,k}^{n+1} - p_{g,i-1,j,k}^{n+1}\right) \\
&\quad + T_{mg2,i,j+1/2,k}^{n+1}\left(p_{g,i,j+1,k}^{n+1} - p_{g,i,j,k}^{n+1}\right) - T_{mg2,i,j-1/2,k}^{n+1}\left(p_{g,i,j,k}^{n+1} - p_{g,i,j-1,k}^{n+1}\right) \\
&\quad + T_{mg3,i,j,k+1/2}^{n+1}\left(p_{g,i,j,k+1}^{n+1} - p_{g,i,j,k}^{n+1}\right) - T_{mg3,i,j,k-1/2}^{n+1}\left(p_{g,i,j,k}^{n+1} - p_{g,i,j,k-1}^{n+1}\right) \\
&\quad - (T_{mg}\gamma_g)_{1,i+1/2,j,k}^{n+1}\left(z_{i+1,j,k} - z_{i,j,k}\right) + \left(T_{mg}\gamma_g\right)_{1,i-1/2,j,k}^{n+1}\left(z_{i,j,k} - z_{i-1,j,k}\right) \\
&\quad - \left(T_{mg}\gamma_g\right)_{2,i,j+1/2,k}^{n+1}\left(z_{i,j+1,k} - z_{i,j,k}\right) + \left(T_{mg}\gamma_g\right)_{2,i,j-1/2,k}^{n+1}\left(z_{i,j,k} - z_{i,j-1,k}\right) \\
&\quad - \left(T_{mg}\gamma_g\right)_{3,i,j,k+1/2}^{n+1}\left(z_{i,j,k+1} - z_{i,j,k}\right) + \left(T_{mg}\gamma_g\right)_{3,i,j,k-1/2}^{n+1}\left(z_{i,j,k} - z_{i,j,k-1}\right) \\
&\quad + (x_{mo}Q_o)_{i,j,k}^{n+1} + (x_{mg}Q_g)_{i,j,k}^{n+1}, \quad m = 1, 2, \ldots, N_c - 1,
\end{aligned} \tag{8.34}$$

$$\begin{aligned}
&\frac{1}{\Delta t}\left(V\left[(\phi F)^{n+1} - (\phi F)^n\right]\right)_{i,j,k} \\
&= T_{o1,i+1/2,j,k}^{n+1}\left(p_{o,i+1,j,k}^{n+1} - p_{o,i,j,k}^{n+1}\right) - T_{o1,i-1/2,j,k}^{n+1}\left(p_{o,i,j,k}^{n+1} - p_{o,i-1,j,k}^{n+1}\right) \\
&\quad + T_{o2,i,j+1/2,k}^{n+1}\left(p_{o,i,j+1,k}^{n+1} - p_{o,i,j,k}^{n+1}\right) - T_{o2,i,j-1/2,k}^{n+1}\left(p_{o,i,j,k}^{n+1} - p_{o,i,j-1,k}^{n+1}\right) \\
&\quad + T_{o3,i,j,k+1/2}^{n+1}\left(p_{o,i,j,k+1}^{n+1} - p_{o,i,j,k}^{n+1}\right) - T_{o3,i,j,k-1/2}^{n+1}\left(p_{o,i,j,k}^{n+1} - p_{o,i,j,k-1}^{n+1}\right) \\
&\quad - (T_o\gamma_o)_{1,i+1/2,j,k}^{n+1}\left(z_{i+1,j,k} - z_{i,j,k}\right) + (T_o\gamma_o)_{1,i-1/2,j,k}^{n+1}\left(z_{i,j,k} - z_{i-1,j,k}\right) \\
&\quad - (T_o\gamma_o)_{2,i,j+1/2,k}^{n+1}\left(z_{i,j+1,k} - z_{i,j,k}\right) + (T_o\gamma_o)_{2,i,j-1/2,k}^{n+1}\left(z_{i,j,k} - z_{i,j-1,k}\right) \\
&\quad - (T_o\gamma_o)_{3,i,j,k+1/2}^{n+1}\left(z_{i,j,k+1} - z_{i,j,k}\right) + (T_o\gamma_o)_{3,i,j,k-1/2}^{n+1}\left(z_{i,j,k} - z_{i,j,k-1}\right) \\
&\quad + T_{g1,i+1/2,j,k}^{n+1}\left(p_{g,i+1,j,k}^{n+1} - p_{g,i,j,k}^{n+1}\right) - T_{g1,i-1/2,j,k}^{n+1}\left(p_{g,i,j,k}^{n+1} - p_{g,i-1,j,k}^{n+1}\right)
\end{aligned}$$

$$
\begin{aligned}
&+ T^{n+1}_{g2,i,j+1/2,k}\left(p^{n+1}_{g,i,j+1,k} - p^{n+1}_{g,i,j,k}\right) - T^{n+1}_{g2,i,j-1/2,k}\left(p^{n+1}_{g,i,j,k} - p^{n+1}_{g,i,j-1,k}\right) \\
&+ T^{n+1}_{g3,i,j,k+1/2}\left(p^{n+1}_{g,i,j,k+1} - p^{n+1}_{g,i,j,k}\right) - T^{n+1}_{g3,i,j,k-1/2}\left(p^{n+1}_{g,i,j,k} - p^{n+1}_{g,i,j,k-1}\right) \qquad (8.35) \\
&- (T_g\gamma_g)^{n+1}_{1,i+1/2,j,k}\left(z_{i+1,j,k} - z_{i,j,k}\right) + (T_g\gamma_g)^{n+1}_{1,i-1/2,j,k}\left(z_{i,j,k} - z_{i-1,j,k}\right) \\
&- (T_g\gamma_g)^{n+1}_{2,i,j+1/2,k}\left(z_{i,j+1,k} - z_{i,j,k}\right) + (T_g\gamma_g)^{n+1}_{2,i,j-1/2,k}\left(z_{i,j,k} - z_{i,j-1,k}\right) \\
&- (T_g\gamma_g)^{n+1}_{3,i,j,k+1/2}\left(z_{i,j,k+1} - z_{i,j,k}\right) + (T_g\gamma_g)^{n+1}_{3,i,j,k-1/2}\left(z_{i,j,k} - z_{i,j,k-1}\right) \\
&+ Q^{n+1}_{o,i,j,k} + Q^{n+1}_{g,i,j,k},
\end{aligned}
$$

$$
\begin{aligned}
&\frac{1}{\Delta t}\left(V\left[(\phi\xi_w S_w)^{n+1} - (\phi\xi_w S_w)^{n}\right]\right)_{i,j,k} \\
&= T^{n+1}_{w1,i+1/2,j,k}\left(p^{n+1}_{w,i+1,j,k} - p^{n+1}_{w,i,j,k}\right) - T^{n+1}_{w1,i-1/2,j,k}\left(p^{n+1}_{w,i,j,k} - p^{n+1}_{w,i-1,j,k}\right) \\
&+ T^{n+1}_{w2,i,j+1/2,k}\left(p^{n+1}_{w,i,j+1,k} - p^{n+1}_{w,i,j,k}\right) - T^{n+1}_{w2,i,j-1/2,k}\left(p^{n+1}_{w,i,j,k} - p^{n+1}_{w,i,j-1,k}\right) \\
&+ T^{n+1}_{w3,i,j,k+1/2}\left(p^{n+1}_{w,i,j,k+1} - p^{n+1}_{w,i,j,k}\right) - T^{n+1}_{w3,i,j,k-1/2}\left(p^{n+1}_{w,i,j,k} - p^{n+1}_{w,i,j,k-1}\right) \qquad (8.36) \\
&- (T_w\gamma_w)^{n+1}_{1,i+1/2,j,k}\left(z_{i+1,j,k} - z_{i,j,k}\right) + (T_w\gamma_w)^{n+1}_{1,i-1/2,j,k}\left(z_{i,j,k} - z_{i-1,j,k}\right) \\
&- (T_w\gamma_w)^{n+1}_{2,i,j+1/2,k}\left(z_{i,j+1,k} - z_{i,j,k}\right) + (T_w\gamma_w)^{n+1}_{2,i,j-1/2,k}\left(z_{i,j,k} - z_{i,j-1,k}\right) \\
&- (T_w\gamma_w)^{n+1}_{3,i,j,k+1/2}\left(z_{i,j,k+1} - z_{i,j,k}\right) + (T_w\gamma_w)^{n+1}_{3,i,j,k-1/2}\left(z_{i,j,k} - z_{i,j,k-1}\right) \\
&+ Q^{n+1}_{w,i,j,k},
\end{aligned}
$$

and

$$
\left[F\left(\frac{L}{\xi_o} + \frac{1-L}{\xi_g}\right) + S\right]^{n+1} = 1, \qquad (8.37)
$$

where $Q_{\alpha,i,j,k} = (Vq_\alpha)_{i,j,k}, \alpha = w, o, g$, and the numerical transmissibilities at the gridblock boundaries

$$
\frac{A_i T_{\alpha i}}{h_i}, \quad \frac{A_i T_{m\alpha i}}{h_i}, \qquad i = 1, 2, 3, \ \alpha = w, o, g, \ m = 1, 2, \ldots, N_c - 1,
$$

are still indicated by $T_{\alpha i}$ and $T_{m\alpha i}$, respectively, where A_i is the cross-sectional area normal to the x_i-direction.

The transmissibility terms in equations (8.33)–(8.37) can be treated as in the black oil. That is, the rock property (i.e., absolute permeability), fluid properties (i.e., viscosities and formation volume factors), and rock/fluid properties (i.e., relative permeabilities and capillary pressures) at internal boundaries of gridblocks should be evaluated using the harmonic averaging, (weighted) arithmetic averaging, and upstream weighting techniques, respectively.

In iterative IMPEC, all the saturation functions k_{rw}, k_{ro}, k_{rg}, p_{cow}, and p_{cgo} are evaluated at the saturation values of the previous time step in a Newton–Raphson iteration, and the fluid formation volume factors and viscosities in the transmissibilities, phase potentials, and well terms are computed using the previous Newton–Raphson iteration values.

Thus we define

$$
\begin{aligned}
p_w^{(l+1)\star} &= p^{l+1} - p_{cow}(S_w^n), \quad p_g^{(l+1)\star} = p^{l+1} + p_{cgo}(S_g^n),\\
\mathbf{T}_\alpha^{(l+1)\star} &= \frac{\xi^l k_{r\alpha}^n}{\mu_\alpha^l}\mathbf{k}, \qquad \alpha = w, o, g,\\
\mathbf{T}_{m\alpha}^{(l+1)\star} &= \frac{x_{m\alpha}^l \xi^l k_{r\alpha}^n}{\mu_\alpha^l}\mathbf{k}, \qquad \alpha = o, g, \quad m = 1, 2, \ldots, N_c.
\end{aligned}
\tag{8.38}
$$

To apply the Newton–Raphson iteration procedure, we must identify the dominance of the gas or oil phase. If the gas phase dominates in the hydrocarbon system (e.g., $L < 0.5$), the primary unknowns will be x_{io}, L, z_i, F, S, and $p, i = 1, 2, \ldots, N_c - 1$, which is the $L - X$ *iteration type* in compositional modeling. If the oil phase dominates (e.g., $L \geq 0.5$), the primary unknowns will be x_{ig}, V, z_i, F, S, and $p, i = 1, 2, \ldots, N_c - 1$, which corresponds to the $V - Y$ *iteration type*. As an example, we use the Newton–Raphson iteration in terms of $\delta x_{io}, \delta L, \delta z_i, \delta F, \delta S$, and $\delta p, i = 1, 2, \ldots, N_c - 1$; a similar calculation can be performed for the $V - Y$ iteration type.

At the Newton–Raphson iteration level $l + 1$, the primary variables are updated as follows:

$$
\begin{aligned}
x_{mo}^{n+1,l+1} &= x_{mo}^{n+1,l} + \delta x_{mo}^{n+1,l+1}, & m &= 1, 2, \ldots, N_c - 1,\\
z_m^{n+1,l+1} &= z_m^{n+1,l} + \delta z_m^{n+1,l+1}, & m &= 1, 2, \ldots, N_c - 1,\\
L^{n+1,l+1} &= L^{n+1,l} + \delta L^{n+1,l+1}, & F^{n+1,l+1} &= F^{n+1,l} + \delta F^{n+1,l+1},\\
S^{n+1,l+1} &= S^{n+1,l} + \delta S^{n+1,l+1}, & p^{n+1,l+1} &= p^{n+1,l} + \delta p^{n+1,l+1}.
\end{aligned}
$$

Below, the superscript $n + 1$ is dropped. Then application of the Newton–Raphson iteration procedure to equations (8.33)–(8.37) gives the following residuals:

$$
\begin{aligned}
R_{fm,i,j,k}^l = f_{mo}(p_o^l, x_{1o}^l, x_{2o}^l, \ldots, x_{N_c o}^l) - f_{mg}(p_g^{l\star}, x_{1g}^l, x_{2g}^l, \ldots, x_{N_c g}^l),\\
m = 1, 2, \ldots, N_c,
\end{aligned}
\tag{8.39}
$$

$$
\begin{aligned}
R_{z_m,i,j,k}^l = {} & \frac{1}{\Delta t}\left(V\left[(\phi F z_m)^l - (\phi F z_m)^n\right]\right)_{i,j,k}\\
& - T_{mo1,i+1/2,j,k}^{l\star}\left(p_{o,i+1,j,k}^l - p_{o,i,j,k}^l\right) + T_{mo1,i-1/2,j,k}^{l\star}\left(p_{o,i,j,k}^l - p_{o,i-1,j,k}^l\right)\\
& - T_{mo2,i,j+1/2,k}^{l\star}\left(p_{o,i,j+1,k}^l - p_{o,i,j,k}^l\right) + T_{mo2,i,j-1/2,k}^{l\star}\left(p_{o,i,j,k}^l - p_{o,i,j-1,k}^l\right)\\
& - T_{mo3,i,j,k+1/2}^{l\star}\left(p_{o,i,j,k+1}^l - p_{o,i,j,k}^l\right) + T_{mo3,i,j,k-1/2}^{l\star}\left(p_{o,i,j,k}^l - p_{o,i,j,k-1}^l\right)\\
& + (T_{mo}\gamma_o)_{1,i+1/2,j,k}^{l\star}\left(z_{i+1,j,k} - z_{i,j,k}\right) - (T_{mo}\gamma_o)_{1,i-1/2,j,k}^{l\star}\left(z_{i,j,k} - z_{i-1,j,k}\right)\\
& + (T_{mo}\gamma_o)_{2,i,j+1/2,k}^{l\star}\left(z_{i,j+1,k} - z_{i,j,k}\right) - (T_{mo}\gamma_o)_{2,i,j-1/2,k}^{l\star}\left(z_{i,j,k} - z_{i,j-1,k}\right)\\
& + (T_{mo}\gamma_o)_{3,i,j,k+1/2}^{l\star}\left(z_{i,j,k+1} - z_{i,j,k}\right) - (T_{mo}\gamma_o)_{3,i,j,k-1/2}^{l\star}\left(z_{i,j,k} - z_{i,j,k-1}\right)\\
& - T_{mg1,i+1/2,j,k}^{l\star}\left(p_{g,i+1,j,k}^{l\star} - p_{g,i,j,k}^{l\star}\right) + T_{mg1,i-1/2,j,k}^{l\star}\left(p_{g,i,j,k}^{l\star} - p_{g,i-1,j,k}^{l\star}\right)\\
& - T_{mg2,i,j+1/2,k}^{l\star}\left(p_{g,i,j+1,k}^{l\star} - p_{g,i,j,k}^{l\star}\right) + T_{mg2,i,j-1/2,k}^{l\star}\left(p_{g,i,j,k}^{l\star} - p_{g,i,j-1,k}^{l\star}\right)\\
& - T_{mg3,i,j,k+1/2}^{l\star}\left(p_{g,i,j,k+1}^{l\star} - p_{g,i,j,k}^{l\star}\right) + T_{mg3,i,j,k-1/2}^{l\star}\left(p_{g,i,j,k}^{l\star} - p_{g,i,j,k-1}^{l\star}\right)\\
& + (T_{mg}\gamma_g)_{1,i+1/2,j,k}^{l\star}\left(z_{i+1,j,k} - z_{i,j,k}\right) - (T_{mg}\gamma_g)_{1,i-1/2,j,k}^{l\star}\left(z_{i,j,k} - z_{i-1,j,k}\right)
\end{aligned}
$$

$$
\begin{aligned}
&+ \left(T_{mg}\gamma_g\right)^{l\star}_{2,i,j+1/2,k} \left(z_{i,j+1,k} - z_{i,j,k}\right) - \left(T_{mg}\gamma_g\right)^{l\star}_{2,i,j-1/2,k} \left(z_{i,j,k} - z_{i,j-1,k}\right) \\
&+ \left(T_{mg}\gamma_g\right)^{l\star}_{3,i,j,k+1/2} \left(z_{i,j,k+1} - z_{i,j,k}\right) - \left(T_{mg}\gamma_g\right)^{l\star}_{3,i,j,k-1/2} \left(z_{i,j,k} - z_{i,j,k-1}\right) \\
&- (x_{mo}Q_o)^l_{i,j,k} - (x_{mg}Q_g)^l_{i,j,k}, \quad m = 1, 2, \ldots, N_c - 1,
\end{aligned}
\tag{8.40}
$$

$$
\begin{aligned}
R^l_{F,i,j,k} = &\frac{1}{\Delta t} \left(V\left[(\phi F)^l - (\phi F)^n\right]\right)_{i,j,k} \\
&- T^{l\star}_{o1,i+1/2,j,k} \left(p^l_{o,i+1,j,k} - p^l_{o,i,j,k}\right) + T^{l\star}_{o1,i-1/2,j,k} \left(p^l_{o,i,j,k} - p^l_{o,i-1,j,k}\right) \\
&- T^{l\star}_{o2,i,j+1/2,k} \left(p^l_{o,i,j+1,k} - p^l_{o,i,j,k}\right) + T^{l\star}_{o2,i,j-1/2,k} \left(p^l_{o,i,j,k} - p^l_{o,i,j-1,k}\right) \\
&- T^{l\star}_{o3,i,j,k+1/2} \left(p^l_{o,i,j,k+1} - p^l_{o,i,j,k}\right) + T^{l\star}_{o3,i,j,k-1/2} \left(p^l_{o,i,j,k} - p^l_{o,i,j,k-1}\right) \\
&+ (T_o\gamma_o)^{l\star}_{1,i+1/2,j,k} \left(z_{i+1,j,k} - z_{i,j,k}\right) - (T_o\gamma_o)^{l\star}_{1,i-1/2,j,k} \left(z_{i,j,k} - z_{i-1,j,k}\right) \\
&+ (T_o\gamma_o)^{l\star}_{2,i,j+1/2,k} \left(z_{i,j+1,k} - z_{i,j,k}\right) - (T_o\gamma_o)^{l\star}_{2,i,j-1/2,k} \left(z_{i,j,k} - z_{i,j-1,k}\right) \\
&+ (T_o\gamma_o)^{l\star}_{3,i,j,k+1/2} \left(z_{i,j,k+1} - z_{i,j,k}\right) - (T_o\gamma_o)^{l\star}_{3,i,j,k-1/2} \left(z_{i,j,k} - z_{i,j,k-1}\right) \\
&- T^{l\star}_{g1,i+1/2,j,k} \left(p^{l\star}_{g,i+1,j,k} - p^{l\star}_{g,i,j,k}\right) + T^{l\star}_{g1,i-1/2,j,k} \left(p^{l\star}_{g,i,j,k} - p^{l\star}_{g,i-1,j,k}\right) \\
&- T^{l\star}_{g2,i,j+1/2,k} \left(p^{l\star}_{g,i,j+1,k} - p^{l\star}_{g,i,j,k}\right) + T^{l\star}_{g2,i,j-1/2,k} \left(p^{l\star}_{g,i,j,k} - p^{l\star}_{g,i,j-1,k}\right) \\
&- T^{l\star}_{g3,i,j,k+1/2} \left(p^{l\star}_{g,i,j,k+1} - p^{l\star}_{g,i,j,k}\right) + T^{l\star}_{g3,i,j,k-1/2} \left(p^{l\star}_{g,i,j,k} - p^{l\star}_{g,i,j,k-1}\right) \\
&+ (T_g\gamma_g)^{l\star}_{1,i+1/2,j,k} \left(z_{i+1,j,k} - z_{i,j,k}\right) - \left(T_g\gamma_g\right)^{l\star}_{1,i-1/2,j,k} \left(z_{i,j,k} - z_{i-1,j,k}\right) \\
&+ \left(T_g\gamma_g\right)^{l\star}_{2,i,j+1/2,k} \left(z_{i,j+1,k} - z_{i,j,k}\right) - \left(T_g\gamma_g\right)^{l\star}_{2,i,j-1/2,k} \left(z_{i,j,k} - z_{i,j-1,k}\right) \\
&+ \left(T_g\gamma_g\right)^{l\star}_{3,i,j,k+1/2} \left(z_{i,j,k+1} - z_{i,j,k}\right) - \left(T_g\gamma_g\right)^{l\star}_{3,i,j,k-1/2} \left(z_{i,j,k} - z_{i,j,k-1}\right) \\
&- Q^l_{o,i,j,k} - Q^l_{g,i,j,k},
\end{aligned}
\tag{8.41}
$$

$$
\begin{aligned}
R^l_{w,i,j,k} = &\frac{1}{\Delta t} \left(V\left[(\phi \xi_w S_w)^l - (\phi \xi_w S_w)^n\right]\right)_{i,j,k} \\
&- T^{l\star}_{w1,i+1/2,j,k} \left(p^{l\star}_{w,i+1,j,k} - p^{l\star}_{w,i,j,k}\right) + T^{l\star}_{w1,i-1/2,j,k} \left(p^{l\star}_{w,i,j,k} - p^{l\star}_{w,i-1,j,k}\right) \\
&- T^{l\star}_{w2,i,j+1/2,k} \left(p^{l\star}_{w,i,j+1,k} - p^{l\star}_{w,i,j,k}\right) + T^{l\star}_{w2,i,j-1/2,k} \left(p^{l\star}_{w,i,j,k} - p^{l\star}_{w,i,j-1,k}\right) \\
&- T^{l\star}_{w3,i,j,k+1/2} \left(p^{l\star}_{w,i,j,k+1} - p^{l\star}_{w,i,j,k}\right) + T^{l\star}_{w3,i,j,k-1/2} \left(p^{l\star}_{w,i,j,k} - p^{l\star}_{w,i,j,k-1}\right) \\
&+ (T_w\gamma_w)^{l\star}_{1,i+1/2,j,k} \left(z_{i+1,j,k} - z_{i,j,k}\right) - (T_w\gamma_w)^{l\star}_{1,i-1/2,j,k} \left(z_{i,j,k} - z_{i-1,j,k}\right) \\
&+ (T_w\gamma_w)^{l\star}_{2,i,j+1/2,k} \left(z_{i,j+1,k} - z_{i,j,k}\right) - (T_w\gamma_w)^{l\star}_{2,i,j-1/2,k} \left(z_{i,j,k} - z_{i,j-1,k}\right) \\
&+ (T_w\gamma_w)^{l\star}_{3,i,j,k+1/2} \left(z_{i,j,k+1} - z_{i,j,k}\right) - (T_w\gamma_w)^{l\star}_{3,i,j,k-1/2} \left(z_{i,j,k} - z_{i,j,k-1}\right) \\
&- Q^l_{w,i,j,k},
\end{aligned}
\tag{8.42}
$$

and

$$
R^l_{p,i,j,k} = \left[F\left(\frac{L}{\xi_o} + \frac{1-L}{\xi_g}\right) + S\right]^l - 1. \tag{8.43}
$$

To find the derivatives of the residuals in the primary variables, we replace the derivatives in x_{ig} by those in these primary variables, $i = 1, 2, \ldots, N_c$. Applying relation (8.21), we see that

$$\frac{\partial x_{ig}}{\partial x_{io}} = \frac{L}{L-1}, \qquad \frac{\partial x_{ig}}{\partial z_i} = \frac{1}{1-L},$$
$$\frac{\partial x_{ig}}{\partial L} = \frac{x_{io} - x_{ig}}{L-1}, \qquad i = 1, 2, \ldots, N_c.$$

Consequently, the chain rule implies

$$\frac{\partial}{\partial x_{io}} = \frac{\partial x_{ig}}{\partial x_{io}} \frac{\partial}{\partial x_{ig}} = \frac{L}{L-1} \frac{\partial}{\partial x_{ig}},$$
$$\frac{\partial}{\partial z_i} = \frac{\partial x_{ig}}{\partial z_i} \frac{\partial}{\partial x_{ig}} = \frac{1}{1-L} \frac{\partial}{\partial x_{ig}},$$
$$\frac{\partial}{\partial L} = \frac{\partial x_{ig}}{\partial L} \frac{\partial}{\partial x_{ig}} = \frac{x_{io} - x_{ig}}{L-1} \frac{\partial}{\partial x_{ig}}.$$

The following relation obtained from (8.12) is also needed for evaluating the derivatives of the residuals:

$$\frac{1}{\xi_\alpha} = \frac{Z_\alpha(p_\alpha, x_{1\alpha}, x_{2\alpha}, \ldots, x_{N_c\alpha}) R\, T}{p_\alpha}, \qquad \alpha = o, g.$$

Finally, equations (8.5) and (8.22) are required to eliminate $x_{N_c o}$ and z_{N_c}.

With all these, application of the Newton–Raphson iteration procedure to equations (8.33)–(8.37) gives the following residual equations:

$$\begin{aligned}
&\sum_{r=1}^{N_c-1} \frac{\partial R^l_{fm,i,j,k}}{\partial (x_{ro})_{i,j,k}} \delta(x_{ro})^{l+1}_{i,j,k} + \frac{\partial R^l_{fm,i,j,k}}{\partial L_{i,j,k}} \delta L^{l+1}_{i,j,k} \\
&\quad + \frac{\partial R^l_{fm,i,j,k}}{\partial p_{i,j,k}} \delta p^{l+1}_{i,j,k} + \sum_{r=1}^{N_c-1} \frac{\partial R^l_{fm,i,j,k}}{\partial (z_r)_{i,j,k}} \delta (z_r)^{l+1}_{i,j,k} \\
&\quad = -R^l_{fm,i,j,k}, \qquad m = 1, 2, \ldots, N_c.
\end{aligned} \tag{8.44}$$

The linear system (8.44) is used to solve for $(\delta x_{1o}, \delta x_{2o}, \ldots, \delta x_{(N_c-1)o}, \delta L)$ in terms of $(\delta z_1, \delta z_2, \ldots, \delta z_{N_c-1}, \delta p)$.

$$\begin{aligned}
&\frac{\partial R^l_{z_m,i,j,k}}{\partial p_{i,j,k-1}} \delta p^{l+1}_{i,j,k-1} + \frac{\partial R^l_{z_m,i,j,k}}{\partial p_{i,j-1,k}} \delta p^{l+1}_{i,j-1,k} + \frac{\partial R^l_{z_m,i,,j,k}}{\partial p_{i-1,j,k}} \delta p^{l+1}_{i-1,j,k} \\
&\quad + \frac{\partial R^l_{z_m,i,j,k}}{\partial p_{i,j,k}} \delta p^{l+1}_{i,j,k} + \frac{\partial R^l_{z_m,i,j,k}}{\partial p_{i+1,j,k}} \delta p^{l+1}_{i+1,j,k} + \frac{\partial R^l_{z_m,i,j,k}}{\partial p_{i,j+1,k}} \delta p^{l+1}_{i,j+1,k} \\
&\quad + \frac{\partial R^l_{z_m,i,j,k}}{\partial p_{i,j,k+1}} \delta p^{l+1}_{i,j,k+1} + \frac{\partial R^l_{z_m,i,j,k}}{\partial F_{i,j,k}} \delta F^{l+1}_{i,j,k} + \frac{\partial R^l_{z_m,i,j,k}}{\partial (z_m)_{i,j,k}} \delta (z_m)^{l+1}_{i,j,k} \\
&\quad + \frac{\partial R^l_{z_m,i,j,k}}{\partial (x_{mo})_{i,j,k}} \delta (x_{mo})^{l+1}_{i,j,k} + \frac{\partial R^l_{z_m,i,j,k}}{\partial (x_{mg})_{i,j,k}} \delta (x_{mg})^{l+1}_{i,j,k} = -R^l_{z_m,i,j,k}.
\end{aligned} \tag{8.45}$$

Equation (8.45) gives δz_m in terms of δp, δF, δx_{mo}, and δx_{mo}.

$$\begin{aligned}&\frac{\partial R^l_{F,i,j,k}}{\partial p_{i,j,k-1}}\delta p^{l+1}_{i,j,k-1}+\frac{\partial R^l_{F,i,j,k}}{\partial p_{i,j-1,k}}\delta p^{l+1}_{i,j-1,k}+\frac{\partial R^l_{F,i,,j,k}}{\partial p_{i-1,j,k}}\delta p^{l+1}_{i-1,j,k}\\&+\frac{\partial R^l_{F,i,j,k}}{\partial p_{i,j,k}}\delta p^{l+1}_{i,j,k}+\frac{\partial R^l_{F,i,j,k}}{\partial p_{i+1,j,k}}\delta p^{l+1}_{i+1,j,k}+\frac{\partial R^l_{F,i,j,k}}{\partial p_{i,j+1,k}}\delta p^{l+1}_{i,j+1,k}\\&+\frac{\partial R^l_{F,i,j,k}}{\partial p_{i,j,k+1}}\delta p^{l+1}_{i,j,k+1}+\frac{\partial R^l_{F,i,j,k}}{\partial F_{i,j,k}}\delta F^{l+1}_{i,j,k}=-R^l_{F,i,j,k}.\end{aligned}\tag{8.46}$$

This equation solves for δF in terms of δp.

$$\begin{aligned}&\frac{\partial R^l_{w,i,j,k}}{\partial p_{i,j,k-1}}\delta p^{l+1}_{i,j,k-1}+\frac{\partial R^l_{w,i,j,k}}{\partial p_{i,j-1,k}}\delta p^{l+1}_{i,j-1,k}+\frac{\partial R^l_{iw,,j,k}}{\partial p_{i-1,j,k}}\delta p^{l+1}_{i-1,j,k}\\&+\frac{\partial R^l_{w,i,j,k}}{\partial p_{i,j,k}}\delta p^{l+1}_{i,j,k}+\frac{\partial R^l_{w,i,j,k}}{\partial p_{i+1,j,k}}\delta p^{l+1}_{i+1,j,k}+\frac{\partial R^l_{w,i,j,k}}{\partial p_{i,j+1,k}}\delta p^{l+1}_{i,j+1,k}\\&+\frac{\partial R^l_{w,i,j,k}}{\partial p_{i,j,k+1}}\delta p^{l+1}_{i,j,k+1}+\frac{\partial R^l_{w,i,j,k}}{\partial S_{w,i,j,k}}\delta S^{l+1}_{w,i,j,k}=-R^l_{w,i,j,k}.\end{aligned}\tag{8.47}$$

Equation (8.47) gives δS_w in terms of δp.

$$\begin{aligned}&\sum_{r=1}^{N_c-1}\frac{\partial R^l_{p,i,j,k}}{\partial (x_{ro})_{i,j,k}}\delta (x_{ro})^{l+1}_{i,j,k}+\frac{\partial R^l_{p,i,j,k}}{\partial L_{i,j,k}}\delta L^{l+1}_{i,j,k}\\&+\sum_{r=1}^{N_c-1}\frac{\partial R^l_{p,i,j,k}}{\partial (z_r)_{i,j,k}}\delta (z_r)^{l+1}_{i,j,k}+\frac{\partial R^l_{p,i,j,k}}{\partial F_{i,j,k}}\delta F^{l+1}_{i,j,k}\\&+\frac{\partial R^l_{p,i,j,k}}{\partial S_{w,i,j,k}}\delta S^{l+1}_{w,i,j,k}+\frac{\partial R^l_{p,i,j,k}}{\partial p_{i,j,k}}\delta p^{l+1}_{i,j,k}=-R^l_{p,i,j,k}.\end{aligned}\tag{8.48}$$

After substituting δx_{ro}, δL, δz_r, δF, and δS, $r = 1, 2, \ldots, N_c - 1$, into (8.48) using equations (8.44)–(8.47), the resulting equation becomes the pressure equation, which, together with the well control equations, is implicitly solved for δp. After δp is obtained, equations (8.47), (8.46), (8.45), and (8.44) are solved explicitly, in turn, for δS, δF, $(\delta z_1, \delta z_2, \ldots, \delta z_{N_c-1})$, and $(\delta x_{1o}, \delta x_{2o}, \ldots, \delta x_{(N_c-1)o}, \delta L)$, respectively.

In summary, iterative IMPEC for the compositional model has the following features:

- The difference between iterative IMPEC and classical IMPEC is that the iterative method is used within each Newton–Raphson iteration loop, while the classical one is utilized outside the Newton–Raphson iteration.
- The saturation constraint equation is used to solve implicitly for pressure p.
- The equilibrium relation is solved for $(x_{1o}, x_{2o}, \ldots, x_{(N_c-1)o}, L)$.
- The hydrocarbon component flow equations are used to obtain explicitly $(z_1, z_2, \ldots, z_{N_c-1})$.

- The global hydrocarbon flow equation is exploited to solve explicitly for F.
- The water flow equation is explicitly solved for S_w.
- Relation (8.21) generates $(x_{1g}, x_{2g}, \ldots, x_{N_c g})$.

8.3 Solution of Equilibrium Relations

We discuss the solution of the thermodynamic equilibrium relation (8.28), which describes the mass distribution of each component in the oil and gas phases. As an example, we concentrate on the PR EOS.

8.3.1 Successive Substitution Method

The *successive substitution method* is often employed to find an initial guess for the computation of the thermodynamic equilibrium relation (8.28) in the Newton–Raphson flash calculation discussed in the next subsection. The *equilibrium flash vaporization ratio* for component i is defined by

$$K_i = \frac{x_{ig}}{x_{io}}, \quad i = 1, 2, \ldots, N_c, \tag{8.49}$$

where the quantity K_i is the K-value of component i. If the iterative IMPEC in the previous section is used (i.e., the capillary pressure p_{cg} is evaluated at the previous time step value of saturations in the Newton–Raphson iteration), it follows from (8.15) that

$$f_{i\alpha} = p x_{i\alpha} \varphi_{i\alpha}, \quad i = 1, 2, \ldots, N_c, \ \alpha = o, g. \tag{8.50}$$

Then, using (8.8), we see that

$$x_{io}\varphi_{io} = x_{ig}\varphi_{ig}, \quad i = 1, 2, \ldots, N_c.$$

Thus, by (8.49), we have

$$K_i = \frac{\varphi_{io}}{\varphi_{ig}}, \quad i = 1, 2, \ldots, N_c, \tag{8.51}$$

where the fugacity coefficients φ_{io} and φ_{ig} are defined in equation (8.14).

A *flash calculation* is an instant phase equilibrium:

Given p, T, and z_i;
Find L (or V), x_{io}, and x_{ig}, $\quad i = 1, 2, \ldots, N_c$.

It follows from (8.21) and (8.49) that

$$\begin{aligned} x_{io} &= \frac{z_i}{L + (1 - L)K_i}, \quad i = 1, 2, \ldots, N_c, \\ \sum_{i=1}^{N_c} &\frac{z_i(1 - K_i)}{L + (1 - L)K_i} = 0. \end{aligned} \tag{8.52}$$

Based on (8.52), we introduce the following successive substitution method for the flash calculation:

Initially, K_i is evaluated by the empirical formula

$$K_i = \frac{1}{p_{ir}} \exp\left(5.3727(1+\omega_i)\left[1 - \frac{1}{T_{ir}}\right]\right),$$

$$p_{ir} = \frac{p}{p_{ic}}, \quad T_{ir} = \frac{T}{T_{ic}};$$

(F1) Given K_i and z_i, find L by

$$\sum_{i=1}^{N_c} \frac{z_i(1-K_i)}{L+(1-L)K_i} = 0;$$

(F2) Find x_{io} and x_{ig} by

$$x_{io} = \frac{z_i}{L+(1-L)K_i}, \quad x_{ig} = K_i x_{io}, \quad i = 1, 2, \ldots, N_c;$$

(F3) Calculate K_i and z_i by

$$K_i = \frac{\varphi_{io}}{\varphi_{ig}}, \quad z_i = L x_{io} + (1-L) x_{ig}, \quad i = 1, 2, \ldots, N_c;$$

Return to (F1) and iterate until the convergence of the values K_i.

In general, convergence of this successive substitution method is very slow. However, it can be used as an initialization for the Newton–Raphson flash iteration discussed below.

8.3.2 Newton–Raphson Flash Calculation

Introduce the notation

$$G_{ij} = \left(\frac{\partial f_{io}}{\partial x_{jo}}\right)^l - \left(\frac{\partial f_{io}}{\partial x_{N_c o}}\right)^l + \frac{L^l}{1-L^l}\left[\left(\frac{\partial f_{ig}}{\partial x_{jg}}\right)^l - \left(\frac{\partial f_{ig}}{\partial x_{N_c g}}\right)^l\right],$$

$$G_{iN_c} = \frac{1}{1-L^l}\sum_{j=1}^{N_c}\left(\frac{\partial f_{ig}}{\partial x_{jg}}\left(x_{jo} - x_{jg}\right)\right)^l,$$

$$H_i(\delta p, \delta z_1, \delta z_2, \ldots, \delta z_{N_c - 1}) = f_{ig}^l - f_{io}^l + \left[\left(\frac{\partial f_{ig}}{\partial p}\right)^l - \left(\frac{\partial f_{io}}{\partial p}\right)^l\right]\delta p$$

$$+ \frac{1}{1-L^l}\sum_{j=1}^{N_c - 1}\left[\left(\frac{\partial f_{ig}}{\partial x_{jg}}\right)^l - \left(\frac{\partial f_{ig}}{\partial x_{N_c g}}\right)^l\right]\delta z_j$$

for $i = 1, 2, \ldots, N_c$, $j = 1, 2, \ldots, N_c - 1$. Then equation (8.44) can be written in matrix form:

$$\begin{pmatrix} G_{11} & G_{12} & \cdots & G_{1,N_c-1} & G_{1,N_c} \\ G_{21} & G_{22} & \cdots & G_{2,N_c-1} & G_{2,N_c} \\ \cdot & \cdot & \cdots & \cdot & \cdot \\ G_{N_c-1,1} & G_{N_c-1,2} & \cdots & G_{N_c-1,N_c-1} & G_{N_c-1,N_c} \\ G_{N_c,1} & G_{N_c,2} & \cdots & G_{N_c,N_c-1} & G_{N_c,N_c} \end{pmatrix} \begin{pmatrix} \delta x_{1o} \\ \delta x_{2o} \\ \cdot \\ \delta x_{(N_c-1)o} \\ \delta L \end{pmatrix} = \begin{pmatrix} H_1 \\ H_2 \\ \cdot \\ H_{N_c-1} \\ H_{N_c} \end{pmatrix}. \tag{8.53}$$

This system gives $(\delta x_{1o}, \delta x_{2o}, \ldots, \delta x_{(N_c-1)o}, \delta L)$ in terms of δz_i, $i = 1, 2, \ldots, N_c - 1$, and δp.

We point out the difference between the successive substitution method and the Newton–Raphson iteration in the flash calculation:

- The former method is easier to implement and is more reliable, even near a critical point. However, its convergence is usually slower; it may take over 1,000 iterations near the critical point.
- The latter method is faster. But it needs a good initial guess for x_{io} and L, $i = 1, 2, \ldots, N_c$; moreover, this method may not converge near a critical point.
- These two methods can be combined. For example, the former is used to find a good initial guess for the latter. Also, in places where the latter is difficult to converge, the former can be utilized instead.

8.3.3 Derivatives of Fugacity Coefficients

We calculate the partial derivatives involved in the Jacobian coefficient matrix of (8.53). First, by (8.50), for $i, j = 1, 2, \ldots, N_c$, $\alpha = o, g$,

$$\frac{\partial f_{i\alpha}}{\partial p} = x_{i\alpha}\varphi_{i\alpha} + p x_{i\alpha}\frac{\partial \varphi_{i\alpha}}{\partial p}, \quad \frac{\partial f_{i\alpha}}{\partial x_{j\alpha}} = p\frac{\partial x_{i\alpha}}{\partial x_{j\alpha}}\varphi_{i\alpha} + p x_{i\alpha}\frac{\partial \varphi_{i\alpha}}{\partial x_{j\alpha}},$$

where

$$\frac{\partial x_{i\alpha}}{\partial x_{j\alpha}} = \begin{cases} 1 & \text{if } i = j, \\ 0 & \text{if } i \neq j. \end{cases}$$

So it suffices to find the derivatives of $\varphi_{i\alpha}$, which is defined by (8.14), $i = 1, 2, \ldots, N_c$, $\alpha = o, g$.

It follows from (8.10) that

$$\frac{\partial A_\alpha}{\partial p} = \frac{a_\alpha}{R^2 T^2}, \quad \frac{\partial B_\alpha}{\partial p} = \frac{b_\alpha}{R T}, \quad \alpha = o, g. \tag{8.54}$$

Differentiating both sides of (8.14) gives

$$\begin{aligned}\frac{1}{\varphi_{i\alpha}}\frac{\partial \varphi_{i\alpha}}{\partial p} &= \frac{b_i}{b_\alpha}\frac{\partial Z_\alpha}{\partial p} - \frac{1}{Z_\alpha - B_\alpha}\left(\frac{\partial Z_\alpha}{\partial p} - \frac{B_\alpha}{p}\right) \\ &\quad - \frac{A_\alpha}{2\sqrt{2}B_\alpha}\left(\frac{2}{a_\alpha}\sum_{j=1}^{N_c} x_{j\alpha}(1-\kappa_{ij})\sqrt{a_i a_j} - \frac{b_i}{b_\alpha}\right) \\ &\quad \cdot 2B_\alpha\left(\frac{Z_\alpha}{p} - \frac{\partial Z_\alpha}{\partial p}\right) \Big/ \left(Z_\alpha^2 + 2\sqrt{2}Z_\alpha B_\alpha + B_\alpha^2\right).\end{aligned} \tag{8.55}$$

Similarly, we can obtain $\partial \varphi_{i\alpha}/\partial x_{j\alpha}$ using the expressions

$$\begin{aligned}&\frac{\partial A_\alpha}{\partial x_{j\alpha}} = \frac{p}{R^2T^2}\frac{\partial a_\alpha}{\partial x_{j\alpha}}, \qquad \frac{\partial B_\alpha}{\partial x_{j\alpha}} = \frac{p}{RT}\frac{\partial b_\alpha}{\partial x_{j\alpha}}, \\ &\frac{\partial a_\alpha}{\partial x_{j\alpha}} = 2\sum_{i=1}^{N_c} x_{i\alpha}(1-\kappa_{ij})\sqrt{a_i a_j}, \qquad \frac{\partial b_\alpha}{\partial x_{j\alpha}} = b_j\end{aligned} \tag{8.56}$$

for $i, j = 1, 2, \ldots, N_c$, $\alpha = o, g$.

The Z-factors, Z_α ($\alpha = o, g$), are determined by equation (8.13), which can be differentiated to find their derivatives. Implicit differentiation on (8.13) yields

$$\begin{aligned}\frac{\partial Z_\alpha}{\partial p} = &-\left\{\frac{\partial B_\alpha}{\partial p}Z_\alpha^2 + \left(\frac{\partial A_\alpha}{\partial p} - 2\left[1+3B_\alpha\right]\frac{\partial B_\alpha}{\partial p}\right)Z_\alpha\right. \\ &\left. - \left(\frac{\partial A_\alpha}{\partial p}B_\alpha + \left[A_\alpha - 2B_\alpha - 3B_\alpha^2\right]\frac{\partial B_\alpha}{\partial p}\right)\right\} \\ &\Big/ \left(3Z_\alpha^2 - 2(1-B_\alpha)Z_\alpha + (A_\alpha - 2B_\alpha - 3B_\alpha^2)\right).\end{aligned} \tag{8.57}$$

Consequently, substituting (8.54) into (8.57) gives $\partial Z_\alpha/\partial p$. A similar argument, together with (8.56), gives the derivatives $\partial Z_\alpha/\partial x_{j\alpha}$, $j = 1, 2, \ldots, N_c$.

8.3.4 Solution of the PR Cubic Equation

The PR cubic equation (8.13) has the form

$$\mathcal{Z}^3 + B\mathcal{Z}^2 + C\mathcal{Z} + D = 0, \tag{8.58}$$

with given inputs B, C, and D. Before discussing the solution of this equation, we consider a simpler cubic equation:

$$X^3 + PX + Q = 0. \tag{8.59}$$

With

$$\Delta = \left(\frac{Q}{2}\right)^2 + \left(\frac{P}{3}\right)^3,$$

equation (8.59) has three roots,

$$X_1 = \sqrt[3]{-\frac{Q}{2} + \sqrt{\Delta}} + \sqrt[3]{-\frac{Q}{2} - \sqrt{\Delta}},$$
$$X_2 = \omega\sqrt[3]{-\frac{Q}{2} + \sqrt{\Delta}} + \omega^2\sqrt[3]{-\frac{Q}{2} - \sqrt{\Delta}},$$
$$X_3 = \omega^2\sqrt[3]{-\frac{Q}{2} + \sqrt{\Delta}} + \omega\sqrt[3]{-\frac{Q}{2} - \sqrt{\Delta}},$$

where

$$\omega = \frac{-1 + i\sqrt{3}}{2}, \quad \omega^2 = \frac{-1 - i\sqrt{3}}{2}, \quad i^2 = -1.$$

Note that

$$X_1 + X_2 + X_3 = 0, \quad \frac{1}{X_1} + \frac{1}{X_2} + \frac{1}{X_3} = -\frac{P}{Q}, \quad X_1X_2X_3 = -Q. \tag{8.60}$$

If $\Delta > 0$, equation (8.59) has only one real root, X_1. If $P = Q = 0$, there is solely the trivial solution $X_1 = X_2 = X_3 = 0$. When $\Delta \leq 0$, there are three real roots given by

$$\begin{aligned} &X_1 = 2\sqrt[3]{\mathcal{R}}\cos\theta, \quad X_2 = 2\sqrt[3]{\mathcal{R}}\cos\left(\frac{2\pi}{3} + \theta\right), \\ &X_3 = 2\sqrt[3]{\mathcal{R}}\cos\left(\frac{4\pi}{3} + \theta\right), \end{aligned} \tag{8.61}$$

where

$$\mathcal{R} = \sqrt{-\left(\frac{P}{3}\right)^3}, \quad \theta = \frac{1}{3}\arccos\left(-\frac{Q}{2\mathcal{R}}\right).$$

To solve equation (8.58), set $\mathcal{Z} = X - \frac{B}{3}$. Then (8.58) is converted into equation (8.59) with

$$P = -\frac{B^2}{3} + C, \quad Q = \frac{2B^3}{27} - \frac{BC}{3} + D.$$

Thus the roots of equation (8.58) are

$$\mathcal{Z}_1 = X_1 - \frac{B}{3}, \quad \mathcal{Z}_2 = X_2 - \frac{B}{3}, \quad \mathcal{Z}_3 = X_3 - \frac{B}{3}. \tag{8.62}$$

If $\mathcal{Z}_1$ is the sole real root, it is selected. In the case where there are three real roots, say,

$$\mathcal{Z}_1 > \mathcal{Z}_2 > \mathcal{Z}_3,$$

we select $\mathcal{Z}_1$ if the vapor (gas) phase dominates. If the liquid (oil) phase dominates, we select $\mathcal{Z}_1$ when $\mathcal{Z}_2 \leq 0$; select $\mathcal{Z}_2$ when $\mathcal{Z}_2 > 0$ and $\mathcal{Z}_3 \leq 0$; select $\mathcal{Z}_3$ when $\mathcal{Z}_3 > 0$. Namely, we select the smallest positive root.

8.3.5 Practical Considerations

We point out a few practical issues in programming the solution of equilibrium relations.

Iteration switch

As noted, depending on the size of L, different variables, either x_{io} and L or x_{ig} and V, should be used in the flash calculation, $i = 1, 2, \ldots, N_c$. If the gas phase dominates in the hydrocarbon system (e.g., $L < 0.5$), the primary unknowns will be x_{io} and L. If the oil phase dominates (e.g., $L \geq 0.5$), the primary unknowns will be x_{ig} and V. This choice can improve solution accuracy and convergence speed. For example, as L gets close to 1, the flash calculation may not converge. In this case, the primary unknown needs to be switched to V. In programming, the switch of iterations should be done automatically.

Determination of bubble points

The following system of $N_c + 1$ equations is solved simultaneously for finding the *bubble point pressure* p and the compositions x_{ig} by a Newton–Raphson iteration ($i = 1, 2, \ldots, N_c$):

$$\begin{aligned} z_i\varphi_{io}(p, x_{1o}, x_{2o}, \ldots, x_{N_c o}) &= x_{ig}\varphi_{ig}(p, x_{1g}, x_{2g}, \ldots, x_{N_c g}), \\ \sum_{i=1}^{N_c} x_{ig} &= 1. \end{aligned} \tag{8.63}$$

In the late steps of the iteration (e.g., after ten iterations), the second equation in (8.63) can be replaced by

$$\sum_{i=1}^{N_c} \frac{\varphi_{io}}{\varphi_{ig}} z_i = 1 \tag{8.64}$$

to speed up convergence. In the Newton–Raphson iteration, if the successive values of pressure change less than a certain value (e.g., 0.01 psi), then this iteration is considered to have converged. We consider that it fails to converge if more than 30 iterations are required or if $|z_i - x_{ig}| < 0.001|z_i|$. In the latter case, the successive substitution method can be used to obtain p and x_{ig}, $i = 1, 2, \ldots, N_c$. A trivial solution occurs when $x_{ig} = z_i$ for any value of p, indicating that a *dew point* occurs (cf. Fig. 8.1).

Determination of dew points

The *dew point pressure* p and the compositions x_{io} satisfy the system of $N_c + 1$ equations ($i = 1, 2, \ldots, N_c$):

$$\begin{aligned} x_{io}\varphi_{io}(p, x_{1o}, x_{2o}, \ldots, x_{N_c o}) &= z_i\varphi_{ig}(p, x_{1g}, x_{2g}, \ldots, x_{N_c g}), \\ \sum_{i=1}^{N_c} x_{io} &= 1. \end{aligned} \tag{8.65}$$

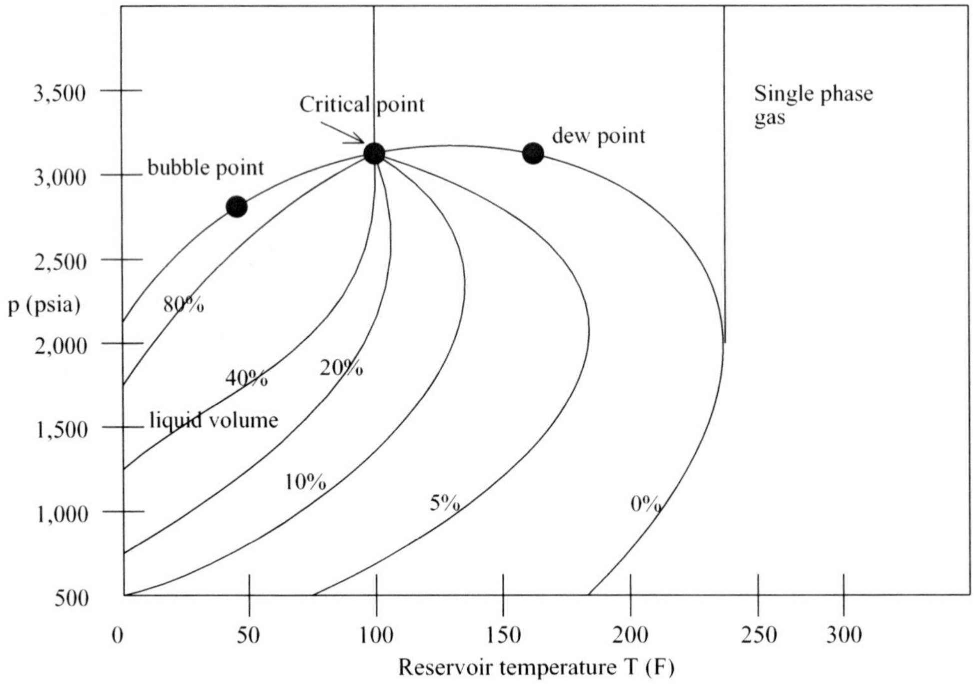

Figure 8.1. *Pressure-temperature phase diagram of a reservoir fluid.*

Again, after about ten Newton–Raphson iterations, the second equation in (8.65) is replaced by

$$\sum_{i=1}^{N_c} \frac{\varphi_{ig}}{\varphi_{io}} z_i = 1. \tag{8.66}$$

Using the same guidelines as in the treatment of bubble points, if the successive values of pressure in the iteration process change less than 0.01 psi, this iteration is considered to have converged. We consider that the convergence fails if more than 30 iterations are required or if $|z_i - x_{io}| < 0.001|z_i|$. In the latter case, the successive substitution method can be used to obtain p and x_{io}, $i = 1, 2, \ldots, N_c$. A trivial solution occurs when $x_{io} = z_i$ for any value of p, indicating that a *bubble point* occurs.

Chapter 9

Nonisothermal Flow and Numerical Solution

The differential equations so far have been developed under the condition that flow is isothermal (nonisothermal flow was briefly touched on in Section 7.6). This condition can be removed by adding an *energy conservation equation*, which introduces an additional dependent variable, temperature, to the system. Unlike the case of mass transport, where the solid itself is assumed impervious to mass flux, the solid matrix does conduct heat. The average temperature of the solid and fluids in a porous medium may not be the same. Furthermore, heat may be exchanged between the phases. For simplicity, we invoke the requirement of *local thermal equilibrium* that the temperature be the same in all phases.

The mass conservation equations and Darcy's laws for nonisothermal flow are the same as for the compositional model discussed in Chapter 8; an additional energy conservation equation is required. For the convenience of the reader, we review these equations. These governing equations are based on the displacement mechanisms of thermal methods: (a) reduction of crude viscosity with increasing temperature, (b) change of relative permeabilities for greater oil displacement, (c) vaporization of connate water and of a portion of crudes for miscible displacement of light components, and (d) high temperatures of fluids and rock to maintain high reservoir pressure. They can model the important physical factors and processes:

- viscosity, gravity, and capillary forces;
- heat conduction and convection processes;
- heat losses to overburden and underburden of a reservoir;
- mass transfer between phases;
- effects of temperature on the physical property parameters of oil, gas, and water;
- rock compression and expansion.

9.1 Basic Differential Equations

9.1.1 Mass Conservation and Darcy's Law

We assume that the chemical components form at most three phases (e.g., water, oil, and gas), N_c chemical components may exist in all three phases, and diffusive effects are neglected.

Let ϕ and $\mathbf{k}$ denote the porosity and permeability of a porous medium $\Omega \subset \mathbb{R}^3$, and let S_α, μ_α, p_α, $\mathbf{u}_\alpha$, and $k_{r\alpha}$ be the saturation, viscosity, pressure, volumetric velocity, and relative permeability, respectively, of the α phase, $\alpha = w, o, g$. Also, let $\xi_{m\alpha}$ represent the molar density of component m in the α phase, $m = 1, 2, \ldots, N_c$, $\alpha = w, o, g$. The molar density of phase α is given by

$$\xi_\alpha = \sum_{m=1}^{N_c} \xi_{m\alpha}, \qquad \alpha = w,\ o,\ g. \tag{9.1}$$

The mole fraction of component m in phase α is then defined by

$$x_{m\alpha} = n_{m\alpha}/n_\alpha, \qquad m = 1, 2, \ldots, N_c,\ \alpha = o, g, \tag{9.2}$$

where $n_{m\alpha}$ is the number of moles of component m in the α phase and n_α is the total moles of this phase. The total mass is conserved for each component:

$$\begin{aligned} &\frac{\partial}{\partial t} \sum_{\alpha=w}^{g} x_{m\alpha}\xi_\alpha S_\alpha + \nabla \cdot \sum_{\alpha=w}^{g} x_{m\alpha}\xi_\alpha \mathbf{u}_\alpha \\ &\qquad = \sum_{\alpha=w}^{g} x_{m\alpha} q_\alpha, \qquad m = 1, \ldots, N_c, \end{aligned} \tag{9.3}$$

where q_α stands for the flow rate of phase α at the wells. In equation (9.3), the volumetric velocity $\mathbf{u}_\alpha$ is given by Darcy's law:

$$\mathbf{u}_\alpha = -\frac{k_{r\alpha}}{\mu_\alpha}\mathbf{k}\left(\nabla p_\alpha - \rho_\alpha \wp \nabla z\right), \qquad \alpha = w,\ o,\ g, \tag{9.4}$$

where ρ_α is the mass density of the α phase, $\wp$ is the magnitude of the gravitational acceleration, and z is the depth.

In addition to the differential equations (9.3) and (9.4), there are also algebraic constraints. The mole fraction balance implies

$$\sum_{m=1}^{N_c} x_{m\alpha} = 1, \qquad \alpha = w,\ o,\ g. \tag{9.5}$$

In the transport process, the saturation constraint reads

$$S_w + S_o + S_g = 1. \tag{9.6}$$

Finally, the phase pressures are related by capillary pressures

$$p_{cow} = p_o - p_w, \qquad p_{cgo} = p_g - p_o. \tag{9.7}$$

9.1.2 Energy Conservation

The energy conservation equation takes the form

$$\frac{\partial}{\partial t}\left(\phi \sum_{\alpha=w}^{g} \rho_\alpha S_\alpha U_\alpha + (1-\phi)\rho_s C_s T\right) + \nabla \cdot \sum_{\alpha=w}^{g} \rho_\alpha \mathbf{u}_\alpha H_\alpha - \nabla \cdot (k_T \nabla T) = q_c - q_L, \tag{9.8}$$

where T is the temperature, U_α and H_α are the specific internal energy and enthalpy of the α phase (per unit mass), ρ_s and C_s are the density and the specific heat capacity of the solid, k_T represents the total thermal conductivity, q_c denotes the heat source item, and q_L indicates the heat loss to overburden and underburden. In (9.8), the *specific internal energy* U_α and *enthalpy* H_α of phase α can be computed from

$$U_\alpha = C_{V\alpha} T, \quad H_\alpha = C_{p\alpha} T, \tag{9.9}$$

where $C_{V\alpha}$ and $C_{p\alpha}$ represent the *heat capacities* of phase α at constant volume and constant pressure, respectively. In addition, the bulk thermal conductivity can be defined by

$$k_T = k_f^{\phi} k_s^{1-\phi},$$

where k_f and k_s are the fluid and rock thermal conductivities, respectively. Substituting relation (9.9) into equation (9.8) gives the energy conservation equation which we will solve:

$$\frac{\partial}{\partial t}\left(\phi \sum_{\alpha=w}^{g} \rho_\alpha S_\alpha C_{V\alpha} T + (1-\phi)\rho_s C_s T\right) + \nabla \cdot \sum_{\alpha=w}^{g} \rho_\alpha \mathbf{u}_\alpha C_{p\alpha} T - \nabla \cdot (k_T \nabla T) = q_c - q_L. \tag{9.10}$$

In thermal methods used in the petroleum industry, heat is lost to the adjacent strata of a reservoir, or the *overburden* and *underburden*, which is included in the term q_L of equation (9.10). We assume that the overburden and underburden extend to infinity along both the positive and negative x_3-axis (the vertical direction); see Fig. 9.1. If the overburden and underburden are impermeable, heat is transferred entirely through conduction. With all

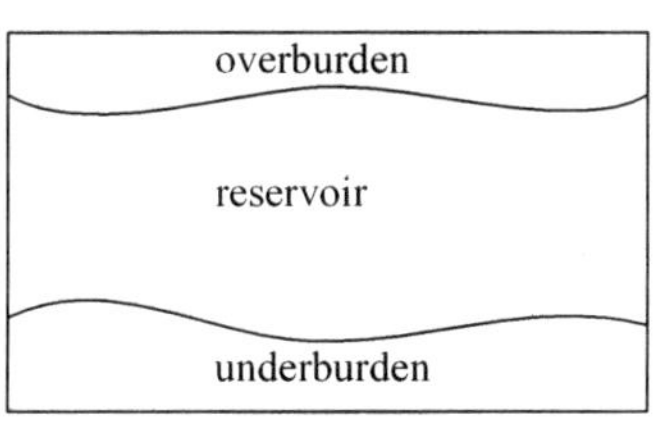

Figure 9.1. *Reservoir, overburden, and underburden.*

fluid velocities and convective fluxes being zero, the energy conservation equation (9.10) reduces to

$$\frac{\partial}{\partial t}\left(\rho_{ob}C_{p,ob}T_{ob}\right) = \nabla \cdot (k_{ob}\nabla T_{ob}), \tag{9.11}$$

where the subscript *ob* indicates that the variables are associated with the overburden and $C_{p,ob}$ is the heat capacity at constant pressure. The initial condition is the original temperature $T_{ob,0}$ of the overburden:

$$T_{ob}(\mathbf{x}, 0) = T_{ob,0}(\mathbf{x}).$$

The boundary condition at the top of the reservoir is

$$T_{ob}(x_1, x_2, x_3, t) = T(x_1, x_2, x_3, t).$$

At $x_3 = \infty$, T_{ob} is fixed:

$$T_{ob}(x_1, x_2, \infty, t) = T_\infty.$$

On other boundaries, we use the impervious boundary condition

$$k_{ob}\nabla T_{ob} \cdot \boldsymbol{\nu} = 0,$$

where $\boldsymbol{\nu}$ represents the outward unit normal to these boundaries. Now, the rate of heat loss to the overburden can be calculated by $k_{ob}\nabla T_{ob} \cdot \boldsymbol{\nu}$, where $\boldsymbol{\nu}$ is the unit normal to the interface between the overburden and reservoir (pointing to the overburden). For the underburden, the heat conduction equation is

$$\frac{\partial}{\partial t}\left(\rho_{ub}C_{p,ub}T_{ub}\right) = \nabla \cdot (k_{ub}\nabla T_{ub}), \tag{9.12}$$

and similar initial and boundary conditions can be developed as for the overburden.

Equations (9.3)–(9.7) and (9.10) provide $N_c + 10$ independent relations, differential or algebraic, for the $3N_c + 10$ dependent variables: $x_{i\alpha}$, $\mathbf{u}_\alpha$, p_α, T, and S_α, $\alpha = w, o, g$, $i = 1, 2, \ldots, N_c$. If equations (9.11) and (9.12) are included, two more unknowns, T_{ob} and T_{ub}, are added. The additional $2N_c$ relations will be provided below (cf. (9.17)). With proper initial and boundary conditions, there is a closed differential system for the unknowns.

9.1.3 Rock Properties

The *rock properties* for nonisothermal flow are similar to those for the isothermal black oil and compositional models; but now these properties depend on temperature. In particular, the capillary pressures are of the form

$$p_{cow}(S_w, T) = p_o - p_w, \quad p_{cgo}(S_g, T) = p_g - p_o. \tag{9.13}$$

Similarly, the relative permeabilities for water, oil, and gas are

$$\begin{aligned} &k_{rw} = k_{rw}(S_w, T), \qquad && k_{row} = k_{row}(S_w, T), \\ &k_{rg} = k_{rg}(S_g, T), && k_{rog} = k_{rog}(S_g, T), \\ &k_{ro} = k_{ro}(S_w, S_g, T). \end{aligned} \tag{9.14}$$

Stone's models (cf. Section 2.5.2) can be adapted for the oil relative permeability k_{ro}, for example.

As an example, the relative permeability functions k_{rw} and k_{row} for a water-oil system can be defined by

$$\begin{aligned} k_{rw} &= k_{rwro}(T)\left(\frac{S_w - S_{wir}(T)}{1 - S_{orw}(T) - S_{wir}(T)}\right)^{nw}, \\ k_{row} &= k_{rocw}(T)\left(\frac{1 - S_w - S_{orw}(T)}{1 - S_{orw}(T) - S_{wc}(T)}\right)^{now}, \end{aligned} \tag{9.15}$$

and for a gas-oil system, k_{rg} and k_{rog} by

$$\begin{aligned} k_{rg} &= k_{rgro}(T)\left(\frac{S_g - S_{gr}^*}{1 - S_{wc}(T) - S_{oinit} - S_{gr}^*}\right)^{ng}, \\ k_{rog} &= k_{rocw}(T)\left(\frac{1 - S_g - S_{wc}(T) - S_{org}(T)}{1 - S_{wc}(T) - S_{org}(T)}\right)^{nog}, \end{aligned} \tag{9.16}$$

where nw, now, ng, and nog are nonnegative real numbers measured in the laboratory; S_{wc}, S_{wir}, S_{orw}, S_{org}, and S_{gr}^* are the connate water saturation, irreducible water saturation; residual oil saturation in the water-oil system, residual oil saturation in the gas-oil system, and residual gas saturation; k_{rwro}, k_{rocw}, and k_{rgro} are the water relative permeability at the residual oil saturation for the water-oil system, the oil relative permeability at the connate water saturation, and the gas relative permeability at the residual oil saturation for the gas-oil system, respectively; and S_{oinit} is the initial oil saturation in the gas-oil system. Finally, for the rock properties, one must consider the thermal conductivity and heat capacity of the reservoir, overburden, and underburden.

9.1.4 Fluid Properties

The thermodynamic equilibrium relation (8.8) discussed in Section 8.1 can be used to determine mass interchange between phases. Because of the complexity of nonisothermal flow, however, an *equilibrium K-value approach* is often used to describe the equilibrium relations (cf. Section 8.1.2):

$$x_{mw} = K_{mw}(p, T)x_{mo}, \quad x_{mg} = K_{mg}(p, T)x_{mo}, \quad m = 1, 2, \ldots, N_c. \tag{9.17}$$

One example of evaluating the K-values $K_{m\alpha}$ uses the empirical formula

$$K_{m\alpha} = \left(\kappa_{m\alpha}^1 + \frac{\kappa_{m\alpha}^2}{p} + \kappa_{m\alpha}^3 p\right)\exp\left(-\frac{\kappa_{m\alpha}^4}{T - \kappa_{m\alpha}^5}\right), \tag{9.18}$$

where the constants $\kappa_{m\alpha}^l$ are obtained in the laboratory, $m = 1, 2, \ldots, N_c$, $l = 1, 2, 3, 4, 5$, $\alpha = w, g$. For the notational convenience, we use $K_{mo} = 1$, $m = 1, 2, \ldots, N_c$.

Water properties

Physical properties of water and steam, such as density, internal energy, enthalpy, and viscosity, can be found from a *water-steam table* (Lake, 1989). Such a table is given in terms of the independent variables: pressure and temperature. In the case where all three phases coexist, a reservoir is in the saturated state. In this case, there is free gas; pressure and temperature are related, and only one of them is employed as an independent variable.

Oil properties

While any number of hydrocarbon components can be treated in the differential system describing the nonisothermal multiphase, multicomponent flow considered in this chapter, computational work and time significantly increase as the number of components increases. It is often computationally convenient (or necessary) to group several similar chemical components into one mathematical component as in the treatment of compositional flow (Chen, Huan, and Ma, 2006). In this way, only a few components (or *pseudocomponents*) are simulated in practical applications.

The oil phase is a mixture of hydrocarbon components, and these components range from the lightest component, methane (CH_4), to the heaviest component, bitumen. One way to reduce the number of components is to introduce pseudocomponents, as noted. According to the compositions of each pseudocomponent, one can deduce its physical properties, such as its pseudomolecular weight (which may not be a constant), critical pressure and temperature, compressibility, density, viscosity, thermal expansion coefficient, and specific heat. These properties are functions of pressure and temperature.

The most important property is the oil and gas phase viscosity dependence on temperature:

$$\mu_{io} = \exp\left(a_1 T^{b_1}\right) + c_1, \qquad \mu_{ig} = a_2 T^{b_2},$$

where T is in absolute degrees, a_1, b_1, c_1, a_2, and b_2 are empirical parameters that can be measured in the laboratory, and μ_{io} and μ_{ig} are the viscosities of the ith component in the oil and gas phases, respectively.

9.2 Numerical Solution of Nonisothermal Flow

In simulation of nonisothermal flow, three parts must be treated: the oil reservoir, overburden, and underburden. Because of the weak coupling between the reservoir and the overburden and underburden, the equations in these three parts can be decoupled; that is, they are solved in a sequential manner. In the reservoir domain, the IMPES, sequential, and SS techniques introduced for the black oil model in Chapter 6 can be applied. For the nonisothermal flow, because there exist strong nonlinearity and coupling in the governing equations, pressure and temperature greatly vary, and mass and energy transfer frequently between the oil and gas phases, the SS technique should be used for the reservoir system. The heat conduction equations for overburden and underburden are simple enough that a fully implicit scheme in time can be employed for their solution.

9.2.1 Choice of Primary Variables

As discussed earlier, equations (9.3)–(9.7), (9.10), and (9.17) form a strongly coupled system of time-dependent, nonlinear differential equations and algebraic constraints for $3N_c + 10$ unknowns. Although there are the same number of equations for these dependent variables, the entire system can be rewritten in terms of certain primary variables, with other variables being obtained from them.

Undersaturated state

As discussed in Section 6.1.4, if all three phases coexist, a reservoir is in the saturated state. When all the gas dissolves into the oil phase (i.e., there is no free gas; $S_g = 0$), the reservoir is in the undersaturated state. The choice of primary unknowns depends on the state of a reservoir.

We define the transmissibilities

$$\begin{aligned}\mathbf{T}_\alpha &= \frac{\rho_\alpha k_{r\alpha}}{\mu_\alpha}\mathbf{k},\\ \mathbf{T}_{m\alpha} &= \frac{x_{m\alpha}\xi_\alpha k_{r\alpha}}{\mu_\alpha}\mathbf{k}, \qquad m = 1, 2, \ldots, N_c,\ \alpha = w, o, g.\end{aligned} \tag{9.19}$$

Moreover, we use the total mole fraction

$$x_m = \sum_{\alpha=w}^{g} x_{m\alpha}, \qquad m = 1, 2, \ldots, N_c. \tag{9.20}$$

Using (9.17), equation (9.20) becomes

$$x_{mo} = \frac{1}{K_{mwog}(p, T)} x_m, \qquad m = 1, 2, \ldots, N_c, \tag{9.21}$$

where $K_{mwog}(p, T) = K_{mw} + 1 + K_{mg}$. As a result, we see that

$$x_{mw} = \frac{K_{mw}}{K_{mwog}} x_m, \quad x_{mg} = \frac{K_{mg}}{K_{mwog}} x_m, \qquad m = 1, 2, \ldots, N_c. \tag{9.22}$$

Thus x_m should be used as a primary unknown, $m = 1, 2, \ldots, N_c$. Due to (9.5), only $N_c - 2$ unknowns are independent. Consequently, in the undersaturated state, $(p, S, x_1, x_2, \ldots, x_{N_c-2}, T)$ are chosen as the primary unknowns, where $p = p_o$ and $S = S_w$. The differential system for these unknowns consists of the N_c component mass conservation equations

$$\frac{\partial(\phi F_m x_m)}{\partial t} = \sum_{\alpha=w}^{g} \nabla \cdot (\mathbf{T}_{m\alpha}[\nabla p_\alpha - \gamma_\alpha \nabla z]) + \sum_{\alpha=w}^{g} x_{m\alpha} q_\alpha, \qquad m = 1, 2, \ldots, N_c, \tag{9.23}$$

and the energy conservation equation

$$\begin{aligned}&\frac{\partial}{\partial t}\left(\phi \sum_{\alpha=w}^{g} \rho_\alpha S_\alpha C_{V\alpha} T + (1-\phi)\rho_s C_s T\right)\\ &\quad - \nabla \cdot \sum_{\alpha=w}^{g} C_{p\alpha} T \mathbf{T}_\alpha (\nabla p_\alpha - \gamma_\alpha \nabla z) - \nabla \cdot (k_T \nabla T) = q_c - q_L,\end{aligned} \tag{9.24}$$

where $\gamma_\alpha = \rho_\alpha \wp$ and

$$F_m = \sum_{\alpha=w}^{g} \frac{K_{i\alpha}}{K_{iwog}} \xi_\alpha S_\alpha .$$

Saturated state

In the saturated state, there is free gas. Pressure p and temperature T are related; their relationship may be given through a saturated steam table. Thus only one can be used as a primary unknown. In this case, we choose the primary unknowns $(p, S_w, S_o, x_1, x_2, \ldots, x_{N_c-2})$, where $p = p_o$. The system of differential equations is composed of the N_c component mass conservation equations (9.23) and the energy conservation equation (9.24).

9.2.2 Finite Difference Equations

Assume that the permeability tensor **k** is diagonal. For $n > 0$ (an integer), a time approximation at the $(n+1)$th level for the system of equations (9.23) and (9.24) is

$$\begin{aligned}
&\frac{1}{\Delta t}\left(V\left[(\phi F_m x_m)^{n+1} - (\phi F_m x_m)^n\right]\right)_{i,j,k} \\
&= \sum_{\alpha=w}^{g}\Big\{ (T_{m\alpha 1})^{n+1}_{i+1/2,j,k}\left(p^{n+1}_{\alpha,i+1,j,k} - p^{n+1}_{\alpha,i,j,k}\right) \\
&\quad - (T_{m\alpha 1})^{n+1}_{i-1/2,j,k}\left(p^{n+1}_{\alpha,i,j,k} - p^{n+1}_{\alpha,i-1,j,k}\right) \\
&\quad + (T_{m\alpha 2})^{n+1}_{i,j+1/2,k}\left(p^{n+1}_{\alpha,i,j+1,k} - p^{n+1}_{\alpha,i,j,k}\right) - (T_{m\alpha 2})^{n+1}_{i,j-1/2,k}\left(p^{n+1}_{\alpha,i,j,k} - p^{n+1}_{\alpha,i,j-1,k}\right) \\
&\quad + (T_{m\alpha 3})^{n+1}_{i,j,k+1/2}\left(p^{n+1}_{\alpha,i,j,k+1} - p^{n+1}_{\alpha,i,j,k}\right) - (T_{m\alpha 3})^{n+1}_{i,j,k-1/2}\left(p^{n+1}_{\alpha,i,j,k} - p^{n+1}_{\alpha,i,j,k-1}\right) \qquad (9.25) \\
&\quad - (T_{m\alpha}\gamma_\alpha)^{n+1}_{1,i+1/2,j,k}\left(z_{i+1,j,k} - z_{i,j,k}\right) + (T_{m\alpha}\gamma_\alpha)^{n+1}_{1,i-1/2,j,k}\left(z_{i,j,k} - z_{i-1,j,k}\right) \\
&\quad - (T_{m\alpha}\gamma_\alpha)^{n+1}_{2,i,j+1/2,k}\left(z_{i,j+1,k} - z_{i,j,k}\right) + (T_{m\alpha}\gamma_\alpha)^{n+1}_{2,i,j-1/2,k}\left(z_{i,j,k} - z_{i,j-1,k}\right) \\
&\quad - (T_{m\alpha}\gamma_\alpha)^{n+1}_{3,i,j,k+1/2}\left(z_{i,j,k+1} - z_{i,j,k}\right) + (T_{m\alpha}\gamma_\alpha)^{n+1}_{3,i,j,k-1/2}\left(z_{i,j,k} - z_{i,j,k-1}\right)\Big\} \\
&\quad + \sum_{\alpha=w}^{g} (x_{m\alpha} Q_\alpha)^{n+1}_{i,j,k}, \qquad m = 1, 2, \ldots, N_c,
\end{aligned}$$

and

$$\begin{aligned}
&\frac{1}{\Delta t}\left(V\left[\left(\phi \sum_{\alpha=w}^{g} \rho_\alpha S_\alpha C_{V\alpha} T + (1-\phi)\rho_s C_s T\right)^{n+1}\right.\right. \\
&\quad \left.\left. - \left(\phi \sum_{\alpha=w}^{g} \rho_\alpha S_\alpha C_{V\alpha} T + (1-\phi)\rho_s C_s T\right)^{n}\right]\right)_{i,j,k}
\end{aligned}$$

$$
\begin{aligned}
&= \sum_{\alpha=w}^{g} \Big\{ (C_{p\alpha} T T_\alpha)^{n+1}_{1,i+1/2,j,k} \left(p^{n+1}_{\alpha,i+1,j,k} - p^{n+1}_{\alpha,i,j,k} \right) \\
&- (C_{p\alpha} T T_\alpha)^{n+1}_{1,i-1/2,j,k} \left(p^{n+1}_{\alpha,i,j,k} - p^{n+1}_{\alpha,i-1,j,k} \right) \\
&+ (C_{p\alpha} T T_\alpha)^{n+1}_{2,i,j+1/2,k} \left(p^{n+1}_{\alpha,i,j+1,k} - p^{n+1}_{\alpha,i,j,k} \right) \\
&- (C_{p\alpha} T T_\alpha)^{n+1}_{2,i,j-1/2,k} \left(p^{n+1}_{\alpha,i,j,k} - p^{n+1}_{\alpha,i,j-1,k} \right) \\
&+ (C_{p\alpha} T T_\alpha)^{n+1}_{3,i,j,k+1/2} \left(p^{n+1}_{\alpha,i,j,k+1} - p^{n+1}_{\alpha,i,j,k} \right) \\
&- (C_{p\alpha} T T_\alpha)^{n+1}_{3,i,j,k-1/2} \left(p^{n+1}_{\alpha,i,j,k} - p^{n+1}_{\alpha,i,j,k-1} \right) \\
&- (C_{p\alpha} T T_\alpha \gamma_\alpha)^{n+1}_{1,i+1/2,j,k} \left(z_{i+1,j,k} - z_{i,j,k} \right) \\
&+ \left(C_{p\alpha} T T_\alpha \gamma_\alpha\right)^{n+1}_{1,i-1/2,j,k} \left(z_{i,j,k} - z_{i-1,j,k} \right) \\
&- \left(C_{p\alpha} T T_\alpha \gamma_\alpha\right)^{n+1}_{2,i,j+1/2,k} \left(z_{i,j+1,k} - z_{i,j,k} \right) \\
&+ \left(C_{p\alpha} T T_\alpha \gamma_\alpha\right)^{n+1}_{2,i,j-1/2,k} \left(z_{i,j,k} - z_{i,j-1,k} \right) \\
&- \left(C_{p\alpha} T T_\alpha \gamma_\alpha\right)^{n+1}_{3,i,j,k+1/2} \left(z_{i,j,k+1} - z_{i,j,k} \right) \\
&+ \left(C_{p\alpha} T T_\alpha \gamma_\alpha\right)^{n+1}_{3,i,j,k-1/2} \left(z_{i,j,k} - z_{i,j,k-1} \right) \Big\} \\
&+ (k_T)^{n+1}_{1,i+1/2,j,k} \left(T^{n+1}_{i+1,j,k} - T^{n+1}_{i,j,k} \right) - (k_T)^{n+1}_{1,i-1/2,j,k} \left(T^{n+1}_{i,j,k} - T^{n+1}_{i-1,j,k} \right) \\
&+ (k_T)^{n+1}_{2,i,j+1/2,k} \left(T^{n+1}_{i,j+1,k} - T^{n+1}_{i,j,k} \right) - (k_T)^{n+1}_{2,i,j-1/2,k} \left(T^{n+1}_{i,j,k} - T^{n+1}_{i,j-1,k} \right) \\
&+ (k_T)^{n+1}_{3,i,j,k+1/2} \left(T^{n+1}_{i,j,k+1} - T^{n+1}_{i,j,k} \right) - (k_T)^{n+1}_{3,i,j,k-1/2} \left(T^{n+1}_{i,j,k} - T^{n+1}_{i,j,k-1} \right) \\
&+ Q^{n+1}_{c,i,j,k} - Q^{n+1}_{L,i,j,k},
\end{aligned}
\tag{9.26}
$$

where $Q_{\alpha,i,j,k} = (Vq_\alpha)_{i,j,k}$, $Q_{c,i,j,k} = (Vq_c)_{i,j,k}$, $Q_{L,i,j,k} = (Vq_L)_{i,j,k}$, and the numerical transmissibilities at the gridblock boundaries

$$
\frac{A_i k_T}{h_i}, \quad \frac{A_i T_{\alpha i}}{h_i}, \quad \frac{A_i T_{m\alpha i}}{h_i}, \quad i = 1, 2, 3, \ \alpha = w, o, g, \ m = 1, 2, \ldots, N_c,
$$

are still indicated by $T_{\alpha i}$, $T_{m\alpha i}$, and k_{Ti}, respectively, where A_i is the cross-sectional area normal to the x_i-direction.

As in the previous chapters, the transmissibility terms in equations (9.25) and (9.26) must be properly averaged. That is, the rock property (e.g., absolute permeability and solid heat capacity), fluid properties (e.g., viscosities and densities), and rock/fluid properties (i.e., relative permeabilities, capillary pressures, and thermal conductivity) at internal boundaries of gridblocks should be evaluated using the harmonic averaging, (weighted) arithmetic averaging, and upstream weighting techniques, respectively.

Undersaturated state

Recall that the capillary pressures p_{cow} and p_{cgo} and relative permeabilities $k_{r\alpha}$ are known functions of saturation and temperature, and the viscosities μ_α, molar densities ξ_α, and mass densities ρ_α are functions of their respective phase pressure, compositions, and temperature, $\alpha = w, o, g$.

Since the system of equations (9.25) and (9.26) is nonlinear in the primary unknowns $(p, S, x_1, x_2, \dots, x_{N_c-2}, T)$, it can be linearized via the Newton–Raphson iteration. For a generic function v of time, set

$$v^{n+1,l+1} = v^{n+1,l} + \delta v^{n+1,l+1},$$

where l refers to the Newton–Raphson iteration number and $\delta v^{n+1,l+1}$ represents the increment in this iteration step. When no ambiguity occurs, we replace $v^{n+1,l+1}$ and $v^{n+1,l}$ by v^{l+1} and v^l, respectively (i.e., the superscript $n+1$ is omitted).

The residuals at the Newton–Raphson iteration level l for equations (9.25) and (9.26) are (recalling that $S_g = 0$ in the undersaturated state)

$$\begin{aligned}
R^l_{m,i,j,k} = {} & \frac{1}{\Delta t}\left(V\left[(\phi F_m x_m)^l - (\phi F_m x_m)^n\right]\right)_{i,j,k} \\
& - \sum_{\alpha=w}^{o}\Big\{(T_{m\alpha 1})^l_{i+1/2,j,k}\left(p^l_{\alpha,i+1,j,k} - p^l_{\alpha,i,j,k}\right) \\
& - (T_{m\alpha 1})^l_{i-1/2,j,k}\left(p^l_{\alpha,i,j,k} - p^l_{\alpha,i-1,j,k}\right) \\
& + (T_{m\alpha 2})^l_{i,j+1/2,k}\left(p^l_{\alpha,i,j+1,k} - p^l_{\alpha,i,j,k}\right) - (T_{m\alpha 2})^l_{i,j-1/2,k}\left(p^l_{\alpha,i,j,k} - p^l_{\alpha,i,j-1,k}\right) \\
& + (T_{m\alpha 3})^l_{i,j,k+1/2}\left(p^l_{\alpha,i,j,k+1} - p^l_{\alpha,i,j,k}\right) - (T_{m\alpha 3})^l_{i,j,k-1/2}\left(p^l_{\alpha,i,j,k} - p^l_{\alpha,i,j,k-1}\right) \\
& - (T_{m\alpha}\gamma_\alpha)^l_{1,i+1/2,j,k}\left(z_{i+1,j,k} - z_{i,j,k}\right) + (T_{m\alpha}\gamma_\alpha)^l_{1,i-1/2,j,k}\left(z_{i,j,k} - z_{i-1,j,k}\right) \\
& - (T_{m\alpha}\gamma_\alpha)^l_{2,i,j+1/2,k}\left(z_{i,j+1,k} - z_{i,j,k}\right) + (T_{m\alpha}\gamma_\alpha)^l_{2,i,j-1/2,k}\left(z_{i,j,k} - z_{i,j-1,k}\right) \\
& - (T_{m\alpha}\gamma_\alpha)^l_{3,i,j,k+1/2}\left(z_{i,j,k+1} - z_{i,j,k}\right) + (T_{m\alpha}\gamma_\alpha)^l_{3,i,j,k-1/2}\left(z_{i,j,k} - z_{i,j,k-1}\right)\Big\} \\
& - \sum_{\alpha=w}^{o}(x_{m\alpha}Q_\alpha)^l_{i,j,k}, \quad m = 1, 2, \dots, N_c,
\end{aligned} \tag{9.27}$$

and

$$\begin{aligned}
R^l_{T,i,j,k} = {} & \frac{1}{\Delta t}\left(V\left[\left(\phi\sum_{\alpha=w}^{o}\rho_\alpha S_\alpha C_{V\alpha}T + (1-\phi)\rho_s C_s T\right)^l\right.\right. \\
& \left.\left. - \left(\phi\sum_{\alpha=w}^{o}\rho_\alpha S_\alpha C_{V\alpha}T + (1-\phi)\rho_s C_s T\right)^n\right]\right)_{i,j,k} \\
& - \sum_{\alpha=w}^{o}\Big\{(C_{p\alpha}TT_\alpha)^l_{1,i+1/2,j,k}\left(p^l_{\alpha,i+1,j,k} - p^l_{\alpha,i,j,k}\right) \\
& \qquad - (C_{p\alpha}TT_\alpha)^l_{1,i-1/2,j,k}\left(p^l_{\alpha,i,j,k} - p^l_{\alpha,i-1,j,k}\right)
\end{aligned}$$

$$
\begin{aligned}
&+(C_{p\alpha}TT_\alpha)^l_{2,i,j+1/2,k}\left(p^l_{\alpha,i,j+1,k}-p^l_{\alpha,i,j,k}\right)\\
&-(C_{p\alpha}TT_\alpha)^l_{2,i,j-1/2,k}\left(p^l_{\alpha,i,j,k}-p^l_{\alpha,i,j-1,k}\right)\\
&+(C_{p\alpha}TT_\alpha)^l_{3,i,j,k+1/2}\left(p^l_{\alpha,i,j,k+1}-p^l_{\alpha,i,j,k}\right)\\
&-(C_{p\alpha}TT_\alpha)^l_{3,i,j,k-1/2}\left(p^l_{\alpha,i,j,k}-p^l_{\alpha,i,j,k-1}\right)\\
&-(C_{p\alpha}TT_\alpha\gamma_\alpha)^l_{1,i+1/2,j,k}\left(z_{i+1,j,k}-z_{i,j,k}\right)\\
&+\left(C_{p\alpha}TT_\alpha\gamma_\alpha\right)^l_{1,i-1/2,j,k}\left(z_{i,j,k}-z_{i-1,j,k}\right)\\
&-\left(C_{p\alpha}TT_\alpha\gamma_\alpha\right)^l_{2,i,j+1/2,k}\left(z_{i,j+1,k}-z_{i,j,k}\right)\\
&+\left(C_{p\alpha}TT_\alpha\gamma_\alpha\right)^l_{2,i,j-1/2,k}\left(z_{i,j,k}-z_{i,j-1,k}\right)\\
&-\left(C_{p\alpha}TT_\alpha\gamma_\alpha\right)^l_{3,i,j,k+1/2}\left(z_{i,j,k+1}-z_{i,j,k}\right)\\
&+\left(C_{p\alpha}TT_\alpha\gamma_\alpha\right)^l_{3,i,j,k-1/2}\left(z_{i,j,k}-z_{i,j,k-1}\right)\Bigg\}\\
&-(k_T)^l_{1,i+1/2,j,k}\left(T^l_{i+1,j,k}-T^l_{i,j,k}\right)+(k_T)^l_{1,i-1/2,j,k}\left(T^l_{i,j,k}-T^l_{i-1,j,k}\right)\\
&-(k_T)^l_{2,i,j+1/2,k}\left(T^l_{i,j+1,k}-T^l_{i,j,k}\right)+(k_T)^l_{2,i,j-1/2,k}\left(T^l_{i,j,k}-T^l_{i,j-1,k}\right)\\
&-(k_T)^l_{3,i,j,k+1/2}\left(T^l_{i,j,k+1}-T^l_{i,j,k}\right)+(k_T)^l_{3,i,j,k-1/2}\left(T^l_{i,j,k}-T^l_{i,j,k-1}\right)\\
&-Q^l_{c,i,j,k}+Q^l_{L,i,j,k}.
\end{aligned}
\tag{9.28}
$$

Accordingly, the residual equations for the increments of the primary unknowns are

$$
\begin{aligned}
&\frac{\partial R^l_{m,i,j,k}}{\partial p_{i,j,k-1}}\delta p^{l+1}_{i,j,k-1}+\frac{\partial R^l_{m,i,j,k}}{\partial p_{i,j-1,k}}\delta p^{l+1}_{i,j-1,k}+\frac{\partial R^l_{m,i,,j,k}}{\partial p_{i-1,j,k}}\delta p^{l+1}_{i-1,j,k}\\
&\quad+\frac{\partial R^l_{m,i,j,k}}{\partial p_{i,j,k}}\delta p^{l+1}_{i,j,k}+\frac{\partial R^l_{m,i,j,k}}{\partial p_{i+1,j,k}}\delta p^{l+1}_{i+1,j,k}+\frac{\partial R^l_{m,i,j,k}}{\partial p_{i,j+1,k}}\delta p^{l+1}_{i,j+1,k}\\
&\quad+\frac{\partial R^l_{m,i,j,k}}{\partial p_{i,j,k+1}}\delta p^{l+1}_{i,j,k+1}+\frac{\partial R^l_{m,i,j,k}}{\partial S_{i,j,k-1}}\delta S^{l+1}_{i,j,k-1}+\frac{\partial R^l_{m,i,j,k}}{\partial S_{i,j-1,k}}\delta S^{l+1}_{i,j-1,k}\\
&\quad+\frac{\partial R^l_{m,i,,j,k}}{\partial S_{i-1,j,k}}\delta S^{l+1}_{i-1,j,k}+\frac{\partial R^l_{m,i,j,k}}{\partial S_{i,j,k}}\delta S^{l+1}_{i,j,k}+\frac{\partial R^l_{m,i,j,k}}{\partial S_{i+1,j,k}}\delta S^{l+1}_{i+1,j,k}\\
&\quad+\frac{\partial R^l_{m,i,j,k}}{\partial S_{i,j+1,k}}\delta S^{l+1}_{i,j+1,k}+\frac{\partial R^l_{m,i,j,k}}{\partial S_{i,j,k+1}}\delta S^{l+1}_{i,j,k+1}\\
&\quad+\sum_{r=1}^{N_c-2}\left\{\frac{\partial R^l_{m,i,j,k}}{\partial (x_r)_{i,j,k}}\delta(x_r)^{l+1}_{i,j,k}+\frac{\partial R^l_{m,i,j,k}}{\partial (x_r)_{i+1,j,k}}\delta(x_r)^{l+1}_{i+1,j,k}+\frac{\partial R^l_{m,i,j,k}}{\partial (x_r)_{i,j+1,k}}\delta(x_r)^{l+1}_{i,j+1,k}\right.\\
&\qquad+\frac{\partial R^l_{m,i,j,k}}{\partial (x_r)_{i,j,k+1}}\delta(x_r)^{l+1}_{i,j,k+1}+\frac{\partial R^l_{m,i,j,k}}{\partial (x_r)_{i,j,k-1}}\delta(x_r)^{l+1}_{i,j,k-1}\\
&\qquad\left.+\frac{\partial R^l_{m,i,j,k}}{\partial (x_r)_{i,j-1,k}}\delta(x_r)^{l+1}_{i,j-1,k}+\frac{\partial R^l_{m,i,,j,k}}{\partial (x_r)_{i-1,j,k}}\delta(x_r)^{l+1}_{i-1,j,k}\right\}
\end{aligned}
$$

$$
\begin{aligned}
&+ \frac{\partial R^l_{m,i,j,k}}{\partial T_{i,j,k-1}} \delta T^{l+1}_{i,j,k-1} + \frac{\partial R^l_{m,i,j,k}}{\partial T_{i,j-1,k}} \delta T^{l+1}_{i,j-1,k} + \frac{\partial R^l_{m,i,,j,k}}{\partial T_{i-1,j,k}} \delta T^{l+1}_{i-1,j,k} \\
&+ \frac{\partial R^l_{m,i,j,k}}{\partial T_{i,j,k}} \delta T^{l+1}_{i,j,k} + \frac{\partial R^l_{m,i,j,k}}{\partial T_{i+1,j,k}} \delta T^{l+1}_{i+1,j,k} \\
&+ \frac{\partial R^l_{m,i,j,k}}{\partial T_{i,j+1,k}} \delta T^{l+1}_{i,j+1,k} + \frac{\partial R^l_{m,i,j,k}}{\partial T_{i,j,k+1}} \delta T^{l+1}_{i,j,k+1} \\
&= -R^l_{m,i,j,k}, \quad m = 1, 2, \ldots, N_c,
\end{aligned}
\tag{9.29}
$$

and

$$
\begin{aligned}
&\frac{\partial R^l_{T,i,j,k}}{\partial p_{i,j,k-1}} \delta p^{l+1}_{i,j,k-1} + \frac{\partial R^l_{T,i,j,k}}{\partial p_{i,j-1,k}} \delta p^{l+1}_{i,j-1,k} + \frac{\partial R^l_{T,i,,j,k}}{\partial p_{i-1,j,k}} \delta p^{l+1}_{i-1,j,k} \\
&+ \frac{\partial R^l_{T,i,j,k}}{\partial p_{i,j,k}} \delta p^{l+1}_{i,j,k} + \frac{\partial R^l_{T,i,j,k}}{\partial p_{i+1,j,k}} \delta p^{l+1}_{i+1,j,k} + \frac{\partial R^l_{T,i,j,k}}{\partial p_{i,j+1,k}} \delta p^{l+1}_{i,j+1,k} \\
&+ \frac{\partial R^l_{T,i,j,k}}{\partial p_{i,j,k+1}} \delta p^{l+1}_{i,j,k+1} + \frac{\partial R^l_{T,i,j,k}}{\partial S_{i,j,k-1}} \delta S^{l+1}_{i,j,k-1} + \frac{\partial R^l_{T,i,j,k}}{\partial S_{i,j-1,k}} \delta S^{l+1}_{i,j-1,k} \\
&+ \frac{\partial R^l_{T,i,,j,k}}{\partial S_{i-1,j,k}} \delta S^{l+1}_{i-1,j,k} + \frac{\partial R^l_{T,i,j,k}}{\partial S_{i,j,k}} \delta S^{l+1}_{i,j,k} + \frac{\partial R^l_{T,i,j,k}}{\partial S_{i+1,j,k}} \delta S^{l+1}_{i+1,j,k} \\
&+ \frac{\partial R^l_{T,i,j,k}}{\partial S_{i,j+1,k}} \delta S^{l+1}_{i,j+1,k} + \frac{\partial R^l_{T,i,j,k}}{\partial S_{i,j,k+1}} \delta S^{l+1}_{i,j,k+1} \\
&+ \sum_{r=1}^{N_c-2} \left\{ \frac{\partial R^l_{T,i,j,k}}{\partial (x_r)_{i,j,k}} \delta (x_r)^{l+1}_{i,j,k} + \frac{\partial R^l_{T,i,j,k}}{\partial (x_r)_{i+1,j,k}} \delta (x_r)^{l+1}_{i+1,j,k} + \frac{\partial R^l_{T,i,j,k}}{\partial (x_r)_{i,j+1,k}} \delta (x_r)^{l+1}_{i,j+1,k} \right. \\
&\qquad + \frac{\partial R^l_{T,i,j,k}}{\partial (x_r)_{i,j,k+1}} \delta (x_r)^{l+1}_{i,j,k+1} + \frac{\partial R^l_{T,i,j,k}}{\partial (x_r)_{i,j,k-1}} \delta (x_r)^{l+1}_{i,j,k-1} \\
&\qquad \left. + \frac{\partial R^l_{T,i,j,k}}{\partial (x_r)_{i,j-1,k}} \delta (x_r)^{l+1}_{i,j-1,k} + \frac{\partial R^l_{T,i,,j,k}}{\partial (x_r)_{i-1,j,k}} \delta (x_r)^{l+1}_{i-1,j,k} \right\} \\
&+ \frac{\partial R^l_{T,i,j,k}}{\partial T_{i,j,k-1}} \delta T^{l+1}_{i,j,k-1} + \frac{\partial R^l_{T,i,j,k}}{\partial T_{i,j-1,k}} \delta T^{l+1}_{i,j-1,k} + \frac{\partial R^l_{T,i,,j,k}}{\partial T_{i-1,j,k}} \delta T^{l+1}_{i-1,j,k} \\
&+ \frac{\partial R^l_{T,i,j,k}}{\partial T_{i,j,k}} \delta T^{l+1}_{i,j,k} + \frac{\partial R^l_{T,i,j,k}}{\partial T_{i+1,j,k}} \delta T^{l+1}_{i+1,j,k} + \frac{\partial R^l_{T,i,j,k}}{\partial T_{i,j+1,k}} \delta T^{l+1}_{i,j+1,k} \\
&+ \frac{\partial R^l_{T,i,j,k}}{\partial T_{i,j,k+1}} \delta T^{l+1}_{i,j,k+1} = -R^l_{T,i,j,k}.
\end{aligned}
\tag{9.30}
$$

The linear system of equations (9.29) and (9.30) is solved for the increments of the primary unknowns, and then the following update is performed until a tolerance in error is reached:

$$
\begin{aligned}
&p^{l+1} = p^l + \delta p^{l+1}, \quad S^{l+1} = S^l + \delta S^{l+1}, \\
&x^{l+1}_m = x^l_m + \delta x^{l+1}_m, \quad T^{l+1} = T^l + \delta T^{l+1}, \quad m = 1, 2, \ldots, N_c.
\end{aligned}
$$

Saturated state

In the saturated state, all three phases coexist, and the primary unknowns are p, S_w, S_o, and x_i, $i = 1, 2, \ldots, N_c - 2$. The residuals are

$$
\begin{aligned}
R^l_{m,i,j,k} = \frac{1}{\Delta t}\left(V\left[(\phi F_m x_m)^l - (\phi F_m x_m)^n\right]\right)_{i,j,k} \\
- \sum_{\alpha=w}^{g}\Big\{ &(T_{m\alpha 1})^l_{i+1/2,j,k}\left(p^l_{\alpha,i+1,j,k} - p^l_{\alpha,i,j,k}\right) - (T_{m\alpha 1})^l_{i-1/2,j,k}\left(p^l_{\alpha,i,j,k} - p^l_{\alpha,i-1,j,k}\right) \\
&+ (T_{m\alpha 2})^l_{i,j+1/2,k}\left(p^l_{\alpha,i,j+1,k} - p^l_{\alpha,i,j,k}\right) - (T_{m\alpha 2})^l_{i,j-1/2,k}\left(p^l_{\alpha,i,j,k} - p^l_{\alpha,i,j-1,k}\right) \\
&+ (T_{m\alpha 3})^l_{i,j,k+1/2}\left(p^l_{\alpha,i,j,k+1} - p^l_{\alpha,i,j,k}\right) - (T_{m\alpha 3})^l_{i,j,k-1/2}\left(p^l_{\alpha,i,j,k} - p^l_{\alpha,i,j,k-1}\right) \\
&- (T_{m\alpha}\gamma_\alpha)^l_{1,i+1/2,j,k}\left(z_{i+1,j,k} - z_{i,j,k}\right) + (T_{m\alpha}\gamma_\alpha)^l_{1,i-1/2,j,k}\left(z_{i,j,k} - z_{i-1,j,k}\right) \\
&- (T_{m\alpha}\gamma_\alpha)^l_{2,i,j+1/2,k}\left(z_{i,j+1,k} - z_{i,j,k}\right) + (T_{m\alpha}\gamma_\alpha)^l_{2,i,j-1/2,k}\left(z_{i,j,k} - z_{i,j-1,k}\right) \\
&- (T_{m\alpha}\gamma_\alpha)^l_{3,i,j,k+1/2}\left(z_{i,j,k+1} - z_{i,j,k}\right) + (T_{m\alpha}\gamma_\alpha)^l_{3,i,j,k-1/2}\left(z_{i,j,k} - z_{i,j,k-1}\right)\Big\} \\
- \sum_{\alpha=w}^{g} (x_{m\alpha} Q_\alpha)^l_{i,j,k}, \quad & m = 1, 2, \ldots, N_c,
\end{aligned}
$$

and

$$
\begin{aligned}
R^l_{T,i,j,k} = \frac{1}{\Delta t}\Bigg(V\Bigg[&\left(\phi\sum_{\alpha=w}^{g}\rho_\alpha S_\alpha C_{V\alpha} T + (1-\phi)\rho_s C_s T\right)^l \\
&- \left(\phi\sum_{\alpha=w}^{g}\rho_\alpha S_\alpha C_{V\alpha} T + (1-\phi)\rho_s C_s T\right)^n\Bigg]\Bigg)_{i,j,k} \\
- \sum_{\alpha=w}^{g}\Big\{ &(C_{p\alpha} T T_\alpha)^l_{1,i+1/2,j,k}\left(p^l_{\alpha,i+1,j,k} - p^l_{\alpha,i,j,k}\right) \\
&- (C_{p\alpha} T T_\alpha)^l_{1,i-1/2,j,k}\left(p^l_{\alpha,i,j,k} - p^l_{\alpha,i-1,j,k}\right) \\
&+ (C_{p\alpha} T T_\alpha)^l_{2,i,j+1/2,k}\left(p^l_{\alpha,i,j+1,k} - p^l_{\alpha,i,j,k}\right) \\
&- (C_{p\alpha} T T_\alpha)^l_{2,i,j-1/2,k}\left(p^l_{\alpha,i,j,k} - p^l_{\alpha,i,j-1,k}\right) \\
&+ (C_{p\alpha} T T_\alpha)^l_{3,i,j,k+1/2}\left(p^l_{\alpha,i,j,k+1} - p^l_{\alpha,i,j,k}\right) \\
&- (C_{p\alpha} T T_\alpha)^l_{3,i,j,k-1/2}\left(p^l_{\alpha,i,j,k} - p^l_{\alpha,i,j,k-1}\right) \\
&- (C_{p\alpha} T T_\alpha \gamma_\alpha)^l_{1,i+1/2,j,k}\left(z_{i+1,j,k} - z_{i,j,k}\right) \\
&+ \left(C_{p\alpha} T T_\alpha \gamma_\alpha\right)^l_{1,i-1/2,j,k}\left(z_{i,j,k} - z_{i-1,j,k}\right) \\
&- \left(C_{p\alpha} T T_\alpha \gamma_\alpha\right)^l_{2,i,j+1/2,k}\left(z_{i,j+1,k} - z_{i,j,k}\right)
\end{aligned}
$$

$$
\begin{aligned}
&+ \left(C_{p\alpha} T T_\alpha \gamma_\alpha\right)^l_{2,i,j-1/2,k} \left(z_{i,j,k} - z_{i,j-1,k}\right) \\
&- \left(C_{p\alpha} T T_\alpha \gamma_\alpha\right)^l_{3,i,j,k+1/2} \left(z_{i,j,k+1} - z_{i,j,k}\right) \\
&+ \left(C_{p\alpha} T T_\alpha \gamma_\alpha\right)^l_{3,i,j,k-1/2} \left(z_{i,j,k} - z_{i,j,k-1}\right) \Big\} \\
&- (k_T)^l_{1,i+1/2,j,k} \left(T^l_{i+1,j,k} - T^l_{i,j,k}\right) + (k_T)^l_{1,i-1/2,j,k} \left(T^l_{i,j,k} - T^l_{i-1,j,k}\right) \\
&- (k_T)^l_{2,i,j+1/2,k} \left(T^l_{i,j+1,k} - T^l_{i,j,k}\right) + (k_T)^l_{2,i,j-1/2,k} \left(T^l_{i,j,k} - T^l_{i,j-1,k}\right) \\
&- (k_T)^l_{3,i,j,k+1/2} \left(T^l_{i,j,k+1} - T^l_{i,j,k}\right) + (k_T)^l_{3,i,j,k-1/2} \left(T^l_{i,j,k} - T^l_{i,j,k-1}\right) \\
&- Q^l_{c,i,j,k} + Q^l_{L,i,j,k}.
\end{aligned}
$$

The corresponding residual equations for the increments of these variables are

$$
\begin{aligned}
&\frac{\partial R^l_{m,i,j,k}}{\partial p_{i,j,k-1}} \delta p^{l+1}_{i,j,k-1} + \frac{\partial R^l_{m,i,j,k}}{\partial p_{i,j-1,k}} \delta p^{l+1}_{i,j-1,k} + \frac{\partial R^l_{m,i,,j,k}}{\partial p_{i-1,j,k}} \delta p^{l+1}_{i-1,j,k} \\
&+ \frac{\partial R^l_{m,i,j,k}}{\partial p_{i,j,k}} \delta p^{l+1}_{i,j,k} + \frac{\partial R^l_{m,i,j,k}}{\partial p_{i+1,j,k}} \delta p^{l+1}_{i+1,j,k} + \frac{\partial R^l_{m,i,j,k}}{\partial p_{i,j+1,k}} \delta p^{l+1}_{i,j+1,k} + \frac{\partial R^l_{m,i,j,k}}{\partial p_{i,j,k+1}} \delta p^{l+1}_{i,j,k+1} \\
&+ \frac{\partial R^l_{m,i,j,k}}{\partial S_{w,i,j,k-1}} \delta S^{l+1}_{w,i,j,k-1} + \frac{\partial R^l_{m,i,j,k}}{\partial S_{w,i,j-1,k}} \delta S^{l+1}_{w,i,j-1,k} + \frac{\partial R^l_{m,i,,j,k}}{\partial S_{w,i-1,j,k}} \delta S^{l+1}_{w,i-1,j,k} \\
&+ \frac{\partial R^l_{m,i,j,k}}{\partial S_{w,i,j,k}} \delta S^{l+1}_{w,i,j,k} + \frac{\partial R^l_{m,i,j,k}}{\partial S_{w,i+1,j,k}} \delta S^{l+1}_{w,i+1,j,k} + \frac{\partial R^l_{m,i,j,k}}{\partial S_{w,i,j+1,k}} \delta S^{l+1}_{w,i,j+1,k} \\
&+ \frac{\partial R^l_{m,i,j,k}}{\partial S_{w,i,j,k+1}} \delta S^{l+1}_{w,i,j,k+1} + \frac{\partial R^l_{m,i,j,k}}{\partial S_{o,i,j,k-1}} \delta S^{l+1}_{o,i,j,k-1} + \frac{\partial R^l_{m,i,j,k}}{\partial S_{o,i,j-1,k}} \delta S^{l+1}_{o,i,j-1,k} \\
&+ \frac{\partial R^l_{m,i,,j,k}}{\partial S_{o,i-1,j,k}} \delta S^{l+1}_{o,i-1,j,k} + \frac{\partial R^l_{m,i,j,k}}{\partial S_{o,i,j,k}} \delta S^{l+1}_{o,i,j,k} + \frac{\partial R^l_{m,i,j,k}}{\partial S_{o,i+1,j,k}} \delta S^{l+1}_{o,i+1,j,k} \\
&+ \frac{\partial R^l_{m,i,j,k}}{\partial S_{o,i,j+1,k}} \delta S^{l+1}_{o,i,j+1,k} + \frac{\partial R^l_{m,i,j,k}}{\partial S_{o,i,j,k+1}} \delta S^{l+1}_{o,i,j,k+1} \\
&+ \sum_{r=1}^{N_c-2} \Bigg\{ \frac{\partial R^l_{m,i,j,k}}{\partial (x_r)_{i,j,k}} \delta (x_r)^{l+1}_{i,j,k} + \frac{\partial R^l_{m,i,j,k}}{\partial (x_r)_{i+1,j,k}} \delta (x_r)^{l+1}_{i+1,j,k} + \frac{\partial R^l_{m,i,j,k}}{\partial (x_r)_{i,j+1,k}} \delta (x_r)^{l+1}_{i,j+1,k} \\
&\qquad + \frac{\partial R^l_{m,i,j,k}}{\partial (x_r)_{i,j,k+1}} \delta (x_r)^{l+1}_{i,j,k+1} + \frac{\partial R^l_{m,i,j,k}}{\partial (x_r)_{i,j,k-1}} \delta (x_r)^{l+1}_{i,j,k-1} \\
&\qquad + \frac{\partial R^l_{m,i,j,k}}{\partial (x_r)_{i,j-1,k}} \delta (x_r)^{l+1}_{i,j-1,k} + \frac{\partial R^l_{m,i,,j,k}}{\partial (x_r)_{i-1,j,k}} \delta (x_r)^{l+1}_{i-1,j,k} \Bigg\} \\
&\qquad = -R^l_{m,i,j,k}, \quad m = 1, 2, \ldots, N_c,
\end{aligned}
\tag{9.31}
$$

and

$$
\begin{aligned}
&\frac{\partial R^l_{T,i,j,k}}{\partial p_{i,j,k-1}}\delta p^{l+1}_{i,j,k-1} + \frac{\partial R^l_{T,i,j,k}}{\partial p_{i,j-1,k}}\delta p^{l+1}_{i,j-1,k} + \frac{\partial R^l_{T,i,,j,k}}{\partial p_{i-1,j,k}}\delta p^{l+1}_{i-1,j,k} \\
&+\frac{\partial R^l_{T,i,j,k}}{\partial p_{i,j,k}}\delta p^{l+1}_{i,j,k} + \frac{\partial R^l_{T,i,j,k}}{\partial p_{i+1,j,k}}\delta p^{l+1}_{i+1,j,k} + \frac{\partial R^l_{T,i,j,k}}{\partial p_{i,j+1,k}}\delta p^{l+1}_{i,j+1,k} + \frac{\partial R^l_{T,i,j,k}}{\partial p_{i,j,k+1}}\delta p^{l+1}_{i,j,k+1} \\
&+\frac{\partial R^l_{T,i,j,k}}{\partial S_{w,i,j,k-1}}\delta S^{l+1}_{w,i,j,k-1} + \frac{\partial R^l_{T,i,j,k}}{\partial S_{w,i,j-1,k}}\delta S^{l+1}_{w,i,j-1,k} + \frac{\partial R^l_{T,i,,j,k}}{\partial S_{w,i-1,j,k}}\delta S^{l+1}_{w,i-1,j,k} \\
&+\frac{\partial R^l_{T,i,j,k}}{\partial S_{w,i,j,k}}\delta S^{l+1}_{w,i,j,k} + \frac{\partial R^l_{T,i,j,k}}{\partial S_{w,i+1,j,k}}\delta S^{l+1}_{w,i+1,j,k} + \frac{\partial R^l_{T,i,j,k}}{\partial S_{w,i,j+1,k}}\delta S^{l+1}_{w,i,j+1,k} \\
&+\frac{\partial R^l_{T,i,j,k}}{\partial S_{w,i,j,k+1}}\delta S^{l+1}_{w,i,j,k+1} + \frac{\partial R^l_{T,i,j,k}}{\partial S_{o,i,j,k-1}}\delta S^{l+1}_{o,i,j,k-1} + \frac{\partial R^l_{T,i,j,k}}{\partial S_{o,i,j-1,k}}\delta S^{l+1}_{o,i,j-1,k} \\
&+\frac{\partial R^l_{T,i,,j,k}}{\partial S_{o,i-1,j,k}}\delta S^{l+1}_{o,i-1,j,k} + \frac{\partial R^l_{T,i,j,k}}{\partial S_{o,i,j,k}}\delta S^{l+1}_{o,i,j,k} + \frac{\partial R^l_{T,i,j,k}}{\partial S_{o,i+1,j,k}}\delta S^{l+1}_{o,i+1,j,k} \\
&+\frac{\partial R^l_{T,i,j,k}}{\partial S_{o,i,j+1,k}}\delta S^{l+1}_{o,i,j+1,k} + \frac{\partial R^l_{T,i,j,k}}{\partial S_{o,i,j,k+1}}\delta S^{l+1}_{o,i,j,k+1} \\
&+\sum_{r=1}^{N_c-2}\Bigg\{\frac{\partial R^l_{T,i,j,k}}{\partial (x_r)_{i,j,k}}\delta (x_r)^{l+1}_{i,j,k} + \frac{\partial R^l_{T,i,j,k}}{\partial (x_r)_{i+1,j,k}}\delta (x_r)^{l+1}_{i+1,j,k} + \frac{\partial R^l_{T,i,j,k}}{\partial (x_r)_{i,j+1,k}}\delta (x_r)^{l+1}_{i,j+1,k} \\
&\quad+\frac{\partial R^l_{T,i,j,k}}{\partial (x_r)_{i,j,k+1}}\delta (x_r)^{l+1}_{i,j,k+1} + \frac{\partial R^l_{T,i,j,k}}{\partial (x_r)_{i,j,k-1}}\delta (x_r)^{l+1}_{i,j,k-1} \\
&\quad+\frac{\partial R^l_{T,i,j,k}}{\partial (x_r)_{i,j-1,k}}\delta (x_r)^{l+1}_{i,j-1,k} + \frac{\partial R^l_{T,i,,j,k}}{\partial (x_r)_{i-1,j,k}}\delta (x_r)^{l+1}_{i-1,j,k}\Bigg\} = -R^l_{T,i,j,k}.
\end{aligned}
\tag{9.32}
$$

Again, equations (9.31) and (9.32) produce a linear system for the increments of the primary variables. Once these increments are obtained, the primary variables are updated until a convergence criterion is achieved.

Chapter 10

Practical Topics in Reservoir Simulation

This book has concentrated on the development of reservoir simulators using finite difference methods. These methods can be replaced by other discretization methods such as finite volume and finite element methods. In this chapter we discuss some practical topics that must be addressed at certain stages in petroleum reservoir simulation, no matter which discretization methods are used. The major steps in a reservoir simulation study are sketched in Fig. 10.1, with more details given in Table 10.1. Here we briefly touch on some of the basic elements of the simulation study: design of study objectives, analysis of reservoir data, development of simulation models, selection of simulation grids, history matching, and prediction of reservoir performance. Other practical topics such as upscaling, recovery optimization, and surface network systems can be found in Chen, Huan, and Ma (2006).

10.1 Design of Study Objectives

The main goal of reservoir simulation is to predict future performance of a reservoir and find ways and means of optimizing the recovery of some of the hydrocarbons under various operating conditions. Its major study objectives involve estimating fluid movement and storage, impact of operating strategies, distribution of fluid and heat, production profiles for profitability decision making, reservoir management, and evaluation and design of novel production technologies. The design of these study objectives is the most crucial and fundamental step in the application of any reservoir simulator. They dictate the scope and type of the simulation model to be employed, the amount of company resources to be allocated, the quality of history matching to be performed, and the type of prediction cases to be made.

The design of study objectives depends on the stage of hydrocarbon recovery of the subject reservoir (the appraisal (preproduction), primary, secondary, or enhanced recovery stage), the quantity and quality of available production and reservoir data, and time frames. For example, (1) in the appraisal stage, the available data from the drillstem tests from exploration and appraisal wells are very limited and can be used solely for a preliminary history matching of the reservoir model; the objectives in this stage may include the determination of the range of reserves and timing of production for the field. (2) As for the data availability, if oil/gas relative permeability data are not available, a simulation study

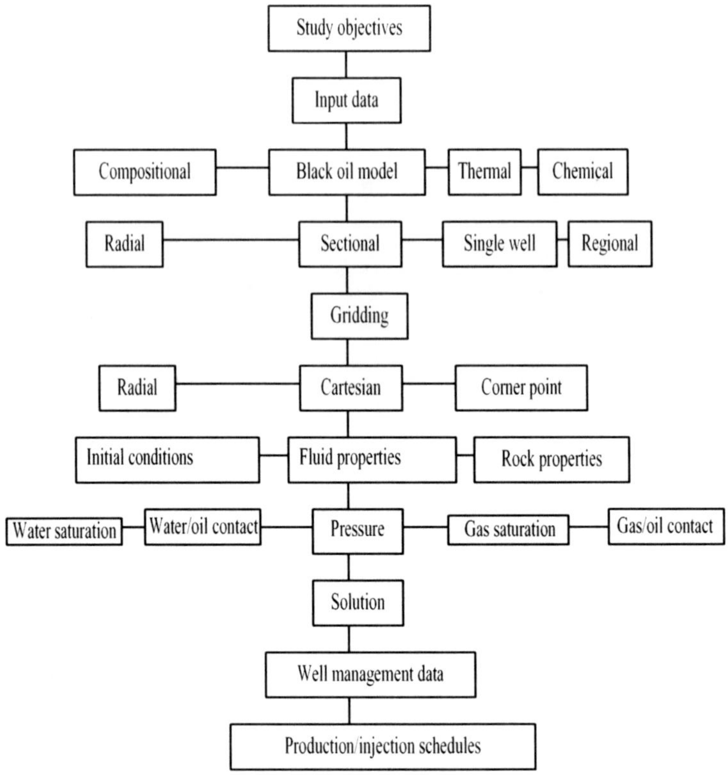

Figure 10.1. *Major steps in reservoir simulation study.*

Table 10.1. *More details on reservoir simulation steps.*

Reservoir description	Input data
Recovery mechanism	Water flood, gravity drainage, ...
Mathematical model	Black oil, compositional, ...
Numerical model	Finite difference, finite elements, ...
Computer model	Fortran, C++, ...
Model validation	Pressure vs. time, ...
History match	Gas/oil ratio, water/oil ratio, ...
Performance prediction	Production vs. time, ...

of potential gas flood will not produce any reliable prediction. (3) Regarding the timing of the study, if simulation results are required in a short period of time, detailed predictions cannot be made.

10.2 Analysis of Reservoir Data

Reservoir simulation is the most powerful predictive tool available to the reservoir engineer; it involves far more geological and reservoir data than any other reservoir performance forecast tool. As a result, it requires much more data than others. A list of data sources available to the reservoir engineer are shown in Table 10.2 (Ertekin, Abou-Kassem, and King, 2001), and data for reservoir simulation are listed in Table 10.3. Here we briefly describe some of the more common data required in reservoir simulation.

Table 10.2. *Reservoir data.*

Depth	Structure area	Gross thickness
Net thickness	Lithology	Production rate
Mechanical properties	Contacts	Pressure
Porosity	Permeability	Fluid saturation
Pore sizes	Production mechanism	Hydrocarbon properties
Water properties		

Table 10.3. *Data for reservoir simulation.*

Static data	Rock
D	Formation top (structure)
h_t	Gross formation thickness
h_n	Net pay thickness
ϕ	Porosity at initial pressure
p_{cow}	Drainage (water/oil) capillary pressure
p_{cgo}	Drainage (gas/oil) capillary pressure
	Fluid
B_w	Water formation volume factor
B_o	Oil formation volume factor
B_g	Gas formation volume factor
ρ_{ws}	Water density at standard conditions
ρ_{os}	Oil density at standard conditions
ρ_{gs}	Gas density at standard conditions
Data for water/oil displacement	Rock
k	Absolute permeability
$k_{rw}(S_w)$	Water relative permeability
$k_{ro}(S_w)$	Oil relative permeability
$p_{cow}(S_w)$	Imbibition (water/oil) capillary pressure
c_R	Rock compressibility
	Fluid
$B_w(p)$	Water formation volume factor
$B_o(p)$	Oil formation volume factor
$\mu_w(p)$	Water viscosity
$\mu_o(p)$	Oil viscosity
c_w	Water compressibility
c_o	Oil compressibility
Data for gas/oil displacement	Rock
$k_{ro}(S_o)$	Oil relative permeability
$k_{rg}(S_o)$	Gas relative permeability
	Fluid
$R_{so}(p)$	Gas solubility
$B_g(p)$	Gas formation volume factor
$\mu_g(p)$	Gas viscosity

10.2.1 Geophysical Data

Geophysical and geological data are required in any reservoir simulator. In general, the geophysical data describe the envelope of a reservoir, and the geological data give the internal anatomy. Seismic data are the most widely used geophysical data in reservoir simulation. These data are generated using acoustic energy at the earth's surface, transmitting this energy toward the subject formation, and measuring and recording the time required for this energy to be reflected back to the surface through subsurface strata (cf. Fig. 10.2 for a 2D seismic process). Fig. 10.3 presents seismic lines for a cross section through the

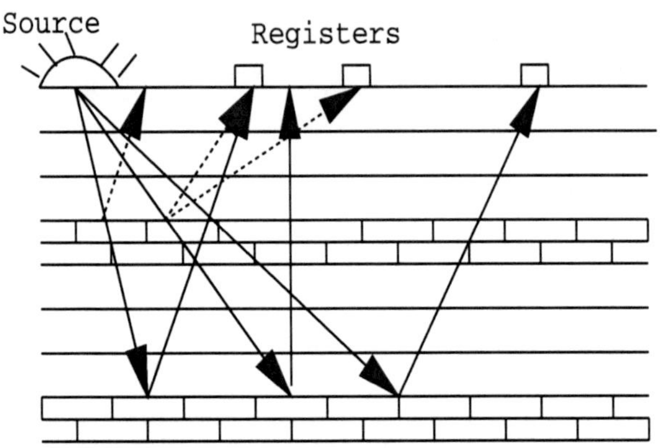

Figure 10.2. *Seismic transmitting and recording.*

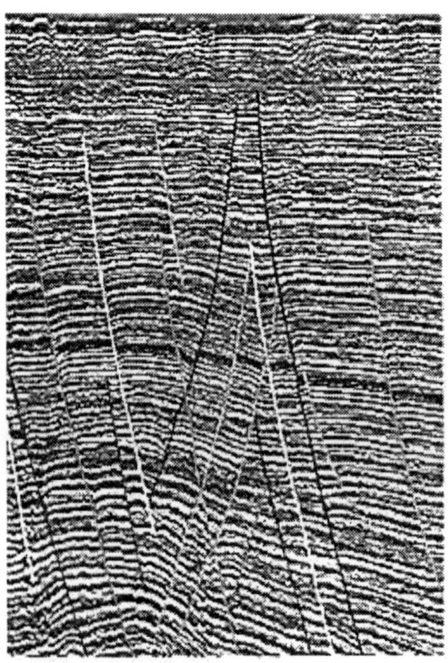

Figure 10.3. *Seismic lines for a cross section.*

reservoir and the adjacent underburden and overburden strata. These cross sections are used to provide the structure top and base and identify faults, formation pinchouts, reservoir continuity, unconformity, and variations in reservoir thickness (Ertekin, Abou-Kassem, and King, 2001). A drawback of seismic data is their relatively poor resolution. Any reservoir property on a scale smaller than the resolution of the seismic survey cannot be detected by the seismic technique.

10.2.2 Geological Data

Reservoir simulation is the only predictive tool that rigorously takes into account reservoir geology. In fact, all available geological data must be properly incorporated into a simulation study to meet the objectives. A geological model determines the distribution of reservoir properties such as porosity, permeability, net pay, and flow barriers and is the basis on which the reservoir simulation model is built. The requirements of the geological model for reservoir simulation are given in Table 10.4. The map requirement can be either a single value for each reservoir layer or a contour map. The geological data sources include stratigraphy, mud logging, geochemistry, paleontology, thin section, scanning electron microscopy, and outcrop studies.

10.2.3 Engineering Data

Geological data are concerned with rock properties and the processes that control the distribution of these properties, while engineering data are concerned with the dynamics and statics of reservoir fluids. The raw data used in the construction of geological models are the same as those in engineering studies: core samples, openhole logs, and pressure-transient data; however, the methods in which the data are processed and the study objectives are distinct.

Core samples provide geologists and engineers with the sole opportunity for directly measuring how actual reservoir rock and fluids behave under constraints. Table 10.5 indicates some of the data obtained from core measurements for reservoir simulation. These data may come from visual inspection, routine core analysis, or special analysis. Visual inspection is a procedure of observing and handling the core material to obtain fundamental

Table 10.4. *Geological data.*

Structure top	Net formation thickness	Gross formation thickness
Porosity	Horizontal permeability	Vertical permeability
Initial saturations	Endpoint saturations	Fluid contacts

Table 10.5. *Engineering data.*

Lithology	Net formation thickness	Gross formation thickness
Porosity	Horizontal permeability	Vertical permeability
Initial saturations	Endpoint saturations	Rock compressibility
Relative permeability	Capillary pressure	

Table 10.6. *Openhole log data in reservoir simulation.*

Lithology	Water saturation	Gas saturation
Porosity	Net formation thickness	Gross formation thickness
Vertical pressure gradient		

Table 10.7. *Pressure-transient data in reservoir simulation.*

Horizontal permeability	Formation damage	Fluid samples
Shut-in well pressure	Static reservoir pressure	

rock properties, including lithology, presence of shale laminations and crossbeddings, gross reservoir thickness, and net- to gross-thickness ratio. Routine core analysis measures basic formation properties such as the porosity, absolute permeability, and initial saturations of core samples. Special analysis deals with more complex reservoir properties such as end-point saturations, relative permeability, capillary pressure, and compressibility, along with many other reservoir properties.

Well log data are critical in measuring formation properties at in situ conditions and at the reservoir scale. As a result, the degree of scaleup for these data is less than core data. Furthermore, openhole logs are run on almost drilled wells. Thus the well log data are the most abundant data available to geologists, petrophysicists, and petroleum engineers. A list of reservoir simulation input data from openhole logs is given in Table 10.6 (Ertekin, Abou-Kassem, and King 2001).

Pressure-transient data are measured at a scale appropriate for numerical reservoir simulation and need little scaleup before use in the simulation as with well log data. These data are concerned with cell permeabilities and well data in the simulation model. A list of pressure-transient data is shown in Table 10.7.

We have so far considered the reservoir-rock properties, i.e., those for the hydrocarbon-bearing rock associated with a subsurface hydrocarbon reservoir. In some petroleum reservoirs, there exist sections of the producing formation that are not hydrocarbon-bearing. The performance of the reservoir can be greatly affected by this nonreservoir rock, such as shales and aquifers. The effect of the nonreservoir rock may need be taken into account in reservoir simulation.

10.3 Development of Simulation Models

10.3.1 Model Selection

The discussion of data sources in the previous section applies to general reservoirs. The development of a reservoir simulation model deals with gathering these data to build an integrated, coherent mathematical representation of the underlying reservoir. Basic components in reservoir modeling are listed in Table 10.8, and some of these parts are briefly discussed.

The two important characteristics of a petroleum reservoir are the nature of the rock and of the fluids filling it. A reservoir is usually *heterogeneous*; its properties heavily depend on the space location. A *fractured reservoir* is heterogeneous, for example. It consists of a

Table 10.8. *Choice of modeling approaches.*

Reservoir type	Single porosity, dual porosity, dual porosity/permeability
Fluid model	Black oil, compositional, thermal, chemical
Recovery processes	Primary, secondary, enhanced oil recovery (miscible, thermal, and chemical)
Model scope	Single well, cross-sectional, window, and full field models
Model dimension	0D (tank type), 1D, 2D, and 3D models
Nonlinear solvers	IMPES (IMPEC), sequential, SS, and adaptive implicit
Linear solvers	Direct, iterative, preconditioned methods

set of blocks of porous media (the *matrix*) and a net of fractures. The rock properties in such a reservoir dramatically change; its permeability may vary from one milli-darcy (md) in the matrix to thousands md in the fractures. While the governing equations for the fractured reservoir are similar to those for an ordinary reservoir, they have additional difficulties that must be overcome. The mathematical models presented in this book take into account the heterogeneity of a porous medium but are limited to heterogeneous reservoirs (i.e., single porosity reservoirs).

The concept of *dual porosity* (and *dual porosity/permeability*) has been utilized to model the flow of fluids through fractured reservoirs (Pirson, 1953; Barenblatt, Zheltov, and Kochina, 1960; Warren and Root, 1963; Kazemi, 1969). In this concept, the fracture system is treated as a porous structure distinct from the usual porous structure of the matrix itself. The fracture system is highly permeable but can store very little fluid, while the matrix has the opposite characteristics. When developing a dual-porosity model, it is critical to treat the *flow transfer terms* between the fracture and matrix systems (Chen, Huan, and Ma, 2006). To some extent, most petroleum reservoirs are naturally fractured.

The nature of the fluids filling a petroleum reservoir strongly depends on the recovery processes occurring in the reservoir. In the very early stage, the reservoir essentially contains a single fluid such as gas or oil (the presence of water can usually be neglected). Often the pressure at this stage is so high that the gas or oil is produced by simple natural decompression without any pumping effort at the wells. This stage is referred to as *primary recovery* (primary depletion), and it ends when a pressure equilibrium between the oil field and the atmosphere occurs. Primary recovery usually leaves 70–85% of hydrocarbons in a conventional oil reservoir. During primary depletion, either the black oil or compositional model can be used in reservoir simulation, depending on the importance of the phase behavior.

To recover part of the remaining oil, for example, a fluid (water or gas) is injected into some wells (*injection wells*) while oil is produced through other wells (*production wells*). This process serves to maintain high reservoir pressure and flow rates. It also displaces some of the oil and pushes it toward the production wells. This stage of oil recovery is called *secondary recovery* (*water* or *gas flood*). For gas-displacement processes in volatile oil reservoirs and pressure maintenance processes in gas condensate reservoirs, mass transfer can significantly influence reservoir behavior. Incorporation of mass transfer effects can be achieved only with the computational model.

Water or gas flood is not very effective, and after this stage 50% or more of hydrocarbons often remain in the reservoir. Due to strong surface tension, a large amount of oil is trapped in small pores and cannot be washed out using this technique. When the oil is heavy and viscous, for example, the water is extremely mobile. If the flow rate is sufficiently high, instead of producing oil, the production wells primarily produce water.

To recover more of the hydrocarbons, several enhanced recovery techniques have been developed. These techniques involve complex chemical and thermal effects and are termed *tertiary recovery* or *enhanced recovery*. Enhanced oil recovery is oil recovery by injecting materials that are not normally present in a petroleum reservoir. There are many different versions of enhanced recovery techniques, but the main objectives of these techniques are to achieve miscibility (and thus eliminate the residual oil saturation) and reduce viscosity. The miscibility is achieved by increasing temperature (e.g., in situ combustion) or by injecting other chemical species like CO_2. Flows of other types involve *thermal methods*, particularly steam drive and soak, and *chemical flooding*, such as alkaline, surfactant, polymer, and foam (ASP+foam) flooding. All flows of these types in petroleum reservoir applications, except chemical flooding process, are considered in this book. The chemical flooding process has special characteristics, such as chemical reaction, adsorption, and non-Newtonian fluids, that must be addressed specifically (Chen, Huan, and Ma, 2006).

10.3.2 Grid Selection

Model discretization involves partitioning of a time interval and a spatial domain into discrete segments. Time discretization results in time steps used to advance the simulation model, while the discretization in space generates a finite difference grid or another type of grid that gives the areal and geologic features to the model.

Gridblock structures

Various gridblock structures used in reservoir simulation are displayed in Fig. 10.4. The issues related to grid selection include the dimension of the reservoir problem (1D, 2D, or 3D), the geometry of the reservoir system (Cartesian, radial, or distorted), and the fineness of the grid (how many gridblocks are required).

The number of dimensions required by simulation depends on the reservoir and study objectives. Zero dimensional (0D) models, or models of tank type, describe only reservoir energy and cannot distinguish flow direction. The material balance method mentioned in Chapter 1 is one such example. The objectives of using 0D models are to determine initial fluid distribution in place, predict field production, estimate water encroachment, and measure average saturation and pressure.

1D models allow for fluids to flow in a single direction and have limited applications for modeling reservoir performance. These models are used to study various flow mechanisms using core and slim tube displacements in laboratory. They can also be used for the study of 1D gravity drainage and calculation of vertical transmissibility.

2D models are historically the most widely used models in reservoir simulation studies since they can be employed for single-well models in an (r, x_3)-coordinate system, cross-sectional models in the x_1- (or x_2-) and x_3-directions, and areal flooding problems. Stacked areal models are composed of several 2D areal models that do not communicate within the reservoir due to vertical transmissibility barriers but may communicate through the wellbore

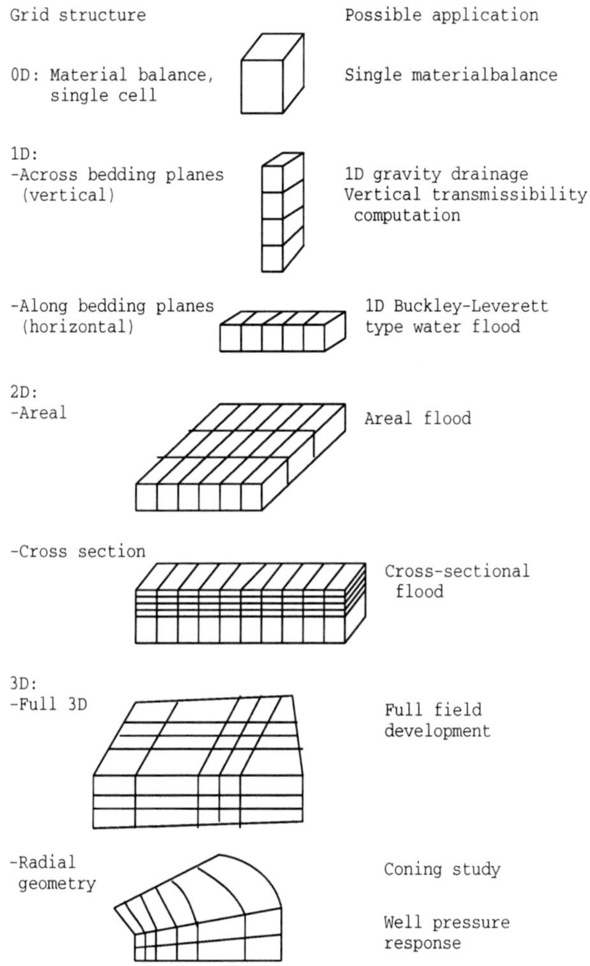

Figure 10.4. *Gridblock structures.*

due to commingled production or through the surface production/injection facilities. They can be used to study commingling and workover/recompletion strategies and tubing/surface-facilities capacities.

With the advent of powerful computer architectures and improved reservoir characterization technologies, many traditional 2D models now become the 3D standard models for full field simulation. 3D models allow for fluid flow in all directions and can be used for most of the simulation studies. A drawback of these models is the requirement of tremendous computational time, particularly for multiphase, multicomponent flow.

In addition to the dependence of the grid selection on the problem dimension, the reservoir flow problem itself also determines the grid geometry. For example, an (r, x_3)-grid should be used near a well, a 2D Cartesian grid is appropriate for a cross-sectional problem, and a complex geology needs a possibly distorted grid.

Grid orientation

As mentioned in Section 3.3.8, the 2D five-point finite difference method and the 3D seven-point finite difference method are sensitive to the spatial orientations of a computational grid. Grid orientation effects can cause a variety of problems, for example, in history matching of individual well performance and in evaluation of sweep efficiency.

To be mathematically and numerically correct, the finite difference grid must be oriented in the directions of the principal permeability axes of a reservoir since the finite difference formulation does not incorporate all cross-derivative terms. In general, however, the directions of these axes are unknown in the reservoir and should be replaced by other directions. An approach is to use the orientation that minimizes the total number of active gridblocks in the model. This approach also minimizes the computational time because the latter depends on the number of the active blocks. Another approach is to select the orientation that maximizes the number of potential wells that fall on or near block centers. Finally, if strong permeability anisotropy is known to exist in the subject reservoir, the finite difference grid must be oriented in the direction of the permeability trend, which is the sole way to handle the anisotropy.

Distorted grids

In reservoir simulation, many important physical and chemical phenomena are sufficiently localized and transient that *local grid refinement* is necessary to resolve them, such as flow behavior near the wellbore, fluid streaking due to stratification, and coning/cusp phenomena. Fig. 10.5 lists various types of local grid refinement techniques used in the finite difference method. In these techniques, the grid is automatically refined or unrefined depending on a local error indicator. They can lead to a very complex data management problem because they involve the dynamic regeneration of a grid, renumbering of nodal points and elements, and element connectivity.

When significant faults are present, the use of a *distorted grid* may be justified to allow for a proper computation of fluid flow across these faults (cf. Fig. 10.6). This is

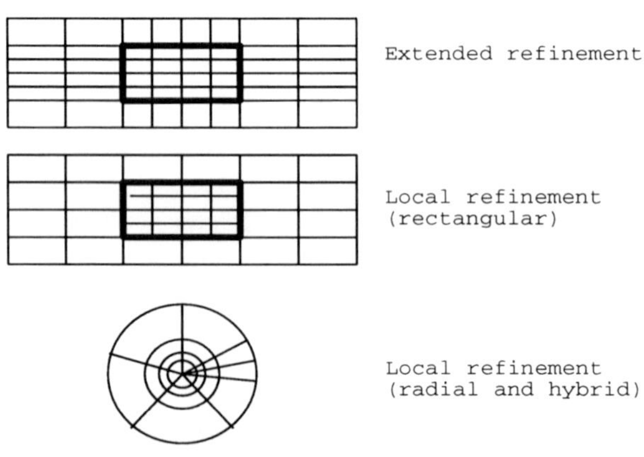

Figure 10.5. *Types of local grid refinement.*

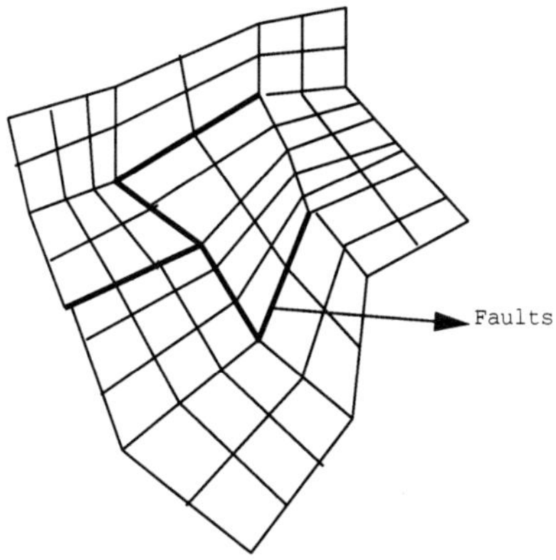

Figure 10.6. *Distorted grid.*

accomplished using *corner point geometry*. However, the reader should keep in mind that the distorted grid stills bears the weakness of grid orientation relative to the locations of wells.

10.4 History Matching

Much of reservoir engineering often involves, first, tuning uncertain parameters of a given reservoir to dynamic data and, second, using the tuned model to predict the performance of the reservoir under different operating scenarios. The tuning process is termed *history matching* and usually requires many simulation runs to determine the sensitivity of the results to the uncertain parameters. Once the sensitivities have been established, the reservoir engineer adjusts the values of the model input parameters to improve the match between the targeted field data and the simulation results. Using the history matched model, different production strategies are tested in which well placement, operating pressure, injectant rate schedules, and other well controls are altered to improve performance of the reservoir. In general, the number of development strategies examined depends on the size of the model, the available computing hardware, and expertise of the engineers.

The main objective of history matching is to improve and validate the reservoir simulation model developed. In practice, for a given production schedule, the matching data usually are (1) observed gas/oil ratios and water/oil ratios; (2) observed average pressures (shut-in pressures) or pressures at observation wells; (3) observed flowing well pressures; and (4) observed oil production rates. The process of history matching is time consuming and extremely difficult. It often represents a large portion of the cost of a petroleum reservoir study.

History matching can be done manually or automatically by adjusting model parameters through a trial-and-error procedure. Manual history matching runs the simulation model for the historical production period and then compares the results with the known field performance. After the comparison is made, the reservoir engineer will adjust the simulation data in an effort to improve the match. The choice of the input data to adjust is made by the engineer and requires knowledge and experience of the engineer on the subject field.

Automatic history matching is the same as manual, except that computer logic is employed to adjust the reservoir data rather than direct intervention of the reservoir engineer. The use of automatic history matching can remove specific knowledge and experience of the field under study. Many approaches are used to do the automatic history matching, and each one attempts to minimize an objective function. This objective function is defined as a function of the difference between the observed reservoir performance and the simulated reservoir behavior during the historical period.

No industrial standard exists as to what constitutes a successfully matched simulation model. The determination of a successful history match varies individually. What is important is that the history match be consistent with the simulation study objectives.

10.5 Prediction of Reservoir Performance

In history matching, a simulator is used to match historical reservoir performance. In the prediction stage, the matched simulator is utilized to predict future performance of a well or a reservoir under different operating strategies. The reservoir engineer studies a variety of scenarios and chooses a strategy that will likely generate the most desirable performance. The engineer also has an opportunity to show the potential benefits of new ideas and give results of high interest to a company or client. Predictions can be made through the use of the black oil, compositional, or thermal simulator described in this book. The prediction process involves the selection of prediction cases, preparation of input data for predictions, proper use of history matching, review and analysis of predicted performance, and evaluation and report of the predicted performance. The reader can refer to Mattax and Dalton (1990) for more information on each of these parts.

Bibliography

J. H. Abou-Kassem and K. Aziz (1985), Analytical well models for reservoir simulation, *SPE Journal* (August), 573–579.

T. Ahmed (2006), *Reservoir Engineering Handbook*, Society of Petroleum Engineers, Richardson, TX.

K. A. Akanni, J. W. Evans, and I. S. Abramson (1987), Effective transport coefficients in heterogeneous media, *Chem. Eng. Sci.* **42**, 1945–1954.

D. Andrews and A. Bennett (1981), Measurements of diffusivity near the sediment-water interface with a fine-scale resistivity probe, *Geochimica et Cosmochimica Acta* **45**, 2169–2175.

S. N. Antontsev (1972), On the solvability of boundary value problems for degenerate two-phase porous flow equations, *Dinamika Splošnoĭ Sredy Vyp.* **10**, 28–53, in Russian.

D. Archer, S. Emerson, and C. Reimers (1989), Dissolution of calcite in deep-sea sediments: pH and O_2 electrode results, *Geochimica et Cosmochimica Acta* **53**, 2831–2845.

G. E. Archie (1942), The electrical resistivity log as an aid in determining some reservoir characteristics, *Petrol. Tech.* **1**, 55–62.

R. Aris (1975), *Mathematical Theory of Diffusion and Reaction in Permeable Catalysts*, Clarendon Press, Oxford.

F. E. Ashford (1969), *Computed Relative Permeability Drainage and Imbibition*, Paper SPE 2582, Presented at the 44th Annual SPE Meeting, Denver, CO.

K. Aziz and A. Settari (1979), *Petroleum Reservoir Simulation*, Applied Science Publishers, London.

D. K. Babu and A. S. Odeh (1989), Productivity of a horizontal well, *SPERE* (November), 417–421.

G. I. Barenblatt, Iu. P. Zheltov, and I. N. Kochina (1960), Basic concepts in the theory of seepage of homogeneous liquids in fissured rocks [strata], *Prikl. Mat. Mekh.* **24**, 852–864.

J. Bear (1972), *Dynamics of Fluids in Porous Media*, Dover, New York.

J. Bear and Y. Bachmat (1991), *Introduction to Modeling of Transport Phenomena in Porous Media*, Kluwer Academic Publishers, Dordrecht, The Netherlands.

P. Bedrikovetsky, D. Marchesin, and P. R. Ballin (1996), *Mathematical Model for Immiscible Displacement Honoring Hysteresis*, Paper SPE 36132, presented at the Fourth Latin American and Caribbean Petroleum Engineering Conference, Port-of-Spain, Trinidad & Tabago, April, 557–573.

J. W. Beekman (1990), Mathematical description of heterogeneous materials, *Chem. Eng. Sci.* **45**, 2603–2610.

R. A. Berner (1980), *Early Diagenesis: A Theoretical Approach*, Princeton University Press, Princeton, NJ.

S. K. Bhatia (1985), Directional autocorrelation and the diffusional tortuosity of capillary porous media, *J. Catalysis* **93**, 192–196.

G. Booch, J. Rumbaugh, and I. Jacobson (1998), *The Unified Modelling Language User Guide*, Addison-Wesley, Boston.

B. P. Boudreau (1996), The diffusive tortuosity of fine-grained unlithified sediments, *Geochimica et Cosmochimica Acta* **60**, 3139–3142.

T. B. Boving and P. Grathwohl (2001), Tracer diffusion coefficients in sedimentary rocks: Correlation to porosity and hydraulic conductivity, *J. Contaminant Hydrology* **53**, 85–100.

R. H. Brooks and A. T. Corey (1964), *Hydraulic Properties of Porous Media*, Hydrology Papers, Colorado State University, Fort Collins, CO, March.

K. E. Brown (1977), *The Technology of Artificial Lift Methods*, Petroleum Publishing Co., Tulsa, OK.

D. A. Bruggemann (1935), Berechnung verschiedener physicalischer konstate von heterogen substanzen, *Ann. Physik* **24**, 636–664.

D. A. Calhoun, Jr., M. Lewis, Jr., and R. C. Newman (1949), Experiments on the capillary properties of porous solids, *Trans. SPE AIME* **186**, 180–196.

G. Chavent and J. Jaffré (1978), *Mathematical Models and Finite Elements for Reservoir Simulation*, North-Holland, Amsterdam.

Z. Chen (2000), Formulations and numerical methods of the black oil model in porous media, *SIAM J. Numer. Anal.* **38**, 489–514.

Z. Chen (2001), Degenerate two-phase incompressible flow I: Existence, uniqueness and regularity of a weak solution, *J. Diff. Equations* **171**, 203–232.

Z. Chen (2002), Degenerate two-phase incompressible flow II: Regularity, stability and stabilization, *J. Diff. Equations* **186**, 345–376.

Z. Chen (2005), *Finite Element Methods and Their Applications*, Springer-Verlag, Heidelberg and New York.

Z. Chen, H. Chen, D. Carruthers, and J. Zhou (2009), Integrated reservoir simulation and basin models: Reservoir charging and fluid mixing, SPE 118833, 2009 SPE Reservoir Simulation Symposium held in The Woodlands, Texas, USA, 2-4 February 2009.

Z. Chen and R. E. Ewing (1997a), Fully discrete finite element analysis of multiphase flow in groundwater hydrology, *SIAM J. Numer. Anal.* **34**, 2228–2253.

Z. Chen and R. E. Ewing (1997b), Comparison of various formulations of three-phase flow in porous media, *J. Comp. Physics* **132**, 362–373.

Z. Chen, G. Huan, and Y. Ma (2006), *Computational Methods for Multiphase Flows in Porous Media*, Computational Science and Engineering Series, Vol. 2, SIAM, Philadelphia, PA.

Z. Chen, G. Qin, and R. E. Ewing (2000), Analysis of a compositional model for fluid flow in porous media, *SIAM J. Appl. Math.* **60**, 747–777.

Z. Chen, G. Zhou, and D. Carruthers (2006), Numerical simulation of compositional flow in porous media under gravity, *Communications in Computational Physics* **1**, 827–846.

M. C. H. Chien, M. L. Wasserman, H. E. Yardumian, and E. Y. Chung (1987), *The Use of Vectorization and Parallel Processing for Reservoir Simulation*, Paper SPE 16025, presented at the Ninth SPE Symposium on Reservoir Simulation, San Antonio, TX, February.

A. T. Corey (1954), The interrelation between gas and oil relative permeabilities, *Producers Monthly* **19**, 38–42.

A. T. Corey (1986), *Mechanics of Immiscible Fluids in Porous Media*, Water Resources Publications, Littleton, CO.

A. T. Corey, C. H. Rathjens, J. H. Henderson, and M. R. Wyllie (1956), Three-phase relative permeability, *Trans. SPE AIME* **207**, 349–351.

H. Darcy (1856), *Les Fontaines Publiques de la Ville de Dijon*, Victor Dalmond.

M. Delshad and G. A. Pope (1989), Comparison of the three-phase oil relative permeability models, *Transport in Porous Media* **4**, 59–83.

J. Douglas, Jr., D. W. Peaceman, and H. H. Rachford, Jr. (1959), A method for calculating multi-dimensional immiscible displacement, *Trans. SPE AIME* **216**, 297–306.

F. A. L. Dullien (1979), *Porous Media, Fluid Transport and Pore Structure*, Academic Press, New York.

E. K. Duursma and C. J. Bosch (1970), Theoretical, experimental, and field studies concerning diffusion of radioisotopes in sediment and suspended particles of the sea, *Netherlands J. Sea Res.* **4**, 395–469.

E. K. Duursma and C. Hoede (1967), Theoretical, experimental, and field studies concerning diffusion of radioisotopes in sediment and suspended particles of the sea. Part A. Theories and mathematical calculations, *Netherlands J. Sea Res.* **3**, 423–457.

R. C. Dykhuizen and W. H. Casey (1989), An analysis of solute diffusion in rocks, *Geochimica et Cosmochimica Acta* **53**, 2797–2805.

T. Ertekin, J. H. Abou-Kassem, and G. R. King (2001), *Basic Applied Reservoir Simulation*, Society of Petroleum Engineers, Richardson, TX.

S. R. Faris, L. S. Gournay, L. B. Lipson, and T. S Webb (1954), Verification of tortuosity equations, *AAPG Bulletin* **38**, 2226–2232.

A. Firoozabadi (1998), *Advanced Thermodynamics of Hydrocarbon Reservoirs*, McGraw-Hill, New York.

L. S. Fung, D. A. Collins, and L. X. Nghiem (1989), An adaptive-implicit switching criterion based on numerical stability analysis, *SPE Reservoir Engineering* (Feb.), 45–51.

K. Ghorayeb and A. Firoozabadi (2000), Molecular, pressure, and thermal diffusion in nonideal multicomponent mixtures, *AIChE Journal* **46**, 883–891.

J. Gibbs (1876), On the equilibrium of heterogeneous substances, *Trans. Conn. Acad.* **3**, 108–248.

M. B. Goldhaber, R. C. Aller, J. K. Cochran, J. Rosenfeld, C. Martens, and R. A. Berner (1977), Sulfate reduction, diffusion, and bioturbation in Long Island Sound sediments: Report of the FOAM Group, *Amer. J. Sci.* **277**, 193–237.

F.-G. Ho and W. Strieder (1981), A variational calculation of the effective surface diffusion coefficient and tortuosity, *Chem. Eng. Sci.* **36**, 253–258.

M. K. Hubbert (1956), Darcy's law and the field equations of the flow of underground fluids, *Trans. SPE AIME* **207**, 222–239.

N. Iversen and B. B. Jorgensen (1993), Diffusion coefficients of sulfate and methane in marine sediments: Influence of porosity, *Geochimica et Cosmochimica Acta* **57**, 571–578.

B. B. Jorgensen (1978), A comparison of methods for the quantification of bacterial sulfate reduction in coastal marine sediments. II, Calculations from mathematical models, *Geomicrobiol. J.* **1**, 29–47.

H. Kazemi (1969), Pressure transient analysis of naturally fractured reservoirs with uniform fracture distribution, *SPE Journal* **9**, 451–462.

J. E. Killough and M. F. Wheeler (1987), *Parallel Iterative Linear Equation Solvers: An Investigation of Domain Decomposition Algorithms for Reservoir Simulation*, Paper SPE 16021, presented at the Ninth SPE Symposium on Reservoir Simulation, San Antonio, TX, February.

M. D. Krom and R. A. Berner (1980), The diffusion coefficients of sulfate, ammonia and phosphorus ions in anoxic marine sediments, *Limnol. Oceanogr.* **25**, 327–337.

J. Kuniansky and J. G. Hillstad (1980), Reservoir simulation using bottom hole pressure boundary conditions, *SPE Journal* (December), 473–486.

L. W. Lake (1989), *Enhanced Oil Recovery*, Prentice-Hall, Englewood Cliffs, NJ.

A. Lerman (1979), *Geochemical Processes: Water and Sediment Environments*, John Wiley and Sons, New York.

M. C. Leverett and W. B. Lewis (1941), Steady flow of gas-oil-water mixtures through unconsolidated sands, *Trans. SPE AIME* **142**, 107–116.

Y.-H. Li and S. Gregory (1974), Diffusion of ions in seawater and in deep-sea sediments, *Geochimica et Cosmochimica Acta* **38**, 703–714.

P. L. Lingen (1974), *Description of Groningen Gas Well Performance Suitable for Medium and Long Term Planning*, Paper SPE 4816, presented at the 1974 SPE European Spring Meeting, Amsterdam, April.

J. Lohrenz, B. Bray, and R. Clark (1964), Calculating viscosities of reservoir fluids from their compositions, *J. Petrol. Tech.* (October), 1171–1176.

P. F. Low (1981), Principles of ion diffusion in clays, in Chemistry in the Soil Environment, *Amer. Soc. Agron. Spec. Publ.* **40**, 31–45.

Y. Ma and Z. Chen (2004), Parallel computation for reservoir thermal simulation of multicomponent and multiphase fluid flow, *J. Comp. Physics* **201**, 224–237.

I. F. MacDonald, M. S. El-Sayed, K. Mow, and F. A. L. Dullien (1979), Flow through porous media: The Ergun equation revisited, *Indust. Chem. Fundam.* **18**, 199–208.

R. C. MacDonald and K. H. Coats (1970), Methods for numerical simulation of water and gas coning, *Trans. SPE AIME* **249**, 425–436.

J. S. Mackie and P. Meares (1955), The diffusion of electrolytes in a cation-exchange resin membrane, *Proc. Royal Soc.* **A232**, 498–509.

M. Maerki, B. Wehrli, C. Dinkel, and B. Muller (2004), The influence of tortuosity on molecular diffusion in freshwater sediments of high porosity, *Geochimica et Cosmochimica Acta* **68**, 1519–1528.

F. T. Manheim and L. S. Waterman (1974), Diffusimetry (diffusion constant estimation) on sediment cores by resistivity probe, in Initial Reports of the Deep Sea Drilling Project, Vol. 22, C. C. von der Borch and G. C. Sclater (eds.), U.S. Printing Office, Washington, D.C., 663–670.

C. C. Mattax and R. L. Dalton (1990), *Reservoir Simulation*, SPE Monograph, Vol. 13, Society of Petroleum Engineers, Richardson, TX.

J. C. Maxwell (1881), *Treatise on Electricity and Magnetism*, 2nd edition, Clarendon Press, Oxford.

D. F. Mayer (1989), *Application of Reservoir Simulation Models to a New Parallel Computing System*, Paper SPE 19121, presented at the SPE Petroleum Computer Conference, San Antonio, TX, June.

R. E. McDuff and R. A. Ellis (1979), Determining diffusion coefficients in marine sediments: A laboratory study of the validity of resistivity techniques, *Amer. J. Sci.* **279**, 666–675.

R. J. Millington (1959), Gas diffusion in porous media, *Science* **130**, 100–102.

N. R. Morrow (1970), Irreducible wetting-phase saturations in porous media, *Chem. Eng. Sci.* **25**, 1799–1816.

Y. Mualem (1976), A new model for predicting the hydraulic conductivity of porous media, *Water Resour. Res.* **12**, 513–522.

J. Naar and J. H. Henderson (1961), An imbibition model—its application to flow behavior and the production of oil recovery, *SPE Journal* (June), 61–67.

J. Naar and R. J. Wygal (1961), Three-phase imbibition relative permeabilities, *SPE Journal* (December), 254–260.

G. H. Neale and W. K. Nader (1973), Prediction of transport processes in porous media, *Amer. Inst. Chem. Eng. J.* **19**, 112–119.

J. S. Nelson and E. C. Simmons (1995), Diffusion of methane and ethane through the reservoir cap rock: Implications for the timing and duration of catagenesis, *AAPG Bulletin* **79**, 1064–1074.

T. E. W. Nind (1981), *Principles of Oil Well Production*, 2nd edition, McGraw-Hill, New York.

J. S. Nolen (1973), Numerical simulation of compositional phenomena in petroleum reservoirs, Reprint Series, SPE, Dallas, **11**, 268–284.

L. Onsager (1931a), Reciprocal relations in irreversible processes: I, *Phys. Rev.* **37**, 405–426.

L. Onsager (1931b), Reciprocal relations in irreversible processes: II, *Phys. Rev.* **38**, 2265–2279.

W. W. Owens and D. L. Archer (1971), The effect of rock wettability on oil-water relative permeability relations, *Trans. SPE AIME* **251**, 873–878.

D. W. Peaceman (1977a), *Fundamentals of Numerical Reservoir Simulation*, Elsevier, New York.

D. W. Peaceman (1977b), *Interpretation of Well-Block Pressures in Numerical Reservoir Simulation*, Paper SPE 6893, presented at the 52nd Annual Fall Technical Conference and Exhibition, Denver, CO.

D. W. Peaceman (1983), Interpretation of well-block pressures in numerical reservoir simulation with non-square grid blocks and anisotropic permeability, *SPE Journal* (June), 531–543.

D. W. Peaceman (1991), *Presentation of a Horizontal Well in Numerical Reservoir Simulation*, Paper SPE 21217, presented at the 11th SPE Symposium on Reservoir Simulation, Anaheim, CA, February.

A. Peneloux, E. Rauzy, and R. Freze (1982), A consistent correction for Redlich-Kwong-Soave volumes, *Fluid Phase Equilibria* **8**, 7–23.

D.-Y. Peng and D. B. Robinson (1976), A new two-constant equation of state, *Industrial and Engineering Chemistry Fundamentals* **15**, 59–64.

E. E. Petersen (1958), Diffusion in a pore of varying cross section, *Amer. Inst. Chem. Eng. J.* **4**, 343–345.

S. J. Pirson (1953), Performance of fractured oil reservoirs, *Bull. Amer. Assoc. Petroleum Geologists* **37**, 232–244.

J. Popovičová and M. L. Brusseau (1997), Dispersion and transport of gas-phase contaminants in dry porous media: Effect of heterogeneity and gas velocity, *J. Contaminant Hydrology* **28**, 157–169.

S. W. Poston, S. C. Ysrael, A. K. Hossain, E. F. Montgomery, and H. J. Ramey, Jr. (1970), The effect of temperature on irreducible water saturation and relative permeability of unconsolidated sands, *Trans. SPE AIME* **249**, 171–180.

L. Rayleigh (1892), On the influence of obstacles arranged in rectangular order upon the properties of a medium, *Phil. Mag.* **34**, 481–489.

R. C. Reid, J. M. Prausnitz, and T. K. Sherwood (1977), *The Properties of Gases and Liquids*, 3rd edition, McGraw-Hill, New York.

J. M. Rutledge, D. R. Jones, W. H. Chen, and E. Y. Chung (1991), *The Use of a Massively Parallel SIMD Computer for Reservoir Simulation*, Paper SPE 21213, presented at the Seventh SPE Symposium on Reservoir Simulation, Anaheim, CA, February.

S. L. Scott, R. L. Wainwright, and R. Raghavan (1987), *Application of Parallel (MIMD) Computers to Reservoir Simulation*, Paper SPE 16020, presented at the Ninth SPE Symposium on Reservoir Simulation, San Antonio, TX, February.

C. D. Shackelford (1991), Laboratory diffusion testing for waste disposal: A review, *J. Contaminant Hydrology* **7**, 177–217.

J. W. Sheldon, B. Zondek, and W. T. Cardwell (1959), One-dimensional, incompressible, non-capillary, two-phase fluid flow in a porous medium, *Trans. SPE AIME* **216**, 290–296.

L. Shen and Z. Chen (2007), Critical review of the impact of tortuosity on diffusion, *Chem. Eng. Sci.* **62**, 3748–3755.

T. G. Sherwood, R. L. Pigford, and C. R. Wilke (1975), *Mass Transfer*, B. J. Clark and J. W. Maisel (eds.), McGraw-Hill, New York, 39–43.

N. D. Shutler (1969), Numerical, three-phase simulation of the linear steamflood process, *Trans. SPE AIME* **246**, 232–246.

B. Smith, P. Bjorstad, and W. Gropp (1996), *Domain Decomposition, Parallel Multilevel Methods for Elliptic Partial Differential Equations*, Cambridge University Press, Cambridge, UK.

R. V. Smith (1983), *Practical Natural Gas Engineering*, Penn Well Publishing Co., Tulsa, OK.

R. W. Snell (1962), Three-phase relative permeability in unconstituted sand, *J. Inst. Petrol.* **84**, 80–88.

G. Soave (1972), Equilibrium constants from a modified Redlich-Kwong equation of state, *Chem. Eng. Sci.* **27**, 1197–1203.

H. L. Stone (1970), Probability model for estimating three-phase relative permeability, *Trans. SPE AIME* **249**, 214–218.

H. L. Stone (1973), Estimation of three-phase relative permeability and residual oil data, *J. Can. Petrol. Technol.* **12**, 53–61.

H. L. Stone and A. O. Garder, Jr. (1961), Analysis of gas-cap or dissolved-gas reservoirs, *Trans. SPE AIME* **222**, 92–104.

J.-P. Sweerts, C. A. Kelly, J. W. Rudd, R. Hesslein, and T. E. Cappenberg (1991), Similarity of whole-sediment molecular diffusion coefficients in freshwater sediments of low and high porosity, *Limnol. Oceanogr.* **36**, 335–342.

G. W. Thomas and D. H. Thurnau (1983), Reservoir simulation using an adaptive implicit method, *SPE Journal* (October), 759–768.

J. W. Thomas (1995), *Numerical Partial Differential Equations, Finite Difference Methods*, Springer-Verlag, New York.

M. M. Thomas (1989), Comments on calculation of diffusion coefficients from hydrocarbon concentration profiles in rocks, *AAPG Bulletin* **73**, 787–791.

M. R. Todd, P. M. O'Dell, and G. J. Hirasaki (1972), Methods for increased accuracy in numerical reservoir simulations, *SPE Journal* (December), 515–530.

M. M. Tomadakis and S. V. Sotirchos (1983), Transport properties of random arrays of freely overlapping cylinders with various orientation distributions, *J. Chem. Physics* **98**, 616–626.

W. J. Ullman and R. C. Aller (1982), Diffusion coefficients in near shore marine sediments, *Limnol. Oceanogr.* **27**, 552–556.

J. van Brakel and P. M. Heertjes (1974), Analysis of diffusion in macroporous media in terms of a porosity, a tortuosity and a constrictivity factor, *Intl. J. Heat Mass Transfer* **17**, 1093–1103.

P. van Cappellen and J. F. Gaillard (1996), Biogeochemical dynamics in aquatic systems, in Reactive Transport in Porous Media, P. C. Lichtner, C. I. Steefel, and E. H. Oelkers (eds.), *Reviews in Mineralogy* **34**, 335–376.

M. Th. van Genuchten (1980), A closed form equation for predicting the hydraulic conductivity in soils, *Soil Sci. Soc. Arm. J.* **44**, 892–898.

H. K. van Poolen, E. A. Breitenback, and D. H. Thurnau (1968), Treatment of individual wells and grids in reservoir modeling, *SPE Journal* **8**, 341–346.

S. Verdière, L. H. Quettier, P. Samier, and A. M. Thompson (1999), *Applications of a Parallel Simulator to Industrial Test Cases*, Paper SPE 51887, presented at the 15th SPE Symposium on Reservoir Simulation, Houston, TX, February.

J. Warren and P. Root (1963), The behavior of naturally fractured reservoirs, *SPE Journal* **3**, 245–255.

H. Weissberg (1963), Effective diffusion coefficients in porous media, *J. Appl. Physics* **34**, 2636–2639.

C. H. Whitson (1982), *Effect of Physical Properties Estimation on Equation of State Predictions*, Paper SPE 11200, presented at the 57th Annual Fall Technical Conference and Exhibition of the Society of Petroleum Engineers of AIME, New Orleans, LA.

C. H. Whitson and P. Belery (1994), *Compositional Gradients in Petroleum Reservoirs*, Paper SPE 28000, presented at the University of Tulsa Centennial Petroleum Engineering Symposium, Tulsa, OK.

A. S. Williamson and J. E. Chappelear (1981), Representing wells in numerical reservoir simulation: Part I—theory; Part II—implementation, *SPE Journal* (June), 323–344.

L. C. Young and R. E. Stephenson (1983), A generalized compositional approach for reservoir simulation, *SPE Journal* **23**, 727–742.

D. Zudkevitch and J. Joffe (1970), Correlation and prediction of vapor-liquid equilibria with the Redlich-Kwong equation of state, *American Institute of Chemical Engineers Journal* **16**, 112–199.

Index